CROWDSOURCING FOR SPEECH PROCESSING

CROWDSOURCING FOR SPEECH PROCESSING

APPLICATIONS TO DATA COLLECTION, TRANSCRIPTION AND ASSESSMENT

Editors

Maxine Eskénazi
Carnegie Mellon University, USA

Gina-Anne Levow
University of Washington, USA

Helen Meng
The Chinese University of Hong Kong, SAR of China

Gabriel Parent
Carnegie Mellon University, USA

David Suendermann
DHBW Stuttgart, Germany

A John Wiley & Sons, Ltd., Publication

Library of Congress Cataloging-in-Publication Data

Eskénazi, Maxine.
    Crowdsourcing for speech processing : applications to data collection, transcription, and assessment / Maxine Eskénazi, Gina-Anne Levow, Helen Meng, Gabriel Parent, David Suendermann.
        pages cm
    Includes bibliographical references and index.
    ISBN 978-1-118-35869-6 (hardback : alk. paper) – ISBN
978-1-118-54127-2 (ebook/epdf) – ISBN 978-1-118-54125-8 (epub) –
ISBN 978-1-118-54126-5 (mobi) – ISBN 978-1-118-54124-1
    1. Speech processing systems–Research.   2. Human computation.   3. Data mining.
    I. Title.   II. Title: Crowd sourcing for speech processing.
    TK7882.S65E85 2013
    006.4′54–dc23

                                                                    2012036598

A catalogue record for this book is available from the British Library.

ISBN: 978-1-118-35869-6

Typeset in 10/12pt Times by Aptara Inc., New Delhi, India

Printed and bound in Malaysia by Vivar Printing Sdn Bhd

Contents

6 Crowdsourcing in Speech Perception 137

Martin Cooke, Jon Barker, and Maria Luisa Garcia Lecumberri

7 Crowdsourced Assessment of Speech Synthesis 173

Sabine Buchholz, Javier Latorre, and Kayoko Yanagisawa

List of Contributors

Gilles Adda
LIMSI-CNRS, France

Jon Barker
University of Sheffield, UK

Laurent Besacier
LIG-CNRS, France

Sabine Buchholz
SynapseWork Ltd, UK

Martin Cooke
Ikerbasque, Spain
University of the Basque Country, Spain

Christoph Draxler
Ludwig-Maximilian University, Germany

Maxine Eskénazi
Carnegie Mellon University, USA

Hadrien Gelas
LIG-CNRS, France
DDL-CNRS, France

Javier Latorre
Toshiba Research Europe Ltd, UK

Gina-Anne Levow
University of Washington, USA

Maria Luisa Garcia Lecumberri
University of the Basque Country, Spain

Joseph J. Mariani
LIMSI-CNRS, France
IMMI-CNRS, France

Ian McGraw
Massachusetts Institute of Technology, USA

Helen Meng
The Chinese University of Hong Kong, China

Gabriel Parent
Carnegie Mellon University, USA

Roberto Pieraccini
ICSI, USA

Joseph Polifroni
Quanta Research, USA

David Suendermann
Synchronoss, USA
DHBW Stuttgart, Germany
ICSI, USA

Kayoko Yanagisawa
Toshiba Research Europe Ltd, UK

Zhaojun Yang
University of Southern California, USA

Preface

This book came about as a result of the standing-room-only special session on crowdsourcing for speech processing at Interspeech 2011. There has been a great amount of interest in this new technique as a means to solve some persistent issues. Some researchers dived in head first and have been using crowdsourcing for a few years by now. Others waited to see if it was reliable, and yet others waited for some service to exist in their country. The first results are very encouraging: crowdsourcing can be a solution that approaches expert results. However, it also comes with warnings: the incoming data must go through quality assessment.

This book is a hands-on, how-to manual that is directed at several groups of readers:

- **Experienced users**: Those who have already used crowdsourcing for speech processing should find a good set of references to the literature as well as some novel approaches that they may not be familiar with.
- **Speech processing users who have not yet used crowdsourcing**: The information in this book should help you get up to speed rapidly and avoid reinventing the wheel for common interface and assessment issues.
- **Users who are not speech processing experts who also need to use crowdsourcing for their speech data**: This book should also help you get started since you will have many of the same issues in dealing with your data.

We start the book with an overview of the principles of crowdsourcing. This is followed by some basic concepts and an overview of research in the area. The following chapters in the book cover most of the present types of speech processing. Chapter 3 covers the acquisition of speech. Chapter 4 covers speech labeling. Chapter 5 covers the variability of crowd speech and how to acquire and label speech in one effort. Chapter 6 explains how to run perception experiments using crowdsourcing. Chapter 7 explains how to use crowdsourcing for speech synthesis. Chapter 8 describes how to use crowdsourcing for assessment of spoken dialog systems. Chapter 9 covers the variety of platforms that are used for crowdsourcing and how they work. Chapter 10 covers industrial applications of crowdsourcing for speech processing. Finally, Chapter 11 covers the legal and ethical issues surrounding the use of crowdsourcing.

We express our wholehearted gratitude to all the contributing authors for their hard work and extremely timely delivery. Their extraordinary support helped us meet the unprecedented deadlines from proposal to production of not much more than a year. Also, the continuous guidance by Wiley's editorial team—Alex King, Liz Wingett, and Richard Davies—was essential for the success of this project.

Maxine Eskénazi, Gina-Anne Levow, Helen Meng,
Gabriel Parent and David Suendermann

1

An Overview

Maxine Eskénazi
Carnegie Mellon University, USA

In the early days of automatic speech processing, researchers dealt with relatively small sets of speech data. They used them mainly to build small automatic systems and to test the systems' validity. The data was often obtained by recording speakers in an anechoic chamber on magnetic tape. It was manually sent to a freestanding spectrogram machine in the 1960s and 1970s or input to a computer in the late 1970s and thereafter. Getting speakers (other than colleagues and students) took time, and labeling the speech that was acquired took much more time. Both endeavors were very costly. These difficulties entered into consideration every time a researcher planned a project, often imposing limitations on the amount of data collected and the scientific goals.

As time went on, one factor that dramatically increased the need for more data was the success of statistically based methods for automatic speech processing. The models for these systems, which quickly became ubiquitous, needed large amounts of data. The expression "more data is better data" was born. As automatic speech processing researchers switched from one application, like Broadcast News, to another, like Communicator, they found that the data from the former application was not very useful for the new one. As data specific to the new application was collected, processed, and fed into speech systems, results improved.

At the same time, other speech research publications (speech synthesis, spoken dialog systems, perception, etc.) also included some assessment. This required increasing numbers of speakers, callers, and judges, and thus a significant investment in data resources. This investment involved researcher time as they found and recorded subjects, as they trained transcribers to write down exactly what had been said, and as they found subjects to try out the resulting systems. Besides the time of the researcher, the investment also included the payment of the speakers and the transcribers, and sometimes a company was engaged to either recruit speakers or to manage transcription, thus adding to the costs.

Crowdsourcing for Speech Processing: Applications to Data Collection, Transcription and Assessment, First Edition.
Edited by Maxine Eskénazi, Gina-Anne Levow, Helen Meng, Gabriel Parent and David Suendermann.
© 2013 John Wiley & Sons, Ltd. Published 2013 by John Wiley & Sons, Ltd.

As Google became a major search engine of choice, it gathered a very large amount of data, larger than any that had ever been used before. Google researchers produced conference papers that demonstrated to their colleagues that some previously unsolved issues in natural language processing became surmountable just by using several orders of magnitude more data (Brants *et al.* 2007). The field became ripe for a solution that would provide more processed data at significantly lower cost. The present estimate of the cost of transcribing 1 hour of speech data by an expert (for ASR training) is 6 hours of transcription time for each actual hour of speech that is processed, at a cost of $90–$150 per hour (Williams *et al.* 2011).

At the same time, linguists and sociolinguists, freed from the use of magnetic tape and the onerous postprocessing that accompanied it, found that recording speech directly on computers enabled them to rapidly obtain large samples of the speech that they wanted to study. Many speakers with more diverse backgrounds could be recorded. Several groups of speakers with varying characteristics could be recorded instead of just one. However, as the need for more speakers and more transcriptions of their speech increased, these communities ran up against the same obstacles that the automatic speech processing community had encountered.

What seems like the answer to these needs has come in the form of crowdsourcing. This technique offers the promise of dramatically lowering the cost of collecting and annotating speech data. Some of the automatic speech processing community has quickly embraced crowdsourcing. This chapter and the next will give a short history and description of crowd-sourcing, some basic guidelines, and then review the uses of crowdsourcing for speech that have been published in the past few years.

1.1   Origins of Crowdsourcing

What may be one of the earliest examples of the use of the crowd is the open call that the Oxford English Dictionary (OED) made to the community in the 1800s for volunteers to index all of the words in the English language and to find example quotations for each of the uses of each word (Wikipedia 2012).

More recently, James Surowiecki's (2004) book, *The Wisdom of Crowds,* gives an explanation of the power of the wisdom of the crowd. It maintains that a diverse collection of opinions from people who are making independent decisions can produce some types of decisions better than obtaining them from experts. Surowiecki sees three advantages to what he terms disorganized decisions: **cognition** (thinking and information processing), **coordination** (optimization of actions), and **cooperation** (forming networks of trust with no central control).

A good example of cooperation in a disorganized decision is the US *Defense Advanced Research Projects Agency* (DARPA) experiment in crowdsourcing to mark the 40th anniversary of the Internet. The goal was to locate 10 balloon markers that had been placed in a variety of locations across the United States. Teams were formed, each vying to be the first to find all 10 markers. This required collaborative efforts with networks of informers in many locations across the country. The team from MIT had the shortest time (under 9 hours). Its groups, comprised friends, and friends of friends, signed up to help locate the balloons. This underlines the observation that, in a crowd situation, where each person is independent and fairly anonymous, an individual will give their knowledge and opinions more freely. The success of the endeavor centered on this generous participation. Indeed, authors of crowdsourcing tasks who ask the members of their crowd if they have suggestions on how to improve a task (without giving them additional remuneration) often find that some of the crowd will take the time to make very insightful and helpful suggestions.

1.2   Operational Definition of Crowdsourcing

The operational basis of crowdsourcing rests on the idea that **a task** is to be done, there is a means to **attract many nonexperts** to accomplish this task, and that some **open call** has gone out to advertise the task to the nonexperts (Wikipedia 2012). The presence of the Internet and cellphones facilitates not only the open call for nonexperts but also the presentation of the task, its accomplishment, and the accumulation of the nonexperts' opinions. The nonexperts also possess some relevant knowledge, be it only that they are native speakers of a given language. From these assumptions, it is believed that the aggregate opinion of many nonexperts will approach the quality of the opinion of an expert. It is also believed that the use of nonexperts in this manner will be less onerous and more rapid than the use of experts. Given this economy of means, it is understandable that the speech and language processing communities have seen crowdsourcing as a possible solution to their large data dilemma. To illustrate the operational aspects of crowdsourcing, consider a task that comes up at barbeques and other social events. A total of 672 jellybeans have been put into a clear jar. Each person attending the barbeque is asked to estimate how many jellybeans are in the jar. They are aware that there is something that the organizer of the barbeque wants them to do (the open call for nonexperts). They are also aware that they are to provide an estimate of the number of jellybeans in the jar (the task), and they know how to count jellybeans or make estimates (have some expertise). They put their answers on a piece of paper and put that into a box. The organizer looks at all of the tickets in the box and finds answers like 300, 575, 807, 653, 678, 599, and 775. The aggregate answer, such as the average (626), or the median (653), is very close to the real number of jellybeans in the jar. Thus, the conditions that characterize crowdsourcing are:

- A task.
- An open call.
- Attracting many nonexperts.

So we will define a crowd as a group of nonexperts who have answered an open call to perform a given task.

1.3   Functional Definition of Crowdsourcing

A functional view of crowdsourcing, from Surowiecki, defines four characteristics of the *wise* crowd. First, the members of any crowd have a **diversity of opinions**. The opinions may be only slightly different from one another, and some may be correct while others are wrong. Second, each member of the crowd has an opinion that is **independent** of all of the other members of the crowd. No member's opinion is influenced by that of any other member. Third, information that the crowd may have is **decentralized**. Everyone has some local information, but no one in the crowd has access to all of the information that may be pertinent to the task. Finally, the opinions of the members of the crowd can be merged to form an **aggregate**, one collaborative solution. To illustrate this, here is an example where the crowd has the task of translating some text. If we have the following text in French,

Je pense qu'il est temps de partir. On prendra congé de ma mère et de ma sœur et on se mettra en route au plus tard à neuf heures.

we can ask a crowd, that is, English and French speaking, for its translation. Some of its members may offer these four solutions:

S1: I think it's time to leave. We will say goodbye to my mother and my sister and get going at 9 a.m. at the latest.
S2: The time has come to leave. Let's say goodbye to my mother and my sister and be on our way by 9 a.m.
S3: We need to go. We'll say goodbye to my mother and my sister and leave by 9 a.m.
S4: Let's go. Take leave of my mother and my sister and be on our way by 9 a.m.

The four aspects of the functional nature of crowdsourcing are illustrated here. We can see that the four solutions offered by members of the crowd (S1–S4) reflect *diverse opinions* on exactly what the right translation is. We also can imagine that these opinions have been arrived at *independently* from one another. Each member of this crowd possesses some *individual pieces of information* that they are using when forming their opinion. S1, for example, may reflect the idea that "il est temps de partir" should be translated literally as "it's time to leave" while S4 may reflect a broader definition, which, in this context, results in the expression "let's go." Finally, we can *merge* these opinions to form one solution by, for example, asking the members of another crowd to vote on which one they like the best, by choosing the one that is most frequently produced, or by using a string-merging algorithm such as Banerjee and Lavie (2005). There has also been work (Kittur *et al.* 2011; CastingWords 2012) on having the crowd collaborate with one another to make the translation evolve into something on which they can all agree.

Thus, according to Surowiecki, the four characteristics of a wise crowd are:

• Has a diversity of opinions.
• Each individual works independently of the others.
• The information is decentralized.
• An aggregate solution can be formed.

1.4 Some Issues

While crowdsourcing seems to be a remarkable solution to the problems plaguing the speech and linguistics communities, it must be approached with care since misleading or incorrect results can also easily be obtained from crowdsourcing. Several issues should be kept in mind to prevent this.

The first issue concerns **the amount of information** given to the crowd. The crowd should be given just enough information to be able to complete the task, but not enough to influence their decisions. For the translation task above, for example, although the creator of the task could ask for a translation that is as literal and close to the original text as possible, this additional information may make the final result less desirable. The workers' opinions should not be influenced by information from the task creator. The second issue is having a crowd that is **too homogeneous**. A crowd that is too homogeneous will not give a superior result. Oinas-Kukkonen (2008) has found that the best decisions come when there is disagreement and contest within the crowd. Note that giving too much information is one way that a crowd may be rendered too homogeneous. He mentions another issue that contributes to homogeneity— **too much communication**. When members of the crowd have less anonymity and the creator of the task has more communication with the crowd, too much information may gradually be transmitted. Linked to the issue of communication is that of **imitation**. If participants are given access to the opinions of other workers, they may be influenced by them and, consciously or not, imitate what they have seen (thus leading us back to a more homogeneous crowd). While some tasks are given to one crowd and then the result is sent to another crowd for verification, some mechanism should be in place to avoid having this influence.

A fifth issue that should be addressed concerns the **prerequisites of the crowd**. We have seen that the members of the crowd are presumed to have some local knowledge. It is not evident that everyone who responds to a call has that knowledge. For example, in the above translation task, the creator of the task will assume that the participants speak both French and English. Since there may be some impostors in the crowd, it is wise to give some sort of pretest. We will discuss this further in Chapter 2. Pretests of performance on the specific task at hand are a reasonable way to winnow out those who may not be able to perform the task. However, creation and checking of the pretest is onerous in itself and it may be more time- and cost-saving to let all who respond complete the task and then eliminate outlier answers later on.

Another issue is **motivation**. Why should someone participate in a task? Are they learning something, playing a game, being remunerated? There should be some reason for an individual to not only sign up to work on a task but also want to continue to work on it. This is linked to a seventh issue, keeping a **reasonable expectation of the work effort**. If members of the crowd are lead to believe that there is less work than what is actually expected, especially in the case of remunerated work, they will quit the task and recommend to others (via worker forums and blogs) that they also avoid this task.

Finally, as we will see in several chapters in this book, it is absolutely necessary to carry out some form of **quality control**. This control can come in many forms and it is meant to weed out the work of poor workers (who have good intentions, but who furnish work that is not of good quality) and malicious workers (those who randomly enter answers or automated bots).

When reading research papers that incorporate crowdsourcing results, it is wise to determine whether these issues have been dealt with since this may affect the wellfoundedness of a paper.

Therefore, before proposing a task, researchers should deal with the following issues:

- Giving the crowd too much information.
- A crowd that is too homogeneous.
- Having too much communication with the crowd.
- Avoiding the possibility of imitation.
- Requesting prerequisites from the crowd.
- Maintaining crowd motivation.
- Presenting a reasonable expectation of workload.
- Conducting quality control.

1.5 Some Terminology

At this point, a short discussion of terminology is useful. The person who is creating the task and who submits it is called (at Amazon Mechanical Turk, MTurk, in this book, a platform that is used for crowdsourcing) the requester. This person may be called the client at other crowdsourcing sites. Herein we will use the term *requester*. The person in the crowd who does the work is appropriately called the *worker* (some also say *turker*) at MTurk and other sites, a *freelancer* at MiniFreelance, and a *contributor* at CrowdFlower and elsewhere. We will use the term *worker*. The individual task itself is called a *Human Intelligence Task* or *HIT* at MTurk, a *mission* at AgentAnything.com, a *microjob* at MicroWorkers, and a *task* at CrowdFlower. We will use the term *task* here, but the reader will also see this term broken down into three types of tasks, according to granularity:

- *Set of tasks* is the complete set of items that the requester wants to have done. For example, the transcription of 2000 hours of speech.
- *Unit task*, for example, transcribing one utterance out of the 2000 hours of speech in the set of tasks above.
- *Assignment* is one piece of work within the unit task; that is, the transcription of the above unit task of one utterance may be assigned to three workers, thus there would be three assignments for one unit task.

Also, when referring to the number of unit tasks completed per hour, we will use the term *throughput*. When referring to when the requester makes a set of tasks available to the workers, we will use the term *submission*. When talking about the agreement between multiple workers in an annotation task, we will use the term *interannotator agreement* (ITA).

1.6 Acknowledgments

This work was supported by National Science Foundation grant IIS0914927. Any opinions, findings, and conclusions and recommendations expressed in this material are those of the author and do not necessarily reflect the views of the NSF.

References

Amazon Mechanical Turk—Artificial Artificial Intelligence. http://mturk.com (accessed 9 July 2012).
Brants T, Popat AC, Xu P, Och FJ and Dean J (2007) Large language models in machine translation. *Proceedings of the Conference on Empirical Methods on Natural Language Processing(EMNLP-2007)*.
Banerjee S and Lavie A (2005) METEOR: an automatic metric for MT evaluation with improved correlation with human judgments. *Proceedings of the ACL 2005 Workshop on Intrinsic and Extrinsic Evaluation Measures for MT and/or Summarization*.
CastingWords. http://castingwords.com (accessed 9 July 2012).
Kittur A, Smus B and Kraut R (2011) CrowdForge: crowdsourcing complex work. *Proceedings of the ACM 2011 Annual Conference on Human Factors in Computing Systems*.
Oinas-Kukkonen H (2008) Network analysis and crowds of people as sources of new organizational knowledge, in *Knowledge Management: Theoretical Foundation* (eds A Koohang *et al.*). Informing Science Press, Santa Rosa, CA, pp. 173–189.

Surowiecki J (2004) *The Wisdom of Crowds: Why the Many Are Smarter Than the Few and How Collective Wisdom Shapes Business, Economies, Societies and Nations*. Doubleday Anchor.

Wikipedia—Crowdsourcing. http://en.wikipedia.org/wiki/Crowdsourcing (accessed 9 July 2012).

Williams JD, Melamed ID, Alonso T, Hollister B and Wilpon J (2011) Crowd-sourcing for difficult transcription of speech. *Proceedings of IEEE Workshop on Automatic Speech Recognition and Understanding (ASRU 2011)*.

2

The Basics

Maxine Eskénazi
Carnegie Mellon University, USA

This chapter contains some basic general information about crowdsourcing that is useful for all speech applications. It uses a review of the literature to show what areas of research have used crowdsourcing and to show what has been done to deal with specific concerns, like quality control. It begins with an overview of the literature on crowdsourcing for speech processing. Then it maps out the types of alternatives to crowdsourcing that the reader might consider and describes the crowdsourcing platforms that exist. From there it discusses how to make task creation easier. Then it gives details about some basic considerations such as getting the audio in and out and payment. Prequalification and native language of the worker are also presented in this section and can serve as an introduction to the issue of quality control that comes later in Section 2.6. The chapter finishes with some useful tips.

2.1 An Overview of the Literature on Crowdsourcing for Speech Processing

Any overview in this area can only serve as an indication of trends, since, by the time that it is published, additional knowledge will have been accumulated. We have found speech-related publications both in the usual automatic speech processing venues and in others not specific to speech processing. Readers should also note that speech crowdsourcing has adapted techniques that have been successful in other areas such as machine translation.

In this section, we will look at what has been published mainly in acquiring, labeling, and transcribing speech and assessing automatic speech processing systems. In the chapters that follow, there will be more specific information in each of these areas. This overview is intended as a census, laying out the land: in what areas of automatic speech processing has crowdsourcing been used, how much effort has there been in each area, what has met with success, and which new areas are emerging.

Crowdsourcing for Speech Processing: Applications to Data Collection, Transcription and Assessment, First Edition.
Edited by Maxine Eskénazi, Gina-Anne Levow, Helen Meng, Gabriel Parent and David Suendermann.
© 2013 John Wiley & Sons, Ltd. Published 2013 by John Wiley & Sons, Ltd.

Papers in the realm of natural language processing that use crowdsourcing date back to about 2007. However, we see the first publications on speech processing appearing about 2009. With this short time-frame, it would be difficult to draw strong conclusions about trends in the publications. However, it is possible to see how greatly the number of publications has grown, where the publications come from, and how the use of quality control has been reported. There are many interesting papers in the slightly more mature area of text/natural language processing that reveal new techniques that can be applied to speech processing. Since that area of research is not reviewed here, the reader may want to look at the overview paper of the 2010 NAACL workshop on Creating Speech and Language Data with Amazon's Mechanical Turk (Callison-Burch and Dredze 2010) for a good general description revealing the breadth of research that was presented at that workshop.

In the few years that crowdsourcing for speech has existed, the venues where papers appear have been varied. The reader can find papers in the following venues:

- The Association for Computational Linguistics (ACL).
- The North American Association for Computational Linguistics (NAACL).
- The European Association for Computational Linguistics (EACL).
- The International Speech Communication Association (ISCA) Interspeech Conference.
- The IEEE International Conference on Acoustics, Speech and Signal Processing (ICASSP).
- The ISCA Speech and Language Technologies for Education (SLaTE) Special Interest Group.
- The Association for the Advancement of Artificial Intelligence (AAAI).
- The Educational Data Mining Workshop (EDM).
- The Empirical Methods for Natural Language Processing Conference (EMNLP).
- The IEEE Automatic Speech Recognition and Understanding Workshop (ASRU).
- The meetings of the Association of Computing Machinery (ACM).
- The International Conference on Language Resources and Evaluation (LREC).
- The Special Interest Group on Discourse and Dialog (SIGDIAL).
- The IEEE Spoken Language Technologies Workshop (SLT).

The papers that have been reviewed for in this chapter come from academia, industry, and government. Of 47 papers, 33 (70%) were from academia, 13 (28%) from industry, and 1 (2%) from government.

2.1.1 Evolution of the Use of Crowdsourcing for Speech

Within the perspective of papers that deal with processing speech in some way, the 47 papers that were found through the end of 2011 range from theses to journal articles, book chapters, and conference and workshop papers. It is possible that a few publications may have been overlooked in our search and multiple publications of the same tasks were not counted. Figure 2.1 shows that the number of publications mentioning the authors' use of crowdsourcing for speech processing has increased from 2007 when there was one paper to 2010 when there were 22 papers. There is a decrease in 2011 (16 papers), but there may be publications that appeared at the end of 2011 that were not included here. It will take several more years of data to see how pervasive crowdsourcing will be. This will in part depend on how well it is embraced outside North America.

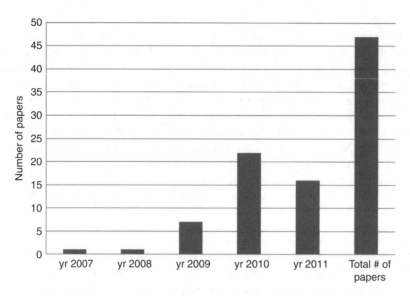

Figure 2.1   Number of publications by year on crowdsourcing for speech processing.

2.1.2   Geographic Locations of Crowdsourcing for Speech

Due to the presence in the United States of the most popular platform, Amazon's Mechanical Turk (MTurk), it is not surprising that the largest number of papers comes from North America. Figure 2.2 shows the distribution of papers by continent.

Of the 47 papers, 31 are from North America, with Europe a distant second at 13 papers. Looking more closely at this phenomenon, Table 2.1 shows the evolution over the past few years of the countries and continents where the papers come from. While there were three papers from Europe and none from Asia in 2007–2009, there is a small increase to seven in 2011 from Europe and two in 2011 from Asia. The very recent appearance of platforms in

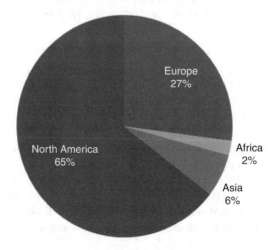

Figure 2.2   Publications 2007–2011 by continent.

Table 2.1 Evolution of publications by country and by continent.

Country	2007–2009	2010	2011
US	7	16	7
Great Britain	1	0	4
Japan	0	0	2
China	0	1	0
France	0	1	1
Portugal	1	0	0
Germany	1	0	0
Ireland	0	2	0
South Africa	0	1	0
Switzerland	0	0	1
Spain	0	0	1
Total	10	21	16
North America	7	16	7
Europe	3	3	7
Asia	0	1	2
Africa	0	1	0

Europe should reinforce the upward trend in future years. It is hoped that the same trend will soon be seen in Asia as well.

The papers surveyed reveal four categories in which the crowd processes speech: to **acquire speech**; to **label and/or transcribe speech**; to **assess speech synthesis, spoken dialog, or other systems**; to **run perception and other studies**.

In order to give the reader a general understanding of how crowdsourcing has been used for speech processing, this section separates papers into the **individual experiments they ran** and labels each task as being one of the four categories above. For example, since McGraw *et al.* (2009) describe both an acquisition and a transcription task, we count each task separately. Yet other papers were overviews or proposals about projects that could be done in the future—for these, of course, there is no data. Of the 38 papers that actually show results, there were 49 individual studies. Section 2.1.3 shows some overall statistics and discusses some trends we observed. Papers on perception and other studies involving human experimentation are often not published in the venues mentioned previously and so the numbers that we have found may not be indicative of all that exists. Table 2.2 shows the breakdown of the 49 experiments into the four categories. In order to provide an overview of the publication rate per category,

Table 2.2 Number of studies per subarea (there may be more than one per paper).

Type	Number of experiments
Labeling/transcription	26 (51%)
Acquisition	12 (25 %)
Studies	6 (12%)
Assessment	6 (12%)

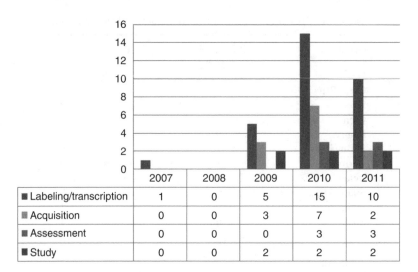

	2007	2008	2009	2010	2011
■ Labeling/transcription	1	0	5	15	10
■ Acquisition	0	0	3	7	2
■ Assessment	0	0	0	3	3
■ Study	0	0	2	2	2

Figure 2.3   The evolution of subareas.

Figure 2.3 provides the same information broken down per year. The only pre-2009 paper we found is Schlaiker's (2007) technical report on the use of speech CAPTCHAs.

2.1.3   Specific Areas of Research

The majority of the 49 experiments were on speech labeling and transcription (51%), with speech acquisition being the second most frequent topic (25%). To our knowledge, only six experiments (12%) have harnessed crowdsourcing for assessment of speech technology, but it should be noted that the latter has seen the greatest increase recently.

Speech Acquisition

There was a bottleneck in real-time acquisition, playback, and archiving of speech over the web that (as we will see in other chapters in this book) was solved by the use of recent web technologies and crowdsourcing platforms. Acquiring speech in this way certainly makes it harder to control microphone type or distance, or the ambient noise, but certain techniques can be used to detect and mitigate these problems. This makes the characteristics of the speech gathered in this way similar to those of the signal that a speech recognizer processes in a web or phone application. In order to acquire speech in a new language, starting with only one speaker and no screen to read from, Ledie *et al.* (2010) have many speakers hear and repeat that person's utterances. Since cellphones can collect speech that is read from the screen, Lane *et al.* (2010) gather training data from speakers reading words from a mobile phone screen. To acquire speech for a targeted domain, Polifroni *et al.* (2010) have the speaker dictate a restaurant review of her own invention. McGraw *et al.* (2010) used MTurk to collect read speech and had workers complete a predefined dialog scenario, thus obtaining more varied and realistic speech data. McGraw *et al.* (2011) created a photo annotation human intelligence task

(HIT) where turkers recorded a spoken description of photos. Others have created interactive games and often use their own interface, not MTurk to do this. To record speech from nonnative speakers while keeping them interested in the task, McGraw *et al.* (2009) and Gruenstein *et al.* (2009) use a language learning game. Chernova *et al.* (2010) also use online games to get material for human–robot dialog research. In order to obtain speech for new synthetic voice models, Freitas *et al.* (2010) use a quiz game with speakers reading text from the screen. Afrikaans, South African English, and Sepedi speech were recorded using an Android-based commercial application, DataHound. This is described in de Vries *et al.* (2011). Finally, Júdice *et al.* (2010) provide the details of an experiment aimed at obtaining elderly speech via a web interface, describing challenges encountered and possible solutions.

Speech Labeling

Speech, or linguistic content, can be annotated using crowdsourcing. At present, the literature reveals more work on speech transcription than on annotation. This trend is changing, as we will see in Chapter 4. This is fueled by the increased interest in detecting sentiment and other nonlinguistic information. Suendermann *et al.* (2010) use an automatic transcription scheme to process very large amounts of speech data from their call center systems. Parent and Eskénazi (2010) have workers transcribe a large amount of spoken dialog turns from real callers. Marge *et al.* (2010a) have workers transcribe meeting data, and Marge *et al.* (2010b) have them transcribe human–robot dialogs. Similarly, Audhkhasi *et al.* (2011b) ask multiple workers to transcribe speech, and use the audio for acoustic model adaptation. Lee and Glass (2011) integrate a series of processing steps in a transcription task in MTurk in order to filter out bad transcripts, as well as to evaluate word level confidence and consequently provide feedback to workers. Williams *et al.* (2011) introduce the idea of using an *incremental redundancy*, where the maximum number of distinct transcriptions obtained from the crowd is set to a certain value N, but utterances stop being submitted if K distinct transcriptions match (such that $K < N$). This variable redundancy attempts to treat easy- and hard-to-transcribe utterances differently, submitting the hard utterances more often, but not infinitely. Another interesting quality control mechanism, called *iterative dual pathway structure,* is described in Liem *et al.* (2011). The approach is iterative: a worker has access to the transcriptions provided by previous workers, which are seeded with an automatic speech recognition (ASR) output, and can provide improvements on the transcriptions. Two independent iteration paths are created, and the iterations stop when four consecutive transcriptions (two from each path) converge.

Akasaka (2009) created a Facebook game where players earn points by transcribing audio while having fun. This idea is also exploited in Luz *et al.* (2010), Novotney and Callison-Burch (2010) (for Korean, Hindi, Tamil), and Audhkhasi *et al.* (2011a) (for Spanish). Gelas *et al.* (2011) (for Swahili and Amharic) obtain transcriptions of speech for low resource languages. Evanini *et al.* (2010) have workers transcribe nonnative speech. Goto and Ogata (2011) present the implementation of a crowdsourcing platform allowing a crowd of volunteers to correct audio transcriptions. Finally, Schlaikjer (2007) looks at audio CAPTCHAs, acquiring transcriptions of audio streams. For annotation, some authors study the identification of accents. Akasaka's (2009) Facebook game collects nonnative speech and then asks workers to identify the accent in each utterance. Kunath and Weinberger (2010) ask listeners to identify accents, and then research the influence of the listeners' backgrounds on their annotations. In Evanini and Zechner (2011), naive annotators are asked to provide prosodic annotation of nonnative speech. McGraw *et al.* (2009) and Parent and Eskénazi (2010) also use crowdsourcing to label whether a text snippet corresponds to the ASR transcription of a specific utterance, thus accomplishing a task similar to ASR output validation.

Assessment of Speech Technology

Six experiments in four papers have used crowdsourcing for some form of assessment. Assessment may cover the ASR alone or some application in which it has been embedded. It can also cover the assessment of spoken dialog systems and computer-assisted language learning systems. Yang *et al.* (2010) have workers review a spoken dialog and then ask them to complete a questionnaire on dialog success. Jurcicek *et al.* (2011) also use crowdsourcing for dialog system assessment and conclude that results provided by MTurk and Cambridge in-house evaluation are indistinguishable. Wolters *et al.* (2010) compared MTurk workers and students at the University of Edinburgh on the task of evaluating speech synthesis. Bucholz and Latorre (2011) report on lessons learned from running 127 crowdsourced speech synthesis preference tests and provide insight on how to detect cheaters. These pioneering studies will lead the way for others to increasingly turn to crowdsourcing to assess the value of a given type of system for a user.

Perception Studies

Cooke *et al.* (2011) study word recognition in noise. Frank *et al.*'s (2010) paper covers two experiments on speech segmentation. Tietze *et al.* (2009) "present two experiments designed to examine the impact of linguistic devices, such as discourse cues and connectives, on comprehension and recall in information presentation for natural language generation (NLG) as used in spoken dialog systems (SDS)." Chapter 6 describes the issues that the requester confronts when creating this type of task. This includes the difficult decision of how to run a study that had only been run before in very well-controlled laboratory conditions and how to obtain informed consent from a worker.

2.2   Alternative Solutions

While this book concerns the use of crowdsourcing as a solution for processing large amounts of speech data, in order to provide some perspective, this section discusses other solutions. One of the prevalent decisions in the past, when faced with large amounts of speech to transcribe, for example, has been to outsource the work to a company that has permanent employees trained as experts in labeling. This is still an onerous solution in both time and money. At the time when this chapter was written, SpeakerText and CastingWords were charging $2 per minute of audio, for example. While this may be high for a research laboratory, the management and training of new personnel to annotate just one corpus is onerous as well. Although several companies have offered this service in the past, only a handful of them have survived. One reason is that the demand for their services is variable over time. In a time of low demand, the company has the dilemma of whether to shoulder the cost of keeping their trained labelers or to lose them and have to hire (and train!) anew when demand picks up.

Another advantage of the use of a professional firm is that the requesters do not need to define (and program) the actual interface that the workers will be using. The company obtains a full description of what the end product is to be and, from past experience in the area, is able to work out the interface issues as well as the instructions to be given to the labelers. This implies that the company has expertise in a given field (such as annotation of speech for ASR training). If researchers have a task that is far from their area of expertise, then the company may not be able to satisfy their need. One goal in creating the Fisher corpus was rapid data collection and annotation for a very large dataset (Cieri *et al.* 2004). With the goal of collecting as much speech as possible in a short amount of time, a robot operator made targeted calls to unique subjects. These speakers had been recruited through newsgroups, print ads, and Google banners related to linguistics and speech technologies. Word of mouth publicity coming from these calls generated even more potential speakers. There were new topics for callers to discuss each day, ensuring some variability in the linguistic content. The system received an average of 54 calls per day after getting 15 per day during its initial testing phase. Although the system was capable of handling even higher call volumes, there were never enough registrants to meet these capabilities. This continuous pipeline of incoming calls was a novel way to obtain large amounts of speech. Another means of obtaining a large volume of calls is to provide a service for the general public. The Let's Go system (Raux *et al.* 2006) has answered calls to the Port Authority of Allegheny County for bus information nightly since March 2005 and has amassed a database of more than 175,000 dialogs with an average of 40 calls per night on weeknights and 90 calls per night on weekends. The nonnegligible cost here is in the maintenance of a working system 365 days a year.

For the Fisher task, as the pipeline of incoming speech data filled up, the transcription had to keep pace. For this, a quick transcription convention was devised that decreased the time to transcribe 1 hour of speech from 20 hours to 6. Two transcription strategies were developed in parallel, one by the Linguistic Data Consortium (LDC) and the other by BBNT (Raytheon BBN Technologies) in conjunction with a commercial service provider, WordWave International (WordWave International 2012). LDC first automatically segmented the speech into utterances that were 2–8 seconds long. The transcribers then made one pass over each utterance to create a verbatim transcription of what they heard. They were not asked to punctuate or to note noises or other nonlinguistic events. BBNT had the entire conversation transcribed with no initial segmentation. They then used forced alignment to furnish timestamps for each word. The result in both cases provided transcriptions that were not as good as an expert, but that were judged to be of sufficient quality to train a speech recognizer.

As indicated above and as we will see later in this book in Chapter 9, the .com crowdsourcing platforms are not the only solution. Other options to consider are **in-house** and **commercial service providers**. Every group's need is different and lack of in-house expertise or limited funds may guide these choices.

2.3 Some Ready-Made Platforms for Crowdsourcing

While there were few crowdsourcing platforms (such as MTurk, Amazon Mechanical Turk 2012, and CrowdFlower, 2012) as late as 2009, since then there has been a significant proliferation of sites. If a .com for crowdsourcing is your vehicle of choice, there are several aspects to consider when selecting which one to go with. For example, tasks proposed to MTurk have little or no intervention on the part of the requester who provides a set of tasks, creates their own tasks, then simply uploads them and collects the results. CrowdFlower, in addition to self-service, has a service that takes over the work of preparing the task by asking the provider for a description of the work to be done. When the work is finished, it remits the results in some easily usable form (just like MTurk). Other characteristics that vary from site to site include having a way for the requester to reject work before the task has finished and giving the requester a means to provide feedback directly to the workers. Later chapters will describe some toolkits that have been developed to make it easier to create and manage a task on some sites.

Some new sites that have sprung up in recent years are: AgentAnything.com (AgentAny-thing.com 2012), microWorkers (microWorkers 2012), MiniFreelance (MiniFreelance 2012), clickworker.com (clickworker 2012), samasource for refugee camps (samasource 2012); and for macrotasks, there are innocentive (innocentive 2012) and Chaordix (Chaordix 2012). A classic example of crowdsourcing on the Internet is Wikipedia (Wikipedia 2012), where readers of the encyclopedic entries can contribute to their content. Facebook (Facebook 2012) also provides a vehicle for a provider to send out an open call and obtain a crowd. Interestingly, Facebook itself benefited from crowdsourcing to translate its site into more than 50+ languages by gathering translations from its own members.

For a more global view of the geography of all of the platforms that have been created to benefit from the wisdom of the crowd, crowdsourcing.org has produced a crowdsourcing landscape, which groups the platforms into eight areas based on the overall goals of each site. They are:

Crowdfunding: Financial contributions from online investors for small business (CROWDFUNDER 2012).

Tools: Applications, platforms and tools that support collaboration (Crowd-Engineering 2012).

Cloud labor: A distributed labor pool that can fulfill a range of tasks (CrowdFlower 2012; Amazon Mechanical Turk 2012).

Civic engagement: Collective actions on issues of public concern (ecycler 2012).

Collective knowledge: Development of knowledge assets (OrganizedWisdom 2012).

Collective creativity: Creative talent used to design original art, media, or content (ADHACK 2012).

Community building: Developing communities through individual active engagement on common passions (CrowdTogether 2012).

Open innovation: Use of sources outside the group to generate and implement ideas (Chaordix 2012).

The reader should note that this geography is very new and thus constantly changing. Some companies may have sprouted up since this chapter was written, and some may have disappeared.

Most of the sites that have sprung up recently are US based. A few non-US sites have been mentioned in later chapters. One successful site in China is Taskcn.com (Taskcn 2012). The globalization of crowdsourcing is essential to its success in such areas as machine translation where access to native speakers of a given language may be limited on some US platforms. As we will see in Section 2.5.3 and also for quality control in Section 2.6, requesters on MTurk have had to rely on geolocation and qualifying tests to get non-English speakers for their tasks. There is always a risk of getting workers who misrepresent themselves. There is also the risk for a task taking a very long time to finish as it awaits workers who speak the required language. As one example, in unpublished work, we have found that it took a task in English about 2 weeks to be finished on MTurk, while the same task in Spanish took 3 weeks to finish. This is another criterion to consider when choosing a platform.

One of the very best ways to become familiar with how to format and present a task as well as to understand how it looks from the worker's point of view is to **sign up to become a worker on one of these sites**. By completing a few tasks yourself, you can clearly see the range of possibilities you have. The ease or difficulty you encounter while completing a task someone else has designed will guide you as you design your own task, helping you make clearer instructions, better layouts and simpler tasks.

2.4   Making Task Creation Easier

Ease in task creation and crowd access has been a critical factor in the general acceptance of crowdsourcing in the research domain. When in 2009, MTurk's online interface was changed to afford batch processing from CSV files (as opposed to command-line arguments), researchers from fields adjacent to speech and natural language, such as linguists, some of whom may have less programming experience, became able to access a crowd. This relative ease of access may influence platform choices.

There have been other efforts to ease the work involved in creating tasks. The advent of TurKit (Little *et al.* 2010) is one example. It allows users to lay out their algorithms in a straightforward manner and to create iterative flows of tasks where, for example, one crowd can verify the preceding crowd's work. Another is CrowdForge by Kittur *et al.* (2011). CrowdForge serves complex tasks, coordinating many workers. It creates and manages subtasks through distributed workflow. It also simplifies the designer's task. The concept of having workers judge one another's production is also used by CastingWords (2012). Here, workers transcribe and improve on earlier workers' transcriptions and also grade the quality of their work.

While many platforms that are used for crowdsourcing are computer based and mainly managed over the web, there are also cellphone-based applications, especially for collecting speech in underdeveloped nations, as we will see in Chapters 5 and 11.

2.5   Getting Down to Brass Tacks

There are six points in this section that, while independent of any specific type of speech application, are common to all of them. First, speech is special in many ways and when approaching a speech crowdsourcing task the requester has to **assure that the audio works**. The other points are: **payment, choice of platform, prequalification, native language of workers**, and **task complexity** and are discussed from a speech perspective here.

2.5.1   Hearing and Being Heard over the Web

All speech processing tasks that are to be accomplished without direct observation of the workers have one common issue. They need to first ensure that the worker can hear the speech signal correctly and/or that the speech signal can be correctly recorded. Getting the speech signal in and out of a web- or telephone-based crowdsourcing application has, in the past, been a major roadblock. Fortunately, this has been addressed in recent years as we will see in Chapters 3 and 4. Another issue had been the type of microphone and headset used. With the advent of small portable devices such as MP3 players and smartphones for playing music, most people own a lightweight headset. Most of the time that headset, as with smartphones, has also a built-in microphone. While this is not the highest quality equipment that could be used, it has been found in many cases to be sufficiently good for speech processing.

With the equipment issues addressed, there are still several possible reasons that audio might not play or be acquired. Some causes of audio problems are:

- Worker not wearing the headset.
- Headset not plugged in.
- Sound levels too high or too low.
- High levels of ambient noise.
- Failure to correctly follow instructions.

Some have argued that poor-quality audio can be dealt with. Novotney and Callison-Burch (2010) found that in the case of very large datasets, poor utterances may only comprise a small part of the whole and do not lower quality significantly for an ASR system trained on this data. However, their expression "good enough" is relevant for applications where a slightly higher word error rate (WER) is acceptable. This may not be an option for other applications, such as for language learning systems, where high precision is paramount. The erroneous part of a large dataset may also be eliminated in part, if not entirely, by some postprocessing. When creating your task, you will save much time in quality control if you check the audio in some way at the beginning of a task.

There are several techniques commonly used to check audio (as we see in Chapters 3 and 4). In the literature, they vary according to whether the task is to acquire speech or to transcribe or label it.

For acquisition, McGraw *et al.* (2010) ask the worker to listen to playback of the utterance and to approve it. Lane *et al.* (2010) similarly asked the worker to do this and then to rerecord it if they felt there were issues with it. In both cases, this action was not mandated or monitored, relying solely on the honesty of the worker. In Lane *et al.*'s case, postprocessing found 10% of the utterances that they collected were not usable. Their later solution to test working audio was to have the worker aurally solve a simple math problem. Jurcicek *et al.* (2011) required the use of a headset and relied on the honesty of the workers for compliance.

Several authors required that workers use a headset microphone although, again, this relies on worker's honesty. Chernova *et al.* (2010) collected dialog data. They recorded the worker and then showed them the ASR transcription of what they had just said, using the WAMI toolkit (Gruenstein *et al.* 2009). Since the workers were participating in a dialog, if they saw that the system had made an error, they simply used their next dialog turn to attempt to correct it. Novotney and Callison-Burch (2010) used an MP3 upload and instructions for the use of the recording software, but they did not report on whether they had checked in some way to see if the worker had complied or if they, too, relied on worker honesty. Hughes *et al.* (2010) took another approach by repeatedly sampling the audio volume to ensure that the phone device was recording properly.

For transcription and labeling, the issue is no longer to determine if a proper audio signal has been obtained, but rather if the worker can hear the signal that is being played. One way to check this is with an audio captcha (captcha.net 2012). Like a visual captcha where the letters in the image must be transcribed, here the audio signal is to be transcribed, thus verifying that the playback is working properly. This also blocks eventual bots. Several papers have the worker transcribe something in this way before they let them work on real data. Munro *et al.* (2010) and Frank *et al.* (2010) asked the worker to correctly transcribe a prechosen word that was played to them and then automatically matched the transcription of the worker with the expected word string. Evanini *et al.* (2010) gave the workers three calibration responses which were known in advance. Then, as for (Frank *et al.* 2010), but for longer strings, the worker responses were matched to the expected correct strings. For transcription, Evanini *et al.* determined that the quality of the worker's output would be greater with this setup since it would help eliminate noise interference and give an overall better quality signal. Chernova *et al.* (2010) also had a transcription task where they asked workers to listen to instructions over the phone. During these instructions, a code was given to the worker. Entering this code, gave access to the transcription mechanism. Roy and Roy (2009) required that the worker wait until the whole audio clip had played before they were allowed to enter any of the transcription. This was designed to keep the worker from simply clicking without listening and typing some random response. They also did not allow the worker to progress from one transcription to the next if they detected an empty response. In a more constantly vigilant approach, Ribeiro *et al.* (2011) asked the worker, at the beginning of each new task (HIT on MTurk here), what type of audio input was being used.

In summary, there are several actions that a requester can take to ensure better audio. In the case of speech acquisition:

- Give instructions as to how to use the microphone and rely on the honesty of the worker.
- Ask the speaker to listen to what was recorded and approve it.
- Sample the recorded signal level in one or two utterances and give the worker feedback.

For speech transcription:

- Use an audio captcha or have the worker transcribe one or more aforeknown utterances.
- Do not let the worker continue if a transcription is empty or was started before the end of the playback.
- At the beginning of each task, remind the worker of the audio input quality standards.

As in other cases, like prequalifications, the above measures may slow down throughput. It is up to each requester to determine the quality trade-off that this will produce for their specific case. They also need to weigh this against the amount of postprocessing that they want to take on.

2.5.2   Prequalification

As mentioned above, with few exceptions it cannot be assumed that every worker is qualified to perform every task. Bots, malicious individuals and nonnative speakers are the most frequent nonqualified workers. It also cannot be assumed that every native speaker of a language is capable of performing any type of linguistic task in their language. One example of this in our past unpublished work has shown that when asking native speakers of French to read ten sentences from the Le Monde journal aloud, only 5 out of 10 speakers could perform that task with 7 or more of the 10 sentences being error free. It is therefore useful to have some sort of prequalification for the workers.

On MTurk, one way to root out malicious workers is to only accept those workers whose task (HIT) acceptance rate is higher than a given threshold. Several papers have relied on this method. Audhkhasi *et al.* (2011a) and Marge *et al.* (2010a) chose an approval rating of at least 95%. Kunath and Weinberger (2010) and Lee and Glass (2011) chose at least 65%. McGraw *et al.* (2009) chose 75% and whether the worker had performed 5 or less assignments in the past hour (to set a pace for the worker). Rayner *et al.* (2011) set a limit of 90% for their first study. On their second study, they did not require a prequalification and noted that most of the workers they gained on the second pass were "scammers." They believe that these individuals were attracted to the task by this lack of a prequalification.

Reliance on past performance has been shown to be effective in eliminating the intentionally malicious workers. Yet there remains the issue that there are well-intentioned individuals who are just not good at performing some tasks. There are two reasons to filter for these workers. The first is to lessen the amount of postprocessing needed. The second is that it is unfair to the worker (and we see their discontent with the requesters in their blogs) to have an honest worker complete a significant amount of work and then not pay them. While, as we will see, there are more sophisticated ways to detect poor work while the tasks are still running and give early feedback to the worker, some find it more appropriate to test the worker's performance on the actual task as a prequalification. Evanini *et al.* (2010) gave the workers three calibration questions, in part, as we have seen, as a check on the audio. However, it also served as a prequalification since if a worker did not have three correct responses, they were not allowed to continue on the task. Ribeiro *et al.* (2011), for the assessment of synthetic speech, chose two samples of synthesis to be scored for quality, one of which had already been assessed by experts to be obviously good and the other to be obviously bad. If the worker's assessments did not match these, they were not accepted for the tasks. Novotney and Callison-Burch (2010) had found that they could reliably estimate a worker's capabilities based on the quality (their ranking relative to other workers) of 15–30 transcriptions and used this to filter workers. de Vries *et al.* (2011) trained field workers to recruit and test potential workers in person.

For speech acquisition, Novotney and Callison-Burch (2010), in order to ensure a "good speaking voice," had workers record a paragraph and asked "public opinion" about the passage. That is, they asked other workers, "Would you listen to a full Wikipedia article read by this person?" They found that most of their speakers had acceptable voices (judged positive by at least 5 out of 10 workers), but there was a minority whose reading received a poor rating.

Another technique that has been used, as we have seen in Section 2.5.1, for checking the audio and as we will see in Section 2.5.3 for checking for native speakers, is to ask the worker about their qualifications and rely on their honesty. For example, Polifroni *et al.* (2010) recruited workers who defined themselves as someone who ate out at restaurants and was familiar with online restaurant review sites (since these workers were going to create their own restaurant reviews after dining out).

As we have seen, there are two types of prequalifications:

- One that is general and is designed to remove malicious workers, **using some sort of past approval rating** (that will only work, of course, for those who use a platform that has this type of service).
- Another, to determine the aptitude of well-intentioned workers for a task, which is to **give the worker a small sample of the task to perform**.

Prequalification is only one form of quality control, a control before the task begins. We will see other means of quality control in Section 2.6.

2.5.3   Native Language of the Workers

Another form of prequalification concerns the native language of the worker and/or the languages that the worker is fluent in. An easy approach to finding out the country of the worker is to use geolocation services such as those offered by MTurk and CrowdFlower. While this may give the location of the worker, it is evident that there are nonnatives working in every country, so just knowing the location does not help filter for the worker's native language. In previous work (Parent and Eskénazi 2010), we used the MTurk country filter, but found that it was ill-adapted for native language detection for two reasons: many US residents aren't native speakers of English and many workers on MTurk are registered as US workers, but aren't working from there (as given by the IP addresses we collected). Knowledge of location does, however, aid in some assessments of work. Recent findings show that posting tasks at noon or midnight local time can produce varying results due to the difference in the countries where workers will respond at these times. Williams *et al.* (2011) show differences in the work obtained when tasks were posted at these two times. Bucholz and Latorre (2011) used CrowdFlower geolocation. Jurcicek *et al.* (2011) got the workers' IP addresses *post hoc* to check honesty since they requested only native speakers for their MTurk task. Tietze *et al.* (2009) and Wolters *et al.* (2010) also relied on the honesty of the workers by simply asking for workers from the United States only. Audhkhasi *et al.* (2011a) disallowed native speakers of English on their Spanish task.

Two papers (de Vries *et al.* 2011; Hughes *et al.* 2010) use employees to recruit their workers and so have some closer knowledge of their demographics. In South Africa, with the intent of recording speech from 11 different native tongues, de Vries *et al.* (2011) use field workers to do the recruiting, training, and rewarding of their workers (who use telephones to enter the data). In a similar manner, Hughes *et al.* (2010) have university students recruit other students to serve as workers. This enables them to obtain demographics such as accent and age. This also lets them store the data on the Android phone that is being used to gather it.

Some tasks require speakers of many different languages. In this case, since there is work for everyone, regardless of their native language, there is no reason for the worker to be dishonest. Lane *et al.* (2010) ask workers to select their native language at the beginning so that they can work on only that language. Akasaka (2009) simply asks for the native language of the worker and records it for use in assessing results concerning the influence of accents. Kunath and Weinberger (2010) asked not only for the native language of the worker but also how well the worker knew other languages.

Most crowdsourced speech processing tasks would benefit from a more effective native language filter. One possibility would be to assess the participant's native linguistic abilities in a pretest such as the translation activities (from English to the native language of the worker) that are required of workers who want to access many text processing tasks. The requester in this case is faced with the negative and positive aspects of this option: the time spent on the pretest would decrease throughput, yet the use of the pretest would improve overall quality.

2.5.4   Payment

One way to attract a crowd is to present a task that people are curious about and want to figure out. This has sometimes been shown to be effective. Most of the time, a worker wants to get something in return for the time spent on the task. This may take the form of enjoyment while playing a game or it may take the form of monetary remuneration. Many issues concerning payment such as the legality and the morality of low wages are discussed in Chapter 11. This section looks at the levels of remuneration in the literature. It also looks at the effect of the level of wages on the time it takes to complete a task, and at the effect of the level of wages on the quality of the output.

The reader can find a wealth of information on this topic in Ipeirotis (2010), which provides data about several aspects of the requester's relationship with the worker. In addressing remuneration, Ipeirotis gave an overall idea of the situation in 2010 concerning price per task. He first looked at the number of tasks at each level of remuneration and found that the majority of the requesters (about 70% of them) paid about $0.05 or less per task item. This does not take into account the relative difficulty of the tasks. He also found that an average task (which can be a HIT on MTurk, comprising several individual items) had an average completion time of 12.5 minutes, which in turn results in an effective average hourly wage of $4.80. These numbers are highly variable from one task to another. In this chapter, we will look at some of these same numbers for speech-related tasks. Ipeirotis also showed the number of tasks completed by day of the week. There is some variation at that level. However, due to MTurk being open at this point to workers in India as well as in the United States, it would have been useful to also see what task completion looked like per hour of the day. McGraw et al. (2010) plotted the data collection rate on MTurk as a function of the worker's local time of day and found low rates of throughput around 4–6 a.m. and high rates from about 1–5 p.m. Yet Williams et al. (2011) found no correlation between submission time and latency, stating that they found workers available around the clock. This issue can be important when working on a platform like MTurk since there are many tasks for the worker to choose from. A worker can scan the list of tasks using the default MTurk ordering, which is newest first. In this case, if many of the type of worker you want are looking for tasks at, say, 2 p.m., you would want to submit your tasks at that time.

Many of the papers dealing with speech processing give information on amount paid per item, duration of task, and amount of data gathered. While the following information has been drawn from those papers, it should be noted that there are several issues when presenting this information. First, all three of the above measures are not mentioned in all of the papers. Next, there are many different types and sizes of tasks, from recording one sentence read aloud from the screen or transcribing one utterance, to assessing a whole dialog or assessing a learning

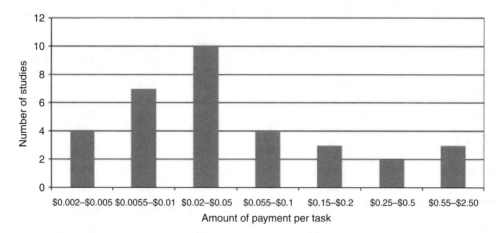

Figure 2.4 Payment per task for speech processing studies (on MTurk).

game. For this reason, the reader will find the measures accompanied by details. The only straight comparison that can be made is for those studies that show the number of utterances transcribed in a task.

Figure 2.4 shows the amount that requesters of speech processing tasks paid per item. This is a very general view and represents only tasks carried out on MTurk. It groups 34 studies, covering many different task types and difficulties and includes such tasks as transcribing an utterance and annotating a whole text. Figure 2.5, on the other hand, only shows tasks involving transcription, where the amount per utterance transcribed was given. In this figure, we see that majority of the payments for these 11 tasks averaged from $0.02 to $0.05.

We can also examine the relation between the amount paid and the time it took to complete the task. Gelas *et al.* (2011) compared the amount of time to obtain work in two relatively rare languages for MTurk workers: Swahili and Amharic. Swahili speakers appear to be more numerous and they show a very large difference in the time it took to accomplish the set of

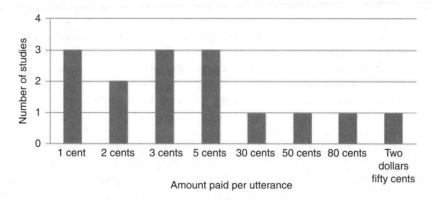

Figure 2.5 Payment per transcription of one utterance on MTurk.

Table 2.3 Payment, time to completion and amount of items processed, per utterance, transcription tasks.

Paid $ per item	Content of item	Hours to completion	Items processed
0.002	1 utterance	22	810 audio snippets
0.004	1 utterance	21	810 audio snippets
0.005	1 utterance	62	52 clips
0.005	1 utterance	46	257,658
0.005	1 utterance	1	200
0.05	1 utterance	13.5	52 clips
0.006	1 utterance	16	10,000
0.01	1 utterance	21.5	52 clips
0.01	1 utterance	20.75	73,643
0.01	1 utterance	17	810
0.02	1 Korean utterance	840	3 h audio

tasks. Fortunately, platforms such as Samasource and txteagle (Eagle 2009) address the issue of how to attract speakers of specific languages. Table 2.3 compares the amount paid per utterance to the time it took to complete the whole task (in hours) on MTurk for transcription tasks, for the cases where the paper broke the task and the cost down per individual utterance.

For more detail on direct comparisons of this nature, a study by McGraw *et al.* (2010) showed the results when the requester paid $0.10 and $0.20 for the same task—the higher rate generated more than twice as many sessions, more than twice as many utterances, and a lower percentage of incomplete sessions. Williams *et al.* (2011) carried out a comparison of the same task at $0.002, $0.004, and $0.01 per utterance transcribed and found that higher pay produced results in less time (22 hours, 21 hours, and 17 hours, respectively). It has also been observed that for tasks with large amounts of data to be processed, workers who are the best performers can be retained by giving them bonuses. This adds slightly to the cost, but also enhances the quality of the work.

One paper addressed whether payment was necessary. Lane *et al.* (2010) compared two groups, one with paid workers and the other with volunteers, to see what portion of the acquired speech was inadequate for acoustic model training. They found that volunteers produced inadequate speech utterances two to five times more often than the paid workers.

Table 2.4 shows all of the sets of tasks where payment was mentioned. Time to completion and amount of items processed are also noted, when available. This gives a more general view of the relationship between the type of task and quantity of data to the amount paid and the time it took to complete the task.

A frequent question related to quality in a microtask market is whether the payment has an effect on the quality of the work. Since it is desirable to retain the best workers, the overall level of payment could affect which workers decide to take on a task. Several papers report investigations of the effect of wage on quality. McGraw *et al.* (2010), Marge *et al.* (2010b), and Novotney and Callison-Burch (2010) all analyzed the productivity per worker at different levels of remuneration. None found a significant difference in quality over different wage levels although Marge *et al.* (2010b) reported the counterintuitive finding that higher pay led to more errors. Williams *et al.* (2011) found similar results. The cheapest pay ($0.002) gave a significantly lower utterance error rate (UER) (at 42.7%) than did the highest pay ($0.01 for

Table 2.4 Amounts paid in 33 studies per item in speech processing tasks.

Paid $ per item	Content of item	Hours to completion	Items processed
0.002	1 utterance transcribed	22	810 audio snippets
0.004	1 utterance transcribed	21	810 audio snippets
0.005	1 utterance transcribed	n/a	10,651 utts
0.005	1 utterance transcribed	62	52 clips
0.005	1 utterance transcribed	46	257,658 utts
0.005	1 utterance transcribed	n/a	n/a
0.005	1 utterance transcribed	1	200 utts
0.006	1 utterance transcribed	16	10,000 utts
0.01	1 utterance transcribed	n/a	10,000 utts
0.01	1 utterance transcribed	21.5	52 clips
0.01	1 utterance transcribed	20.75	73,643 utts
0.01	1 utterance transcribed	17	810 utts
0.02	1 utterance transcribed	n/a	20 h audio
0.02	1 Korean utt transcribed	840	3 h audio
0.025	1 utterance transcribed	n/a	200
0.03	1 audio clip annotated	336	15,911 audio clips
0.03	1 scenario written	3	273
0.05	1 dialog assessed	1,080	11,000
0.05	1 utterance transcribed	13.5	52 clips
0.05	1 audio clip transcribed Swahili	288	1,183 clips
0.05	1 audio clip transcribed Amharic	1752	1,183 clips
0.05	1 description recorded	48	1,099
0.1	1 utterance transcribed	n/a	n/a
0.1	1 description recorded	48	995
0.1	Assess 1 synthetic audio clip	156	8,307
0.1	Assess 1 synthetic audio clip	72	3,831
0.13	1 utterance transcribed	n/a	n/a
0.2	Assess 1 synthetic audio clip	36	4,410 clips
0.2	Recorded 1 dialog	240	1,000
0.3	1 audio clip transcribed	n/a	40
0.5	Assess 1 synthetic audio clip	n/a	n/a
0.8	1 minute audio transcription	n/a	45.1 min of speech
2	1 assessment session	168	26 sessions
2.5	Annotate 1 whole text	2	14 texts
n/a	1 listening test item	504	1,093

a UER of 50.8%). Furthermore, they found that submissions at midnight GMT had a UER of 50.9%, but those submitted at noon had a significantly lower UER (42.8%). They proposed that the latter tasks may have been done by North American workers who were more familiar with the business names in the task (as opposed to workers in India).

2.5.5   Choice of Platform in the Literature

As mentioned previously, there are many types of platforms for crowdsourcing. The 40 papers we examined have a variety of platforms. Table 2.5 shows their relative popularity. There are

Table 2.5   Number of studies per platform.

Platform	Number of studies
MTurk	23 (53.5%)
In-house-web and phone	15 (34.9%)
In-house-phone	5 (11.6%)
CrowdFlower	2 (0.5%)
Offshore outsourcing	1 (0.25%)
Quizlet.com	1 (0.25%)
Facebook	1 (0.25%)

43 studies in total reported in this table due to several papers reporting more than one study on more than one platform.

Most of the studies (53.5%) used the MTurk microtask market. While this platform provides access to a large quantity of workers, it can also become relatively expensive for large amounts of data. Fifteen of the 43 studies (34.9%) had some in-house created platform that ran either over the web or on a phone. Of those, 5 (11.6%) were phone-based applications, and 5 (11.6%) reported using some other platform. It should be noted that some solutions use a game interface, and workers are sometimes paid (as is always the case with MTurk) and sometimes volunteers. The diversity of the tasks in these studies makes for insufficient data for any one type of platform to compare the quality of the results from the different sources of crowds.

Recently, both workers and requesters at MTurk have been required to furnish a Social Security number for tax purposes. Also, at the time this chapter was written, requesters still had to come from the United States and in order to be paid in cash, workers had to have a bank account in the United States or in India. Workers from outside these countries are offered payment in Amazon gift card credit. This attracts some workers, but direct payment would be much more attractive. The move to create platforms for workers outside the United States and India has started. This will enable researchers to publish work on languages other than English with significant amounts of data.

The Source of the Crowd

It would seem that MTurk would make it easier to reach a larger number of workers due to its widespread reputation and also due to remuneration. Figure 2.6 shows the numbers of workers for the studies in the literature. It should be noted that not all papers give the number of distinct workers in their tasks, nor do they always give the amount of data they collected. Figure 2.6 illustrates the fact that platforms other than MTurk may provide a large number of workers. This is important for tasks where a variety of opinions is necessary, such as speech acquisition, where a wide sampling of speech from many workers produces training material for more reliable acoustic models for automatic speech processing.

At the same time, it is interesting to see if the same distribution holds for the amounts of data that have been processed. Again, the papers do not always reveal the total number of items processed. The use of a chart would not be appropriate here since the items that are

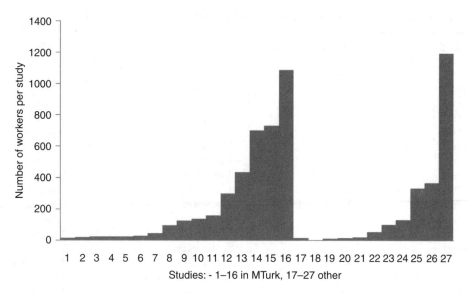

Figure 2.6   Number of workers per study—MTurk vs. other platforms.

collected are different in size (ranging from whole dialogs to individual utterances) and in type (covering tasks that range from acquisition to labeling to assessment and to studies). Table 2.6 shows in detail the amounts of data obtained according to whether the task was run on MTurk or another platform and the type of data that was obtained. Although the numbers of items mentioned in the literature are in descending order, the amounts of data that these numbers correspond to are not all the same. This table also clearly shows that large numbers of items can be processed on platforms other than MTurk and that other platforms afford just as much diversity of purpose and type of task as MTurk.

2.5.6   The Complexity of the Task

There has been consistent evidence in the literature that breaking large tasks up into several smaller subtasks will optimize accuracy. Work such as that by Kittur *et al.* (2011) and Ledie *et al.* (2010) is based on this principle. It harks back to the concept of cognitive load. The more that a worker has to attend to in a given task, the less well that worker will attend to any one aspect of the task.

Here is an example from the realm of automatic speech processing, and more specifically of speech labeling. Given some data, say the spontaneous telephone speech of two siblings talking to one another, we need to come up with a transcription that will reflect all of what is happening in each segment. During this conversation, many things may have occurred. The speakers interrupted one another, they coughed, one turned on the tap in the kitchen, a dog barked, and at one point, they became very upset with one another. We could ask a worker to transcribe the speech with the barge-ins, the extraneous noises (at least indicating that there had been some noise at a certain time), and the emotions that were manifested. If this is to be done in one pass, the worker would have to annotate: the words that were pronounced; parts

Table 2.6   Types and amounts of studies carried out on MTurk and other platforms.

Platform	Number of items	Type of items	Purpose
MTurk	100,000	Utterances	Acquisition
MTurk	257,658	Utterances	Labeling
MTurk	15,911	Audio clips	Labeling
MTurk	11,000	Dialogs	Assessment
MTurk	9,372	Dialog turns	Acquisition
MTurk	5,870	Utterances	Labeling
MTurk	3,052	Utterances	Labeling
MTurk	2,473	Utterances	Labeling
MTurk	1,183	Utterances	Labeling
MTurk	1,183	Utterances	Labeling
MTurk	1,000	Paragraphs	Acquisition
MTurk	910	Dialog calls	Assessment
MTurk	900	Audio snippets	Labeling
MTurk	445	Utterances	Acquisition
MTurk	169	Synthetic snippets	Assessment
MTurk	127	Preference sets	Assessment
MTurk	52	Audio clips	Labeling
MTurk	30	Paragraphs read aloud	Labeling
MTurk	28	Audio clips	Labeling
MTurk	20	Hours of speech	Labeling
Other	157,150	Items	Listening study
Other	103,458	Utterances	Labeling
Other	55,152	Utterances	Acquisition
Other	11,814	Dialogs	Labeling
Other	10,000	Utterances	Labeling
Other	2,094	Utterances	Acquisition
Other	1,229	Utterances	Labeling
Other	549	Utterances	Labeling
Other	421	System calls	Assessment
Other	350	Dialogs	Labeling
Other	50	Paragraphs read aloud	Labeling
Other	27	Hours of speech	Acquisition

of words; each emotion that is heard; and nonlinguistic sounds, thus attending to four subtasks at the same time. It is generally accepted in the speech community that this approach will provide a poorer transcription than if there is a separate task/pass for each of the four types of transcription. The worker who is trained for this labeling task will know four types of labels and a lot of information about how to label, probably too much to remember. So it is probable that that worker will forget some of the training and be slowed down considerably by having to go back and forth to the labeling instructions while working. With so many things to attend to at the same time, the worker will probably attend to each with less concentration than if there had been just one subtask. Although task simplification may seem evident here, the reader can still find that a few papers in the literature involved one-pass complex labeling tasks. As we will see in Chapter 4, it is fairly easy, no matter which platform you choose, to create several subtasks, and this may be designed so that it does not quadruple the work, time, or cost.

As a task designer, you should take a disinterested look at the task you are creating in order to determine its complexity:

- What are the decisions that the worker must make to accomplish this task?
- If there is more than one decision involved, is it possible to divide them into separate tasks?
- What would the instructions for each of these tasks be?
- Can these tasks be cascaded or should they be offered independently of one another?

2.6   Quality Control

As we have seen, there are both malicious workers and poor workers. This necessitates some quality control beyond the above-mentioned prequalification.

2.6.1   Was That Worker a Bot?

Later chapters will also discuss quality control as another criterion for the success of a task. While there are many willing and helpful members in a crowd, there are some who have more pernicious reasons to participate. Some of them are real and some are robotic agents, bots, designed to make simple decisions that imitate what a worker would do. Those decisions are often wrong. This issue is most often found in the case of remunerated tasks. Bots are often found working on the larger tasks, since a smaller task could be finished before there was enough time to program the bot. There are also workers who just want to get paid and will provide any random answer (e.g., pressing "A" all of the time on a multiple choice task) very rapidly. There are also workers who may be well intentioned, but whose work is poor. Many crowdsourcing papers have dealt with quality assurance issues. Novel methods of detecting outliers and fraud appear regularly. Some compare the crowd to experts, some compare workers in the crowd to one another, and yet others compare the work of one crowd to that of another. CrowdFlower and MTurk have a novel system that relies on the programmatic creation of a gold standard, a base of opinions that are known to be true, often from experts, which are compared to the opinions of the workers (Oleson *et al.* 2011). This standard is used both to give workers targeted training feedback and to prevent dishonesty. The authors show that it requires less manual quality management and improves overall results. Quality control has been discussed in detail in most of the chapters in this book.

2.6.2   Quality Control in the Literature

Discussions of quality control often compare the performance of the crowd to that of the experts. The first question that comes up is whether the nonexpert worker can perform as well as an expert. The literature in speech processing has shown thus far that the quality of the work of crowd is beginning to approach that of the experts. However, in fields that have a year or two more experience in the use of crowdsourcing, like machine translation, we see the results of the crowd have now attained the level of the experts. Zaidan and Callison-Burch (2011) used a variety of translation techniques and features that model both the translations and the translators. These features enabled them to discriminate good from poor work. When they used only the acceptable translations, the results were equal to those of professionals and came at significantly lower cost. In the speech processing papers that compared the performance of the crowd to an expert baseline, although the crowd sometimes approached the level of the experts, it has yet to surpass it. This is true for a classification task (in McGraw *et al.* (2010) and Parent and Eskénazi (2010), where experts show a slightly higher kappa), a transcription task (in McGraw *et al.* (2010), Gruenstein *et al.* (2009), and Evanini *et al.* (2010), where experts have slightly lower WER), or an acquisition task (in Lane *et al.* (2010), where experts produce more valid spoken utterances). The major way that has been found to reduce the gap between the crowd and the experts is through quality control. Thus, quality control should be central to the design of every task.

How can you implement quality control for the data you are collecting? Quality control involves the use of a set of measures that can, for some tasks, positively influence the quality of the data obtained. You can perform quality control at several different stages in the crowdsourcing process. **Before** the worker starts on the task, as we have seen, you can implement prequalification requirements such as work history, native language, and/or success on a prequalification task. Online filtering can be used **during** the task to assess the quality of the workers' production. Finally, quality control can be carried out **after** a worker has submitted all of their work.

Given the obvious importance of quality control, it is surprising to see that, of 31 papers that actually described studies where some sort of quality control could have been used, 8, or 26%, **did not use any quality control**. In some cases, this can be explained by the fact that the requester was running a "proof of concept."

Figure 2.7 shows, for the 23 papers where there was quality control, at which point in the task the requesters chose to implement it: before the task began, while the task was running ("during"), or after the task had been completed. The overwhelming majority of the

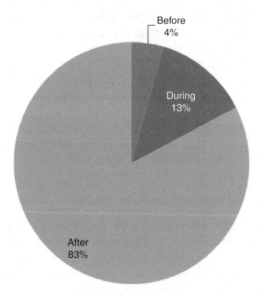

Figure 2.7   Point at which the requesters implemented some type of quality assessment (before, during or after the task).

tasks (83%) were assessed for quality after they had been completed. It should be noted that assessment during the task using MTurk has not been easy and that the papers that describe quality assessment during the task are among the most recently published. We believe that the tendency to assess during the task will increase as time goes on.

The paper that used pretask quality control simply used the performance on several items of the task itself as a filtering mechanism. Only workers who performed well on these items were accepted to go on to work on the whole set of tasks. The authors reasoned that if a worker can perform well on one subset, then they will continue to perform well and so no further control is needed. Others have looked to the data for the answer to this. They determined the number of items that would have to be answered correctly before they found a correlation with good-quality work.

Of the 19 papers that used posttask quality assessment, some used a gold standard. This means that there are known answers for some of the items that the workers complete that were obtained from experts. The experts commonly do only about 10% of the tasks in the whole dataset. The results of each member of the crowd are then compared to those of the experts. This can be done during the task, when the expert-answered questions are evenly distributed in the data (say one out of every 10 items). In this way, after about 30 items, a decision could be made about the quality of the worker based on agreement with the expert. Gold standard can also be used after the task is finished. Eight (25%) of all the papers (42% of the papers that used posttask assessment) described some gold standard. Another technique that has often been used in the literature is interworker comparison. This is where the variability of the work submitted by multiple workers is monitored for discrepancy (e.g., with a voting scheme, where, e.g., all workers agree but one). Table 2.7 shows that this was used in 36% of the tasks. Compared to the gold standard, interworker comparisons have the advantage of eliminating an extra set of human interventions. This type of approach assumes that most workers perform correctly, and so the aggregation of the workers' judgments will provide a good estimate of the true value, which can in turn be used to identify poor workers or cheaters. This unsupervised approach to quality control was, for example, used by McGraw et al. (2009) and Parent and Eskénazi (2010) with a voting scheme on a labeling task. Gruenstein et al. (2009), Parent and Eskénazi (2010), Marge et al. (2010a), Marge et al. (2010b), Novotney and Callison-Burch (2010), and Evanini et al. (2010) used string merging algorithms such as ROVER to sort the data, but not always for quality assessment. Interworker agreement was considered in some cases to be sufficiently accurate that it could be used alone to measure the quality of individual performance. The biggest issue in doing this is that the majority of the crowd could be wrong, in which case the final solution would also not only be wrong, but you could be penalizing

Table 2.7 Types of quality assessment used in tasks that did not use gold standard alone.

Technique	During the task	After the task
Intraworker	1	0
Interworker	0	8
Consistency of response	0	3
Compare to ASR output	1	4
Compare to another crowd	0	4
Confidence in response	0	2

good workers. For this reason, some of these studies either built new gold-standard datasets, or reused existing labeled datasets (e.g., from LDC, such as in Novotney and Callison-Burch (2010) and Audhkhasi *et al.* (2011a)).

Five of the tasks in Table 2.7 compare the work of the crowd to ASR output, assuming that they have sufficient quality output from the latter. In some cases, the ASR output is considered as one of the members of the crowd in an interworker comparison (e.g., Williams *et al.* 2011). This ASR output can be used to filter out poor work while the task is still running. This is advantageous for workers who are planning to spend a lot of time on your task. If they can be told early on that their work is not going to be accepted, they will save a lot of unpaid time (and you will have a better reputation among the workers). Table 2.7 shows that so far only two of the tasks used something other than gold-standard assessed workers during the task (although the prequalification might also be seen to meet this criterion). There are natively supported mechanisms on MTurk and on CrowdFlower to help with this. An example of this type of quality control is the "gold-unit" in CrowdFlower: a requester defines gold-units, which CrowdFlower inserts in the task that the workers complete. The worker receives feedback about whether they are completing the task properly (thus reinforcing their understanding of the task). This online quality control also shows the crowd that quality matters and that they are being monitored. It may be that the predominant use of MTurk, where it is not as easy to automatically insert gold-unit type checks, is the reason that this process has not yet been widely adopted by the speech community.

Four of the tasks assessed the work of a first crowd by passing their work to a second crowd that judges the quality of the work of the former as a separate task (like Yang *et al.* 2010). This has also been shown to increase the quality of the data.

One task, Ledie *et al.* (2010) used an intraworker comparison, and two tasks asked the worker for an opinion on an item, such as "How confident are you in your response?" (Parent and Eskénazi 2010), which was shown to be useful in detecting faulty items. Finally, for some tasks, it is possible to observe how consistent the worker was in their responses. This could entail logic between several of their responses (e.g., if they said yes to question 3, then they should also say no to 5), or giving the same answer to the same question when it is asked later in the same task.

Summing up, you can introduce quality control before, during, or after the task. You can use experts' judgments to make a gold standard, you can compare workers' output to one another, or you can use a second crowd to judge the output of the first. We have seen that the quality control mechanism introduces an extra cost, both in the preparation and processing of the tasks and in slowing down throughput. However, although even gold-standard datasets can contain errors and ambiguous labels, this extra measure ensures that the workers are assessed on a valid basis.

2.7   Judging the Quality of the Literature

From the information presented in this chapter, the reader can see that it is relatively easy to propose a study, gather some data and publish results. Yet if there is no quality control the results may be misleading. Although in short papers there may be little room for details, you should always look for evidence of

- Checking that audio input/output is working.
- Assessing worker qualification.
- Assessing the quality of the work.
- Making an effort to simplify each task.

Low kappa and other agreement scores and absence of the mention of the above points may cast doubt on the validity of the results.

2.8 Some Quick Tips

Here are a few quick tips gained from experience in using MTurk. They are intended to get your set of tasks noticed and make them more desirable than others.

First, **the name of the task** should be appealing. That is, instead of naming a task "Labeling speech," you could, for example, call it "Labels for calls to a bus information system," which tells more about what it is about. This not only lets the workers know more about the nature of the task but it also makes them curious about the application.

Second, **make it pretty**. Make the background, the font, the color, and other graphic elements look nice. This is just another subtle way to make your task more attractive than the others so that it is the one the workers choose to do.

Finally, have each task (HIT in this case) be **contained on one screen**. This means that the worker should not have to scroll down or to the side to complete the task. To do this, you can have a button that hides/presents the instructions. It can hide them after your worker has read them the first time and then recall them if they want to go back and read them again. You can change fonts and placement on the page as well. Why do this? The worker, when deciding what set of tasks to do, is taking into account the time it takes to do each task. Having to scroll takes up much more time and it makes the task look bigger. Making the task look as agreeable as possible will help attract more workers. Now that you are armed with some of the basics, you will find details on how to implement your tasks in the chapters that follow.

2.9 Acknowledgments

This work was supported by National Science Foundation grant IIS0914927. Any opinions, findings, and conclusions and recommendations expressed in this material are those of the author and do not necessarily reflect the views of the NSF.

References

Adda G and Mariani J (2010) Language resources & Amazon Mechanical Turk: ethical, legal and other issues in *Proceedings of LISLR2010 LREC2010*.

ADHACK. http://www.adhack.com/ (accessed 9 July 2012).

AgentAnything.com. http://www.agentanything.com/ (accessed 9 July 2012).

Akasaka R (2009) Foreign accented speech transcription and accent recognition using a game-based approach. Masters thesis, Swarthmore College, Swarthmore, PA.

Amazon Mechanical Turk—Artificial Intelligence. http://mturk.com (accessed 9 July 2012).

Audhkhasi K, Georgiou P and Narayanan S (2011a) Accurate transcription of broadcast news speech using multiple noisy transcribers and unsupervised reliability metrics. *Proceedings of International Conference on Acoustics, Speech and Signal Processing (ICASSP-2011)*.

Audhkhasi K, Georgiou P and Narayanan S (2011b) Reliability-weighted acoustic model adaptation using crowd sourced transcriptions. *Proceedings of Interspeech 2011*.

Bucholz S and Latorre J (2011) Crowdsourcing preference tests and how to detect cheating. *Proceedings of Interspeech 2011*.

Callison-Burch C and Dredze M (2010) Creating speech and language data with Amazon's Mechanical Turk. *Proceedings of NAACL-2010 Workshop on Creating Speech and Language Data With Amazon's Mechanical Turk*.

The Official CAPTCHA Site. http://www.captcha.net (accessed 8 July 2012).

CastingWords. http://www.castingwords.com (accessed 9 July 2012).

Chaordix. http://www.chaordix.com/ (accessed 9 July 2012).

Chernova S, Orkin J and Breazeal C (2010) Crowdsourcing HRI through online multiplayer games. *Proceedings of Dialog with Robots: Papers from the AAAI Fall Symposium.*

Cieri C, Miller D and Walker K (2004) The Fisher corpus: a resource for the next generations of speech-to-text. *Proceedings of 4th International Conference on Language Resources and Evaluation (LREC-2004).*

clickworker. http://www.clickworker.com (accessed 9 July 2012).

Cooke M, Barker J, Lecumberri ML and Wasilewski K (2011) Crowdsourcing for word recognition in noise. *Proceedings of Interspeech 2011.*

CrowdFlower. http://www.crowdflower.com/ (accessed 9 July 2012).

CrowdEngineering. http://www.crowdengineering.com (accessed 9 July 2012).

CROWDFUNDER. http://www.crowdfunder.com (accessed 9 July 2012).

CrowdTogether. http://www.crowdtogether.com/ (accessed 9 July 2012).

de Vries NJ, Badenhorst J, Davel MH, Barnard E and de Waal A (2011) Woefzela—an open-source platform for ASR data collection in the developing world. *Proceedings of Interspeech 2011.*

Eagle N (2009) txteagle: Mobile crowdsourcing, internationalization, design and global development. Lecture Notes in Computer Science, 2009, Volume 5623/2009, pp. 447–456.

ecycler. www.ecycler.com (accessed 9 July 2012).

Evanini K, Higgins D and Zechner K (2010) Using Amazon Mechanical Turk for transcription of non-native speech. *Proceedings of Conference of the North American Chapter of the Association of Computational Linguistics Workshop: Creating Speech and Language Data with Amazon's Mechanical Turk (NAACL-2010).*

Evanini K and Zechner K (2011) Using crowdsourcing to provide prosodic annotations for non-native speech. *Proceedings of Interspeech 2011.*

Facebook. www.facebook.com (accessed 9 July 2012).

Frank MC, Tily H, Arnon I and Goldwater S (2010) Beyond transitional probabilities: human learners apply a parsimony bias in statistical word segmentation. *Proceedings of the 32nd Annual Meeting of the Cognitive Science Society.*

Gelas H, Abate ST, Besacier L and Pellegrino F (2011) Quality assessment of crowdsourcing transcriptions for African languages. *Proceedings of Interspeech 2011.*

Gruenstein A, McGraw I and Sutherland A (2009) A self-transcribing speech corpus: collecting continuous speech with an online educational game. *Proceedings of ISCA Workshop on Speech and Language Technology in Education (SLATE-2009).*

Hughes T, Nakajima K, Ha L, Vasu A, Moreno P and LeBeau M (2010) Building transcribed speech corpora quickly and cheaply for many languages. *Proceedings of Interspeech 2010.*

innocentive. http://www.innocentive.com/ (accessed 9 July 2012).

Ipeirotis PG (2010) Analyzing the Amazon Mechanical Turk marketplace. *ACM XRDS: Crossroads* **17**(2), 16–21.

Jurcicek F, Keizer S, Gasic M, Mairesse F, Thomson B, Yu K and Young S (2011) Real user evaluation of spoken dialogue systems using Amazon Mechanical Turk. *Proceedings of Interspeech 2011*

Kittur A, Smus B and Kraut R (2011) CrowdForge: crowdsourcing complex work. *Proceedings of the ACM 2011 Annual Conference on Human Factors in Computing Systems.*

Kunath SA and Weinberger SH (2010) The wisdom of the crowd's ear: speech accent rating and annotation with Amazon Mechanical Turk. *Proceedings of the NAACL HLT 2010 Workshop on Creating Speech and Language Data with Amazon's Mechanical Turk.*

Lane I, Waibel A, Eck M and Rottmann K (2010) Tools for collecting speech corpora via Mechanical-Turk. *Proceedings of the NAACL HLT 2010 Workshop on Creating Speech and Language Data with Amazon's Mechanical Turk.*

Ledie J, Odero B, Minkov E, Kiss I and Polifroni J (2010) Crowd translator: on building localized speech recognizers through micropayments. *SIGOPS Operating Systems Review* **43**(4), 84–89.

Lee C and Glass J (2011) A transcription task for crowdsourcing with automatic quality control in *Proceedings of Interspeech 2011.*

Little G, Chilton LB, Goldman M and Miller RC (2010) Turkit: human computation algorithms on mechanical turk. *Proceedings of the 23rd Annual ACM Symposium on User Interface Software and Technology (UIST '10).*

Marge M, Banerjee S and Rudnicky A (2010a) Using the Amazon Mechanical Turk for transcription of spoken language. *Proceedings of International Conference on Acoustics, Speech and Signal Processing (ICASSP-2010).*

Marge M, Banerjee S and Rudnicky A (2010b) Using the Amazon Mechanical Turk to transcribe and annotate meeting speech for extractive summarization. *Proceedings of the NAACL HLT 2010 Workshop on Creating Speech and Language Data with Amazon's Mechanical Turk.*

McGraw I, Lee C-y, Hetherington L, Seneff S and Glass J (2010) Collecting voices from the cloud. *Proceedings of LREC 2010.*

McGraw I, Glass J and Seneff S (2011) Growing a spoken language interface on Amazon Mechanical Turk. *Proceedings of Interspeech 2011.*

McGraw I, Gruenstein A and Sutherland A (2009) A self-labeling speech corpus: collecting spoken words with an online educational game. *Proceedings of Interspeech 2009.*

microWorkers. http://www.microworkers.com/ (accessed 9 July 2012).

MiniFreelance. http://www.minifreelance.com/ (accessed 9 July 2012).

Munro R, Bethard S, Kuperman V, Lai VT, Melnick R, Potts C, Schnoebelen T and Tily H (2010) Crowdsourcing and language studies: the new generation of linguistic data. *Proceedings of the NAACL HLT 2010 Workshop on Creating Speech and Language Data with Amazon's Mechanical Turk.*

Novotney S and Callison-Burch C (2010) Cheap, fast and good enough: automatic speech recognition with non-expert transcription. *Proceedings of Conference of the North American Chapter of the Association of Computational Linguistics (NAACL-2010).*

Oinas-Kukkonen H (2008) Network analysis and crowds of people as sources of new organizational knowledge, in *Knowledge Management: Theoretical Foundation* (eds A Koohang *et al.*). Informing Science Press, Santa Rosa, CA, pp. 173–189.

Oleson D, Sorokin A, Laughlin G, Hester V, Le J and Biewald L (2011) Programmatic gold: targeted and scalable quality assurance in crowdsourcing human computation. *Proceedings of 2011 AAAI Workshop.*

OrganizedWisdom. http://www.organizedwisdom.com (accessed 9 July 2012).

Parent G and Eskenazi M (2010) Toward better crowdsourced transcription: transcription of a year of the Let's Go Bus information system data. *Proceedings of IEEE Workshop on Spoken Language Technology (SLT-2010).*

Polifroni J, Kiss I and Seneff S (2010) Speech for content creation. *Proceedings of SiMPE 2010.*

Raux A, Bohus D, Langner B, Black A and Eskenazi M (2006) Doing research on a deployed spoken dialogue system: one year of Let's Go! experience. *Proceedings of Interspeech 2006.*

Rayner M, Frank I, Chua C, Tsourakis N and Bouillon P (2011) For a fistful of dollars: using crowd-sourcing to evaluate a spoken language CALL application. *Proceedings of the SLaTE Workshop 2011.*

Ribeiro R, Florencio D, Zhang C and Seltzer M (2011) CrowdMOS: an approach for crowdsourcing mean opinion score studies. *Proceedings of ICASSP 2011.*

Roy BC and Roy D (2009) Fast transcription of unstructured audio recordings. *Proceedings of Interspeech 2009.*

samasource. http://www.samasource.org (accessed 9 July 2012).

Taskcn.com. http://www.taskcn.com (accessed 9 July 2012).

Tietze MI, Winterboer A and Moore JD (2009) The effect of linguistic devices in information presentation messages on comprehension and recall. *Proceedings of the 12th European Workshop on Natural Language Generation.*

Yang Z, Li B, Zhu Y, King I, Levow G and Meng H (2010) Collection of user judgments on spoken dialog system with crowdsourcing. *Proceedings of the IEEE Workshop on Speech and Language Technologies 2010.*

Wikipedia. http://www.wikipedia.com/ (accessed 9 July 2012).

Williams JD, Melamed ID, Alonso T, Hollister B and Wilpon J (2011) Crowd-sourcing for difficult transcription of speech. *Proceedings of IEEE Workshop on Automatic Speech Recognition and Understanding (ASRU 2011).*

Wolters M, Isaac K and Renals S (2010) Evaluating speech synthesis intelligibility using Amazon Mechanical Turk. *Proceedings of 7th Speech Synthesis Workshop.*

WordWave International. www.wordwave.co.uk (accessed 9 July 2012).

Zaidan O and Callison-Burch C (2011) Crowdsourcing translation: professional quality from non-professionals. *Proceedings of ACL-2011.*

Further reading

Black AW, Bunnell HT, Dou Y, Muthukumar PK, Metze F, Perry D, Polzehl T, Prahallad K, Steidl S and Vaughn C (2012) Articulatory features for expressive speech synthesis. *Proceedings of IEEE ICASSP2012.*

Dredze M, Jansen A, Coppersmith G and Church K (2010) NLP on spoken documents without ASR in *Proceedings of the 2010 Conference on Empirical Methods in Natural Language Processing.*

Freitas J, Calado A, Braga D, Silva P and Dias M (2010) Crowd-sourcing platform for large-scale speech data collection. In*Proceedings of FALA 2010*.

Goto M and Ogata J (2011) PodCastle: recent advances of a spoken document retrieval service improved by anonymous user contributions. *Proceedings of Interspeech 2011*.

Júdice A, Freitas J, Braga D, Calado A, Dias MS, Teixeira A and Oliveira C (2010) Elderly speech collection for speech recognition based on crowd sourcing. *Proceedings of DSAI'2010 DSAI Software Development for Enhancing Accessibility and Fighting Info-exclusion*.

Howe J (2006) The Rise of Crowdsourcing. Wired Magazine, Issue 14, 06 June 2006.

Liem B, Zhang H and Chen Y (2011) An iterative dual pathway structure for speech-to-text transcription. *Proceedings of Association for the Advancement of Artificial Intelligence Workshop* (AAAI 2011).

Luz S, Masoodian M and Rogers B (2010) Supporting collaborative transcription of recorded speech with a 3D game interface, in *Knowledge-Based and Intelligent Information and Engineering Systems*, Volume 6279 of Lecture Notes in Computer Science, pp. 394–401. Springer, Berlin.

Roy BC, Vosoughi S and Roy D (2010) Automatic estimation of transcription accuracy and difficulty. *Proceedings of Interspeech 2010*.

Schlaikjer A (2007) A dual-use speech CAPTCHA: Aiding visually impaired web users while providing transcriptions of audio streams. *Technical Report 07-014*, Language Technologies Institute, Carnegie Mellon University, Pittsburgh, PA.

Suendermann D, Liscombe J and Pieraccini R (2010) How to drink from a fire hose: one person can annoscribe 693 thousand utterances in one month. *Proceedings of Special Interest Group on Discourse and Dialogue (SIGDIAL-2010)*.

Wald M (2011) Crowdsourcing correction of speech recognition captioning errors. *Proceedings of W42011 Microsoft Challenge*.

Worsley M and Blikstein P (2011) What's an expert? Using learning analytics to identify mergent markers of expertise through automated speech, sentiment and sketch analysis. *Proceedings of EDM 2011*.

3

Collecting Speech from Crowds

Ian McGraw
Massachusetts Institute of Technology, USA

The collection of large quantities of speech data typically requires a considerable amount of time and expertise. Moving speech interfaces to the cloud to enable the crowdsourcing paradigm can accelerate the collection process by making such interfaces widely accessible; unfortunately, this advantage often comes at the cost of additional complexity that must be managed by the collector. This tension, between the resources of the collector to create cloud-based speech interfaces and those of speakers to actually access them, steers the growth of crowdsourcing as a tool for speech collection.

For some time, researchers in the speech-related sciences relied on monolithic solutions to collect speech directly in the laboratory. Bringing the speakers to the recording device has the distinct advantage of enabling the collector to control the acoustic environment and the microphone setup, and to monitor every aspect of the collection itself. Clearly, however, this solution does not scale.

Hungry for more data, researchers resorted to phone-based collection methods. Initially, this technique of speech collection required expensive, customized hardware and software, or perhaps even the assistance of a telephone company. Even so, the benefits of this technology sometimes out-weighed the complications, and large-scale speech-corpus collection was born.

Still, phone-based audio collection is only available to institutions that have the resources to configure and deploy this technology. Moreover, it limits the types of interactions that can be collected to those consisting entirely of audio, spurring some researchers to develop hybrid systems using both a website and a phone call for each session. Ideally, however, an all-in-one solution could be found.

For this reason, some researchers have begun turning to the web for their audio collection needs. Although the technology to collect speech over the web is still immature, it has the advantage of being widely accessible, and directly integrable into crowdsourcing platforms such as Amazon Mechanical Turk (MTurk).

Crowdsourcing for Speech Processing: Applications to Data Collection, Transcription and Assessment, First Edition.
Edited by Maxine Eskénazi, Gina-Anne Levow, Helen Meng, Gabriel Parent and David Suendermann.
© 2013 John Wiley & Sons, Ltd. Published 2013 by John Wiley & Sons, Ltd.

This chapter begins in Section 3.1 by giving a short history of speech collection, describing how the field has progressed to date. We assert that web-based speech collection, while not yet well explored, has the potential to vastly simplify audio collection efforts. We describe the current technology for microphone-enabled websites in Section 3.2 and point to future technologies, such as HTML5, which promise to simplify the task even further.

Ultimately, the goal of this chapter is to give the reader the basic tools necessary to perform their own speech collection tasks. We believe the web-based approach to be accessible to the largest number of interested parties, and so, in Sections 3.3 and 3.4, we provide an example audio recorder and server. Then, in Section 3.5, we outline the steps necessary to deploy this audio collection interface to MTurk, to crowdsource the collection of speech. We conclude this chapter with a look to the future in Section 3.7. We describe advanced crowdsourcing techniques that hint at the possibility of producing human-in-the-loop applications that learn speech-related tasks in a new domain.

3.1   A Short History of Speech Collection

If one defines the term broadly enough, the speech research community has been *crowdsourcing* the collection of speech for many years and in many different contexts. This section describes speech collection efforts for a diverse set of applications. Explicit corpus collections have been performed for a myriad of speech-related tasks: speaker verification, language identification, large-vocabulary speech recognition, and so on. In other cases, such as with widely deployed spoken dialog systems, the collection of audio is simply a by-product of using the system.

We describe what we believe to be a representative sample of this work to give the reader a feel for the collection techniques typically used for speech. For each collection effort, we also describe the technology used to make data collection possible and give a broad overview of the protocol used during collection. Speech collection can either occur on-site or remotely. The remote methods of data collection applicable to crowdsourcing include collection over a standard telephone, through the web, or via smartphones.

3.1.1   Speech Corpora

We begin with speech corpus collection, which is a task of central importance in speech research. These corpora provide a standard for comparison for speech-related research in addition to their utility for training various speech-related algorithms. Their very collection, however, is a feat in and of itself. A catalog containing many influential speech corpora is available through the Linguistic Data Consortium (LDC).

For certain purposes, it is desirable to collect audio in the laboratory using multiple microphones under ideal conditions. The *TIMIT* corpus of read speech, sponsored by DARPA, was one such effort (Fisher *et al.* 1986). A collaboration between SRI, Texas Instruments (TI), MIT, the TIMIT corpus was designed to contain phonetically varied read speech. Speech was collected from 630 English speakers using two microphones, one close-talking and the other a free-field microphone. Recording was carried out at Texas Instruments, and after some postprocessing, 16 kHz audio was shipped to MIT for phonetic transcription (Zue *et al.* 1990). To this day, *TIMIT* remains the most popular corpus carried by the LDC.

In subsequent decades, the general trend in speech data collection efforts has been toward collecting data remotely, rather than requiring the speakers to be on-site. This method comes with advantages and disadvantages. First, it can be more difficult, for instance, to ensure channel uniformity and a clean acoustic environment across all of the speakers. Second, remote data collection brings new technological challenges, since the recording software must be set up to handle simultaneous speakers. With such an infrastructure in place, however, it is easier for the collectors to connect with a larger number of participants and ensure greater diversity thereof.

Texas Instruments launched a wave of automated corpus collection with its voice across America (VAA) project (Wheatley and Picone 1991). In order to interface with the telephone network, the protocol required customized hardware and software to connect a PC to a T1 line. Typical of telephone speech, the audio was transmitted and recorded with an 8-bit μ-law encoding scheme. Speech from a normal telephone call could then be recorded to disk. The initial corpus collected contained 3700 speakers producing 50,000 utterances by reading prompted sentences given to them in a unique mailed letter. This collection technique was quickly brought to a larger scale.

The original *SWITCHBOARD* corpus followed in the footsteps of VAA with a similar, entirely automated collection protocol (Godfrey *et al.* 1992). This corpus was meant to contain spontaneous speech. Thus, when a subject called in, a database of participants was queried and an outgoing call was made in an attempt to find a conversational partner. A topic of conversation was suggested, and the participants were given the power to signal when recording should begin. A total of 2430 conversations were collected, which, on average, lasted around 6 minutes each, resulting in over 240 hours of audio. The participants self-rated the naturalness of their conversations on a Likert scale where (1) was very natural and (5) indicated conversation that felt forced. The average rating was 1.48, indicating that the collection methodology was successful in capturing natural, spontaneous speech.

In the mid-90s, countries from across the world were submitting their contributions to a set of corpora collected under the *PolyPhone* project (Bernstein *et al.* 1994; Kudo *et al.* 1996; Wang 1997). These projects also used PCs outfitted to collect audio over standard phone lines. In the *Macrophone* project in the United States, prompt sheets such as the one in Figure 3.1 were mailed to participants. Over the course of a 6-week period, the project was able to collect around 200,000 utterances from 5000 speakers. In Japan, a corresponding project later collected 8866 speakers and validated 122,570 utterances, discarding those that contained noise or hesitations.

Once the research community overcame the technological hurdle of how to automate the collection of large speech corpora, there remained the question of how to recruit participants affordably. Many collection efforts, such as for the NYNEX *PhoneBook*, found success using outside market-research firms to gather participants. The *PhoneBook* corpus now contains more than 1300 speakers reading around 75 isolated words each (Pitrelli *et al.* 1995). The LDC's *CALLHOME* corpus (Canavan *et al.* 1997), applied an innovative technique to the recruitment problem. For this collection, participants were given the ability to make free long-distance phone calls in exchange for having their voices recorded. When the *Fisher* corpus was collected (David *et al.* 2004), a more typical Internet marketing strategy was employed. In this case, however, after participants were registered, the system actively called them to request participation. This system-initiated protocol has been continued more recently in the Mixer corpora for speaker recognition (Cieri *et al.* 2007).

It is clear from these example corpora that the speech community has been crowdsourcing for quite some time. One novelty, however, is that we have begun to automate the verification

Figure 3.1 These are sample prompts from the American and Japanese contribution to the *PolyPhone* project. Countries were tasked with funding the collection of their own speech corpus. Many employed market-research firms to identify potential participants and mailed them prompts to be read aloud over the phone.

process in addition to the collection itself. A group from Nokia, for example, describe the *CrowdTranslator*, for which corpora in a target language are collected using prompts over mobile phones for low-resource languages (Ledlie *et al.* 2009). Verification was performed by comparing answers to redundant prompts given to workers within individual sessions.

3.1.2   Spoken Language Systems

While in some cases corpus collection is performed as an independent step from the speech systems built in the target domain, with many spoken language systems there is a chicken-and-egg problem in which data are needed to construct a system, but the system is also needed to provide proper context for data collection. With such applications, a bootstrapping method is often applied to iteratively edge closer to natural, in-domain data.

The Air Travel Information System (ATIS) domain exemplifies this paradigm well. Initially, a wizard-of-oz style collection procedure was used, whereby subjects brought into the laboratory were led to believe that they were speaking with a computer, when in fact a human transcriber was handling their flight-related queries (Hemphill *et al.* 1990). Sessions were collected at a slow pace of one per day for a period of about 8 weeks, yielding 41 sessions containing 1041 utterances. Later, the collection effort was distributed to multiple sites, including AT&T, BBN, CMU, MIT, and SRI (Hirschman 1992). By 1994, most sites were automating their collection procedure using the spoken dialog systems built for the domain (Dahl *et al.* 1994).

Around the same time, in Europe, similar efforts were in place (Peckham 1993). There was also interest in another travel domain. A system developed in Germany allowed members of the public to call in to get train time-table information (Aust *et al.* 1994). Described in this work are a number of pitfalls associated with making such systems widely available. First,

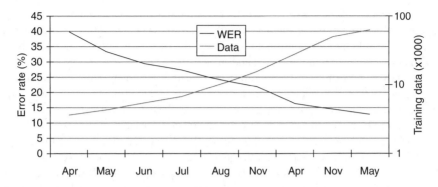

Figure 3.2   The plot, borrowed from Zue *et al.* (2000), depicts 2 years of development early in the life of the Jupiter weather information dialog system. As more training data were collected, new recognizers were released, causing a steady improvement in performance. Note the nonlinear *x*-axis corresponding to the release dates of updated systems. Reproduced by permission of Jim Glass.

there was a struggle to receive enough calls for development purposes before releasing to the general public. Second, technical difficulties with the telephone interface only allowed a single caller through at a time. Finally, the calls themselves varied enormously in almost every aspect. Some users would ask nonsensical questions of the system, background noise was a common phenomenon, and the variation in natural language, even for a response as simple as an affirmation, was unpredictable (e.g., "yes," "great," "sure," and "perfect," "okay").

In subsequent years, an increasing number of institutions began to tackle the task of making spoken language systems widely accessible. The DARPA *Communicator* project renewed interest in the air travel domain (Walker *et al.* 2001). With nine groups participating, this time a common architecture for dialog system development was used and a telephone interface was deployed for short-term data collection (Seneff *et al.* 1998). Examples of longer term deployments can be seen in AT&T's *How may I help you* customer care dialog system (Gorin *et al.* 1997), MIT's Jupiter weather information system (Zue *et al.* 2000), and CMU's Let's Go Bus information system (Raux *et al.* 2006).

The complexity involved in releasing a spoken language system into the wild is difficult to overstate. Maintaining a constant running suite of software that incorporates cutting-edge research is a tricky balance. Still, the benefits to deploying systems early and iterating on development are clear. Figure 3.2 shows the performance of the Jupiter weather information system over a number of development cycles and release dates. Adjusting the axis, it is clear that the word error rate decreases logarithmically with increasing amounts of data, reinforcing the old adage: "There's no data like more data."

3.1.3   User-Configured Recording Environments

In the last decade, researchers have begun to explore a new possibility for speech collection via lightweight client-configurable recording tools. Whether through the web or on a mobile device, these user-driven solutions come with a new set of advantages and disadvantages. Whereas the phone provides a somewhat consistent channel over which the voice is transmitted, collecting speech directly from a user's device requires individuals to record from the equipment they have at their disposal, whether it is a built-in microphone or a cheap headset microphone. Misconfigured microphones can also be a source of variability. An inexperienced user may not be able find the controls to turn down the recording volume, let alone have the wherewithal to recognize clipping in the audio.

Still, the advantages of recording audio in this fashion are many. First and foremost, the tools necessary to perform collection are easier to deploy and load-balance than those required for phone-based collection. An early example of web-based speech data collection comes from Germany, where a team gathered read speech from adolescents across the country using a downloadable audio recording tool called SpeechRecorder (Draxler 2006). SPICE (Schultz *et al.* 2007) and the web-accessible multimodal interface (WAMI) Toolkit (Gruenstein *et al.* 2008) were two early examples that also used Java to record audio directly from a web interface and upload it to a server behind-the-scenes. WAMI was even able to simulate streaming via the chunking protocol already available in HTTP 1.1. More advanced streaming, with data compression, can be performed through a Flash Media Server (FMS).

Other platforms rely on mobile applications (apps) to collect audio. Lane *et al.* for example, developed a suite of tools specifically for data collection through crowdsourcing venues such as MTurk (Lane *et al.* 2010). AT&T developed a platform for speech mashups for mobile devices (Di Fabbrizio *et al.* 2009). The development kit contained a native client for the iPhone and a plug-in for Safari that allowed speech applications to connect to a speech recognizer running remotely.

By attaching a remote speech recognizer to the client, users can interact with full-fledged spoken dialog systems from their personal computers or mobile devices. In 2007, *CityBrowser* was made publicly available using the WAMI Toolkit and e-mail lists were used to recruit subjects for a user study (Gruenstein and Seneff 2007). An iPhone app was later developed for the WAMI Toolkit, and *FlightBrowser*, the latest incarnation of the ATIS and Communicator projects, was given a mobile interface.

While mobile interfaces can be attractive, they can require a considerable amount of developer time and energy. This is compounded by the fact that the popular mobile phone operating systems of today are entirely incompatible with one another, making it difficult to minimize the effort of reaching a large audience through smartphones. The approach we have adopted at MIT is to build a generic webkit-based app capable of streaming speech to a server for iOS and Android. We then implement prototypes in JavaScript and HTML and run them on a desktop browser, or in one of these speech-enabled mobile browsers. This gives us the luxury of a write-once-and-run-anywhere deployment protocol.

The use of web technologies in crowdsourcing for speech is not a new phenomenon. In the mid-1990s, a team from MIT combined their telephone-based collection protocol with a web prompting interface to collect read speech (Hurley *et al.* 1996). This phone-web hybrid strategy has continued today in other groups with full-fledged spoken dialog systems. In Jurčíček *et al.* (2011) and Gašić *et al.* (2011), workers were given a number to call and a task to follow via the MTurk web interface.

The rise of Web 2.0 technologies allows for more dynamic interaction, however, when the speech recording software is embedded directly within the web page. YourSpeech from the Microsoft Language Development Center offers both a Quiz game and a text-to-speech (TTS) generator to incentivize users to participate. The TTS generator prompts users for speech and produces a concatenative speech synthesizer once enough data are received (Freitas *et al.* 2010). In McGraw *et al.* (2010), our flight information system was evaluated and more than

1000 dialog sessions were collected by deploying a unified graphical user interface (GUI) to Amazon Mechanical Turk, combining speech input with visual feedback. Another group subsequently used MTurk to evaluate a somewhat simpler interface for language learning (Rayner *et al.* 2011). For an overview of some of the speech acquisition tasks performed through modern-day crowdsourcing (see Chapter 2).

In the remainder of this chapter, we will focus on technology and applications of web-based crowdsourcing. In our view, this form of speech data collection, when supplemented with the worked examples provided in subsequent sections, has a relatively gentle learning curve. Our goal in this work is to make the process of setting up a speech collection interface for crowdsourcing accessible to even the novice programmer. Perhaps, by providing these tools, we can even help spread these types of large-scale speech collection efforts to other fields of study. We begin with an overview of the available technology.

3.2   Technology for Web-Based Audio Collection

As previously described, one clear advantage of collecting audio through a website rather than over the phone is that the collection task can be supplemented with visual information. To elicit read speech, static prompts can be formatted and displayed using HTML. Basic interactivity, such as the ability to click a button to skip a prompt, can be performed using JavaScript. Web 2.0 technologies, such as Asynchronous JavaScript and XML (AJAX), allow a website to communicate with a server to integrate a database or code that would otherwise be difficult to implement in the browser.

While most of the aforementioned technologies are relatively mature, it is surprisingly difficult to implement a web-based framework to enable the transmission of speech captured from a microphone to a remote server. This is because browsers currently do not offer native support for the microphone. While mobile browsers do not even support a plug-in architecture that allows third-party developers to incorporate native code, desktop browsers can be extended to allow for microphone access. Developing a plug-in, while perhaps the easiest way to control the look-and-feel of the interaction, requires the user to download and install code that they may be hesitant to trust. Fortunately, there are a number of plug-ins (e.g., Silverlight, Java, and Flash) developed by reputable third parties that make microphone access possible and which the user is likely to already have installed. Each, however, comes with its own set of caveats.

At first glance, it would seem that the server-side technology should offer greater freedom of choice. After all, the collection of audio need only be supported by a single server, whereas on the client side, a multitude of browser configurations must be supported for transmission. Unfortunately, there are times when the choice of client-side technology dictates that of the server side. Adobe Flash, for instance, can stream Speex-encoded audio from any of the popular desktop browsers. Currently, however, this comes at the high cost of requiring the installation of Adobe's proprietary FMS. This dilemma comes with its own work-arounds, again with their own costs.

In the remainder of this section, we will delineate the current state of technology regarding audio collection from the web. We begin with the three desktop browser plug-ins that currently enjoy the greatest global market penetration: Silverlight, Java, and Flash. In an ideal world, none of these plug-ins would be required, and browser vendors themselves would implement a standard for microphone access, and perhaps even audio streaming. Thus, we next discuss the current state of the HTML5 specification with respect to these developments, but ultimately, we come to the conclusion that native microphone support in all of the major browsers is unlikely in the near future. Since the plug-in solution does not work for most smartphones, we end this section with a work-around involving the development of native apps.

3.2.1   Silverlight

Microsoft Silverlight was first released in 2007 as a means of creating rich Internet applications using a subset of the .NET framework. There are official plug-ins for all of the major browsers on both Windows and Macs. There is even an open-source version available for Linux, although there is no guarantee that it contains the features necessary for audio collection given that microphone support itself was only recently added to Silverlight in its fourth release in 2010.

Microsoft claims that more than 60% of the browsers accessing the web today have Silverlight installed. This may be due to the success of Netflix, which uses Silverlight to stream movies to consumers over the web, in addition to its movies-by-mail service. It is also quite likely that this statistic varies by country. Silverlight currently does not have quite the name recognition of Adobe's Flash framework, but from the user's perspective, it is just as easy to install and provides many of the same capabilities. Note that, while it was at one point supported on Windows smartphones, Silverlight's future in the mobile arena is far from certain.

In addition to providing microphone access, Silverlight can communicate with the containing web page via JavaScript, or to a remote server using HTTP requests. HTTP stands for Hypertext Transfer Protocol and serves as the web's standard request–response protocol. An HTTP POST in particular can be a useful way to transmit audio data across the wire to an HTTP server. A user interface to control starting or stopping the recording process can be created within the Silverlight application itself, or using JavaScript and communicating the relevant actions directly to Silverlight. Regardless of the look and feel of the rest of the interface, microphone access comes with one aspect that third-party developers have no control over. The alert message below pops up the first time a website requests the microphone through Silverlight:

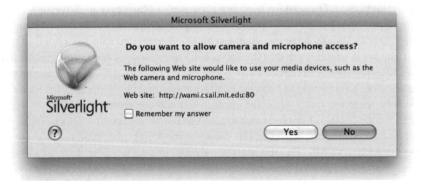

Silverlight application development is only well supported on Windows machines for which Microsoft's development tools are available. The minimum set of tools necessary to develop a Silverlight application are free, but support is greatest in Microsoft's premium integrated development environment (IDE), Visual Studio. For other platforms, *ad hoc* solutions do exist, including an open source .NET development framework called Mono, which appears to have earned a sizable following. Such solutions, however, inevitably lag behind the officially supported software development kit (SDK).

3.2.2 Java

Java Applet technology has been around since the mid to late nineties, when applets quickly became the most popular way to add dynamic content to a website. It was not long before the larger websites of the day were incorporating Java in high-traffic applications. For example, most of the original games on *Yahoo!* were originally implemented in applet form and executed in the java runtime environment via the Java browser plug-in.

Even today Java has retained a surprisingly large portion of the plug-in market share. Currently, Java's coverage on desktop browsers is on par with, if not slightly better than, that of Silverlight. Although somewhat more cumbersome for the user to install and operate, once up and running, a Java applet provides the microphone access necessary for recording-related tasks. Indeed, Java has been explored by a number of researchers as a means of collecting speech from the web (Gruenstein *et al.* 2008; Lane *et al.* 2010).

Java runs in a sandbox environment, meaning that it does not have access to certain system functionality by default. Of interest to us is that the microphone falls under these special security considerations, requiring us to take additional action. Just as in Silverlight, the solution is to present the user with a pop-up message that requests permission to access this feature. Unfortunately, as can be seen in the dialog below, there is no way to specify that only the microphone is of interest, and the message can thus appear ominous to some users.

To be more precise, even presenting the security warning above is not quite so simple. Applets requesting access to features outside the restricted sandbox are required to *sign* their code. Typically, this involves sending a certificate signing request (CSR) to a certificate authority such as VeriSign. For between $500 and $1000, the certificate authority will then return a public key, which can be incorporated into the Java Archive file that contains the applet. When the applet is accessed by a user, the certificate authority's servers are contacted to verify the identity of the code's signer.

Understandably, the signing process may be prohibitively expensive to some. Fortunately, there is a free work-around that comes at the cost of a slightly more daunting security message.

In particular, it is also possible to *self-sign* an applet. In this scenario, no third party is required to generate a pop-up message asking the user for microphone permissions. The downside, however, is that, with self-signing, the already ominous message becomes even more daunting:

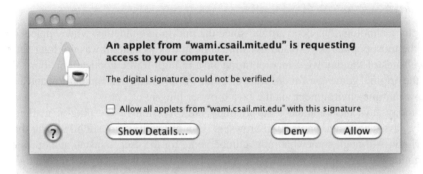

Aside from the aforementioned code-signing issue, Java is arguably the most developer-friendly of the plug-ins we explore here. The Java SDK can be installed on any operating system, and advanced developer tools are both free and crossplatform. The Eclipse IDE, for instance, offers on-the-fly compilation and full-featured development tools for Java. With the application of a Java-based server-side technology, such as Apache Tomcat, most of the development for a recording application can be written in a single language.

3.2.3 Flash

Through much of the last decade, rich Internet applications transitioned away from the heavy-weight Java applet architecture and (where JavaScript alone would not suffice) settled on Adobe's relatively lightweight Flash solution. Currently, Flash can be found on the web in many forms including games, video players, and advertisements. Some popular sites, most notably YouTube, have begun making the switch to HTML5; however, others, including Hulu, have continued to rely on Flash.

Flash enjoys the largest market of desktop browser plug-ins, with more than 95% coverage across the most popular browsers. In fact, Google's Chrome browser comes with Flash pre-installed. For this reason, Flash is arguably the audio collection method that offers the fewest impediments from the user's perspective. As with the Java and Silverlight solutions, Flash can communicate with JavaScript and can even be hidden from view, allowing the website to appear plug-in free, save for the initial microphone permissions security panel. Even the microphone permission settings are relatively unobtrusive:

Unlike Java or Silverlight, the security settings dialog box for Flash appears embedded directly in the web page. While this is convenient for the client, this can make development difficult unless the width and height of the Flash content is kept at least as large as the 214 × 137 pixel security window. Resizing or moving the Flash object will work in some browsers, but not in others, where such actions seem to reload the content, losing the security settings. One solution to these issues is to force the user to check the *Remember* box, which ensures that microphone permissions are retained through page refreshes or the resizing of the Flash content.

Once microphone permissions are obtained, getting the audio from the browser to a server is fraught with additional difficulties. Flash was designed to stream content using the real-time messaging protocol (RTMP) to the proprietary FMS. With the purchase of an FMS license for around $1000, one can stream Speex-encoded audio over the web using the Adobe-sanctioned method. For audio collection, perhaps this meets the requirements. For an interactive application, however, one still needs to figure out how to access the audio stream once it gets to the FMS, in addition to decoding it and passing it along for further processing.

Less expensive media server options do exist. Wowza Media Server is a Java-based server that can be used to stream content to a variety of devices. However, audio collection still requires that the client have access to the microphone. Although it has since expanded, Wowza was originally built to support RTMP and can thus be used to handle Speex audio data collected through Flash. Licenses for Wowza are generally more flexible than those of FMS. There even exist daily or monthly licenses that can be used in conjunction with Amazon's Elastic Compute Cloud (EC2).

Finally, there is an open-source media server called Red5 that offers a free alternative to Wowza or Adobe's FMS. Red5 is also built on Java and provides support for RTMP. While Red5 supports the functionality necessary for collecting audio, the support and documentation are likely to be lacking in some respects relative to those of the licensed media servers.

Regardless of pricing or licensing issues, media servers supporting RTMP are necessarily far more complicated than necessary for what ought to be a relatively simple task of transporting audio from client to server. One of the reasons is that RTMP was designed to transport audio in real time, dropping frames if necessary. Whereas applications like voice chat have strict real-time constraints, many applications for audio do not. Certainly, collecting audio to be processed off-line need not impose latency restrictions on incoming audio. Even a speech recognizer that is run on-the-fly is arguably easier to configure if one does not have to consider arbitrarily dropped frames from a protocol such as RTMP.

A simpler approach, albeit with the potential for a certain amount of latency, is to use an HTTP POST to submit audio data from the Flash client to a server. This technique eliminates the need for a media server altogether, allowing any HTTP compliant web server to collect the audio. Adobe made this possible in late 2008 with the release of Flash 10. Previous versions of Flash did not give direct access to audio samples on the client side, forcing developers to open a `NetStream` that piped audio from the microphone directly to an RTMP compliant media server. Now, however, the samples can be buffered on the client side and sent to a web server via a standard HTTP POST.

Later, we will describe the WAMI recorder, which is an open-source project that makes use of this technique to transport the audio. There do exist a few caveats to using the HTTP POST approach; however, the largest of which is probably that Speex-encoded audio data has *not* yet been made available to the client-side code. In other words, ActionScript, the scripting language used to program Flash content, only has access to raw audio samples. While it is easy to wrap uncompressed audio into a container, such as WAV, there is not yet a client-side solution to compressing the audio using a codec such as Speex, unless one is willing to use RTMP. Other lossy compression techniques, such as the μ-law encoding often used in telecommunications, might be straightforward to implement in ActionScript, but only cut file sizes roughly in half.

Development in Flash is not as straightforward as in Java, but fortunately, it is not quite as platform dependent as Silverlight. Adobe's full-fledged IDE, Flash Professional, costs a few hundred dollars, but is equipped with the functionality necessary to create complex animations as well as to write, compile, and debug ActionScript code. Fortunately, a free alternative exists for those of us primarily interested in application development rather than fancy animations.

Flex is a free open-source application framework also created by Adobe. Unlike Flash Professional, it is not particularly well suited to creating movies and animations, but suffices for compiling ActionScript and generating a SWF file, which can be embedded in a site as a Flash object. The Flex SDK is command-line driven; however, Adobe does release a modified version of Eclipse called Adobe Flash Builder that provides all the amenities of a typical IDE. An education license for Adobe Flash Builder is free, but the general public must pay a hefty fee, and for this reason may prefer to stick with the command-line tools (CLT).

3.2.4   HTML and JavaScript

Most computers have audio recording software preinstalled. Thus, one solution to collecting audio over the web is to require the user to figure out how to record the audio themselves, and then provide a means of uploading the audio through an HTML form. Clearly, this greatly restricts the types of applications into which one can incorporate audio recording. Without a plug-in, however, there is currently no way to access the microphone on any of the major browsers. Still, the future of microphone access in the web browser lies in JavaScript and HTML itself. However, it is unclear how far away this future lies.

The World Wide Web Consortium (W3C), founded and headed by Tim Berners-Lee in 1994, was created in part to guide the development of web standards such as HTML and to ensure compatibility across browsers. A specification goes through a number of draft stages before being implemented, reviewed, and finally ratified in a W3C recommendation. It is within these documents that the fate of microphone access within HTML and JavaScript will be settled. Ultimately, however, it is up to the browser vendors to implement the specifications.

In early 2004, discontent with the W3C's handling of the HTML specification motivated the formation of the Web Hypertext Application Technology Working Group (WHATWG). Their standardization efforts have since been adopted back into the W3C's HTML working group and renamed HTML5. These two working groups continue to develop their mostly overlapping specifications in parallel. However, the WHATWG has dropped the term HTML5 and now refers to its specification as a *living* standard for HTML, noting that pieces of the specification (e.g., the `<canvas>`, `<audio>`, and `<video>` tags) have already been implemented in a number of browsers. The W3C, on the other hand, still intends to issue an official HTML5 recommendation; however, recent official estimates put the release date of such a recommendation sometime in 2014.

Ideally, HTML or JavaScript would provide a standard mechanism for accessing the microphone regardless of the particular browser, mobile or desktop, and this would entirely alleviate the need for plug-ins. Perhaps the standard would be developed along lines similar to the way that browsers are beginning to support the $<$ audio $>$ tag for *playing* audio. However, standardization is a lengthy, complicated process involving the often competing, rarely spoken motives of several interested parties, especially the well-known browser vendors: Microsoft, Apple, Google, Mozilla, and Opera. Indeed, it is most often employees of these companies who serve as members of the aforementioned working groups.

To provide a concrete example of the difficulty in standards development, one need only to look at the working draft of the $<$ audio $>$ tag for playback. As of this writing, there is no single audio format supported across the five most popular browsers. MP3, for instance, is supported by the latest versions of Internet Explorer, Chrome, and Safari, but is not supported in Firefox and Opera, presumably for reasons having to do with the license. Surprisingly, uncompressed audio does not fare much better. The WAV format, a simple container for linear PCM audio data developed in part by Microsoft in the early nineties, is supported by four out of five major browser vendors. The lone holdout is Microsoft Internet Explorer. Hopefully, as the working draft reaches maturity, at least one format will be supported by all of the major browsers.

Currently, the proposal on the table for microphone access, the getUserMedia() function, falls under the purview of the W3C WebRTC working group, whose stated mission is "to enable rich, high-quality, real-time communications (RTC) applications to be developed in the browser via simple Javascript application programming interfaces (APIs) and HTML5." The project is supported by Google, Mozilla, and Opera, which thus represents buy-in from a majority of the browsers accessing the web today. In theory, this API would allow one to collect audio in JavaScript and submit it to a server using a simple XMLHttpRequest, which would appear on the server side as an HTTP POST containing the audio data. To date, however, this portion of the specification is still in a very early draft, and has not been implemented within any officially released browsers. For the time being, it is clear that we will have to rely on plug-ins for our web-based audio recording needs.

3.3   Example: WAMI Recorder

In this section, we describe the WAMI recorder, an open-source Flash recording tool that is wrapped in a simple JavaScript API. The original WAMI toolkit (Gruenstein *et al.* 2008) served primarily as a means of transporting audio from the browser to a speech recognizer running remotely. Initially, a Java technology, it has since been rewritten in Flash and the recorder itself, has been placed into a small open-source project, making it easy to adapt it for a variety of uses. The source code for the WAMI recorder can be checked out from a Google Code repository found at the following URL:

```
https://wami-recorder.googlecode.com
```

3.3.1   The JavaScript API

The simplest WAMI recorder example might consist of just three files. First, an HTML file would define a few buttons to start and stop recording. The HTML file would then include a JavaScript file to set up the recorder, which resides in the third SWF file. The JavaScript would also be responsible for linking the button clicks to actions in the recorder.

The example project we have created has a few more basic features. To ease the process of embedding the Flash into the web page, the project relies on an external dependency called SWFObject. Fortunately, this code is just a single, well-maintained JavaScript file, which is reliably hosted by third parties, meaning it can be conveniently included by linking to those resources directly.

We also provide a somewhat more appealing GUI than can be created using HTML alone. With a single PNG image containing possible button backgrounds and foregrounds, one can duplicate and manipulate the master image in JavaScript to create attractive buttons that even boast a special effect. In particular, when audio is recorded through the microphone, a few JavaScript tricks can be used to display an audio level meter by adding some red to the microphone's silhouette. The same can be done in green for the playback button. When a button is disabled its silhouette becomes gray:

Before this GUI is put into place, however, we must ensure that the proper microphone permissions have been granted. To do this, we need to check whether WAMI already has the appropriate permissions, and, if not, show the privacy settings panel. The security settings are the only piece of the recorder for which the user must interact directly with the Flash itself. It may be worthwhile at this stage to encourage users to click the *Remember* button in the privacy panel in order to avoid going through this procedure every time the page is refreshed.

While granting permissions in the privacy panel is essential, other panels in the settings might be of interest as well. The microphone panel in particular can be useful for selecting an alternative input source, adjusting the record volume, or reducing echo. The following is a screenshot of this panel:

Once the microphone permissions have been granted, the application has full access to the core WAMI recording API though JavaScript. This API consists of six functions that lie in the `Wami` namespace and control actions related to both recording and playing audio:

```
Wami.startRecording(myRecordURL);
Wami.stopRecording();

Wami.startPlaying(myAudioURL);
Wami.stopPlaying();

Wami.getRecordingLevel();
Wami.getPlayingLevel();
```

WAMI's `startRecording` function will begin recording and prepares to perform an HTTP POST of the audio data to the URL provided. When `stopRecording` is called, recording ends and the data are sent. Note that this means audio is buffered on the client during recording, which may make long recordings or continuous listening unfeasible. In a later section, we will describe a work-around that we have implemented to simulate streaming, and list the inevitable caveats that come with trying to perform an action Flash was not designed to handle without RTMP.

The `startPlaying` function also takes a URL, but in this case, the HTTP request made is a GET. The server is then expected to place a WAV file in the body of the response. This is precisely what a browser does when one types the URL of a WAV into the address bar and hits enter. The Flash downloads the WAV to the client and plays it back. The `stopPlaying` function can be used to stop the audio in the middle of playback.

The remainder of the API just supplements the recording and playing functions with a few additional features. The remaining two functions in the list above provide access to the microphone activity level. In our sample GUI, these two functions help animate the buttons when recording and playing so that the silhouettes in the buttons can act as audio meters. There are also a few optional parameters not shown above. The `startRecording` and `startPlaying` functions also accept arguments that specify callbacks that will fire when an action starts, stops, or if an error occurs. This can be useful, for instance, when an application needs to play two short audio files in a row, or simply change in the visual content of the page after audio playing has stopped.

3.3.2   Audio Formats

Audio processing vocabulary (e.g., "format," "encoding," and "container") tends to cause a fair amount of confusion. An *encoding* refers to the way the bits represent the audio itself, whereas the *container* is a wrapper for the audio bits that specifies metainformation. Confusing everything is the word *format*, which can be used to refer to the container, the encoding, or both together.

An audio container is usually capable of holding more than one audio encoding. Encodings can be lossless, in which case the original high-fidelity audio data can be recovered, or lossy, which generally discards some of the information in return for a better compression. To be completely clear what audio format one is referring to, it's best to specify both a container and an encoding.

The primary container used in the WAMI recorder is WAV. A second, less well-known, container that we support is called AU. Both can be used to store uncompressed audio in the linear pulse code modulation (PCM) representation, as well as other encodings such as G.711. Released in 1972, G.711 is a standard for audio companding commonly used in telecommunications. In North America, μ-law encoding is used to allow a sample to be encoded in 8-bits rather than 16, resulting in lossy compression. Although the μ-law algorithm would not be difficult to implement in ActionScript, WAMI currently leaves the PCM samples unaltered.

Uncompressed PCM audio generates files that are larger than one might wish, making bandwidth consumption a limitation of this recorder. Lossy encodings generally solve problems with network constraints; however, as mentioned previously, Flash currently makes compression in ActionScript difficult for all but the most basic algorithms. MP3 encoding, for example, is a computationally intensive task that would likely require the ability to run native code, assuming that the licensing constraints were not a problem. Speex, on the other hand, is an open-source codec that is well suited to the task of the compression of voice data. Often found in the OGG container format, Speex is also supported in Flash using the FLV container. Unfortunately, Speex has only been made available for recording to a FMS, and currently cannot be used to send audio via a POST request.

Adobe has given developers a preview release of an architecture called Alchemy that may offer a solution to the encoding problem. Alchemy allows developers to compile C and C++ code targeted to run in the sandboxed ActionScript virtual machine. It would be feasible, then, to port an efficient audio encoder using Alchemy to be efficiently executed on the client side. Alchemy has been released as a preview, but may not be officially supported until late in 2012.

With these caveats in mind, there is still a way to control the file size of the audio data in the WAMI recorder by manipulating the sample rate used. The following sample rates are supported by Flash's `flash.media.Microphone` object: 8 kHz, 11.025 kHz, 16 kHz, 22.05 kHz, and 44.1 kHz. We recommend a minimum of 16-bits per sample, unless a lossy compression algorithm such as μ-law is employed.

Playing the audio back to the speaker is often a requirement of an audio collection application, if only to let the speaker know that their microphone configuration is working. Unfortunately, for playback, Flash only accepts uncompressed audio at 44.1 kHz, making resampling necessary for the lower rates. Resampling a rate that divides evenly into 44.1 kHz can be approximated easily in ActionScript, but resampling 8 kHz and 16 kHz sound is more problematic. With additional server-side code, the resampling task can be performed remotely. Alternatively, the playback task could be left to the browser using HTML5's `<audio>` tag. Recall that WAV playback is currently supported by all the major browsers except for Internet explorer.

3.4   Example: The WAMI Server

One of the most appealing features of the WAMI project is that it can be used with a variety of server-side technology. We have experimented with implementations that use Java in Apache Tomcat, PHP, pure python as well as python within the Google App Engine (GAE). In this section, we describe two of these configurations in detail. First, we show a simple PHP script that can run from any PHP-capable web server. Since there are countless ways to set up such a server, we leave the general configuration to the reader. For those who have no experience with setting up a server whatsoever, we provide a second example that makes use of the GAE. In this example, the server-side configuration is standardized by Google, and friendly tools are available for debugging and deploying the web service. Finally, we conclude this section with some additional considerations that may be important for some nonstandard use cases of WAMI.

3.4.1   PHP Script

As a simple case-study of how one might use WAMI in practice, we will take a look at a simple PHP implementation of the necessary server-side code. Suppose that we have a file called `collect.php` with the following contents:

```php
<?php
parse_str($_SERVER['QUERY_STRING'], $params);
$name = isset($params['name']) ? $params['name'] : 'output.wav';
$content = file_get_contents('php://input');
$fh = fopen($name, 'w') or die("can't open file");
fwrite($fh, $content);
fclose($fh);
?>
```

This six-line PHP script is enough to implement a basic back-end for WAMI. PHP is a server-side scripting language typically used to generate dynamic web pages. To allow the WAMI recorder to POST data to the script, it must be made available at a particular URL using a PHP-enabled web server. Apache is one such server that can be outfitted with PHP. Assuming such a web server was running on port 80 (the port used most often for HTTP requests) and that a file called `collect.php` was hosted at the root, the following line of code in JavaScript would initiate recording to it:

```
recorder.startRecording('http://localhost/collect.php');
```

Assuming that the proper file-writing permissions were in place, the PHP code would generate a file called `output.wav`. The audio can be made available easily for playback by ensuring that its location is accessible from the web using the same server. With a typical PHP installation under Apache, the audio in our first example would become available at `http://localhost/output.wav`. Of course, in this example, if we were to record again, the audio would be overwritten. For that reason, it can be useful to define a mechanism for naming the files. The PHP code above handles this by parsing the URL's query parameters, which are separated from the rest of the URL using a "?":

```
var name = userID + "." + sessionID + "." +
utteranceID + ".wav";
recorder.startRecording('http://localhost/collect.php?name=' + name);
```

Note that this description is just for illustrative purposes, and does not take into concern the security risks of setting up such a server. Suppose for instance, that a malicious user performed a recording targeted to a URL with `name=collect.php`, overwriting the script with their own content. In practice, it might be wise to ensure on the server side that the file name follows a prespecified format, and ends up in a predetermined directory.

One can use the file name as a simple means of storing metainformation about the audio. The example above depicts a possible file name format that includes a user ID, a session ID, and an utterance ID, all of which are defined and maintained in the JavaScript. The user ID might be stored in a browser cookie, which saves information across page reloads. The session might remain the same for the duration of the time that a particular WAMI-enabled page is loaded. Finally, the utterance ID would be incremented each time the user recorded new audio.

3.4.2   Google App Engine

To show the flexibility and simplicity of implementing WAMI's server side, here we provide
a second example of a WAMI server. Unlike the PHP example, however, we provide a single
standard setup procedure, streamlined by Google, so that the novice web programmer can
begin collecting audio without struggling with the difficulties of hosting a server. The server-
side code is then hosted using Google's cloud technology called GAE. We have open sourced
the necessary code and placed it in a Mercurial repository accessible on Google code:

```
https://wami-gapp.googlecode.com
```

Signing up for a GAE account is almost as easy as creating a Google account. However,
there is a verification process that requires the creator to provide a cell number to which a
verification code will be sent. Once the account has been created, the account holder can create
up to 10 applications. Each application must be given a unique name and descriptive title.

There are two main commands that one uses when developing on GAE in python:
`dev_appserver.py` and `appcfg.py`. However, on Mac and Windows machines it can
be more convenient to download the SDK, which abstracts these commands away, providing a
convenient GUI for development. The launcher can run the application locally to ease debugging, and the SDK console provides an interface to data stored locally so that one's quota is
not wasted during development.

For small amounts of audio collection, use of GAE is free. Each GAE account begins with
5 GB of free storage space, which comes to over 6 hours of audio recorded at 22,050 Hz.
Another 5 GB is less than $1 per month. When collecting audio on MTurk, the subject of the
next section's tutorial, paying the workers will undoubtedly dominate the cost of collection,
relative to the small amount of money one might have to spend to use Google's cloud service.

There are actually two types of data storage on GAE that will be of interest for our
application. The first is the *blobstore*. The term BLOB stands for Binary Large OBject; perfect
for the audio data we intend to collect from the client. The second storage mechanism on GAE
is called the *datastore*. The datastore typically contains smaller, queryable information such
as integers, strings, and dates. This type of storage can be useful for saving metainformation
about the audio.

Web services hosted on GAE can be implemented in one of three programming languages.
However, the data storage APIs are currently only available in two: Java and Python. This
example will make use of Python, since it is typically less verbose. We begin with the simplest
possible service capable of collecting and replaying a single audio recording:

```python
from google.appengine.ext import webapp
from google.appengine.ext.webapp import util

class WamiHandler(webapp.RequestHandler):
    type = ""
    data = []

    def get(self):
        self.response.headers['Content-Type'] = WamiHandler.type
        self.response.out.write(WamiHandler.data);
```

```
def post(self):
    WamiHandler.type = self.request.headers['Content-Type']
    WamiHandler.data = self.request.body
```

Overriding the `RequestHandler` superclass allows us to handle HTTP requests such as POST and GET. Note that there is nothing specific to audio in the `WamiHandler` class shown above. We are simply storing the bits of the request body along with the content type in the POST handler. In the GET handler we set the content-type in the response header and serve the data by writing it to the response. Obviously, this simple servlet does not yet store the data to a permanent location, but it does provide a feel for the ease of working within the GAE framework.

The next step is to deploy our simple script to test it out. If we suppose that the `RequestHandler` above is implemented in a file called `simple.py` along with some scaffolding to expect requests at the path `/audio`, we can set up the server in just a few quick steps. YAML, a human-readable serialization standard, is used to specify mappings between URLs and scripts or static content such as HTML and JavaScript. In fact, we can host the WAMI recorder, including the JavaScript, SWF file, GUI images, and any HTML content, using the GAE:

```
application: NAME
version: 1
runtime: python
api_version: 1

handlers:
- url: /audio
  script: python/simple.py

- url: /client
  static_dir: public/client
```

The `.yaml` file above declares a web service called `wami-recorder` along with a few URL mappings. `http://wami-recorder.appspot.com/audio` points to the python and `http://wami-recorder.appspot.com/client` points to a folder containing the WAMI recorder. No additional configuration is necessary to host this simple example. In general, the application is deployed to a Google-managed address `NAME.appspot.com`, where NAME is the unique application identifier chosen upon its creation. In addition to the `simple.py` script described above, we have also included `sessions.py`, which is a blobstore-capable script. With a few more lines of code, it is relatively easy to create a blob and place it into more permanent storage. Finally, a small change must be made to the `app.yaml` file to change the handler for the `/audio` path:

```
- url: /audio
  script: python/sessions.py
```

One of the advantages of using a well-packaged service like the GAE is that there are developer features that make it easy to visualize the back-end of the application. The dashboard of GAE graphs is the number of requests per second an application receives over the course of its deployment.

A *Datastore Viewer* and a *Blob Viewer* give the developer a peek into what is being stored. Perhaps most importantly for someone collecting audio using this service, clicking on the link to a (moderately sized) file in the Blob Viewer will allow the developer to play the corresponding audio without explicitly downloading it. Chrome, for instance, recognizes the `audio/x-wav` type that the WAMI recorder sends and that the GAE saves. Thus, clicking a link will play the audio directly in the browser using a simple user interface: . This feature can be quite useful for keeping track of a remote collection task. Below is an image of the BlobViewer with a few utterances in it:

File Name	Content Type	Size	Creation Date
20ff869a9bb-4	audio/x-wav	61.4 KBytes	2012-02-12 20:46:14
20ff869a9bb-3	audio/x-wav	81.4 KBytes	2012-02-12 20:46:05
20ff869a9bb-2	audio/x-wav	87.4 KBytes	2012-02-12 20:46:00
20ff869a9bb-1	audio/x-wav	61.4 KBytes	2012-02-12 20:45:54
20ff869a9bb-0	audio/x-wav	143.4 KBytes	2012-02-12 20:45:42
90c0bd315d4-4	audio/x-wav	61.4 KBytes	2012-02-12 20:45:27
90c0bd315d4-3	audio/x-wav	75.4 KBytes	2012-02-12 20:45:22
90c0bd315d4-2	audio/x-wav	79.4 KBytes	2012-02-12 20:45:16
90c0bd315d4-1	audio/x-wav	63.4 KBytes	2012-02-12 20:45:10
90c0bd315d4-0	audio/x-wav	139.4 KBytes	2012-02-12 20:44:38

Sometimes, it can be useful to store additional information about a blob. In this case, the datastore is the more appropriate storage mechanism. In our `sessions.py` example, we show how one can store the retrieval URL of the audio along with a reference to its blob in the BlobStore. Our script accomplishes this with a simple data model containing just two properties:

```
from google.appengine.ext import db

# A simple database model to store a URL and an associated blob.
class DataModel(db.Model):
    url = db.StringProperty(required=True)
    blob = blobstore.BlobReferenceProperty(required=True)
```

It is not hard to imagine storing other types of information, such as a user ID, session ID, or utterance ID, by explicitly giving them their own columns in the datastore. GQL, a subset of SQL, can then be used to create queries over the datastore. Here, however, our goal is to provide a simple example of datastore usage. The resulting Datastore Viewer interface looks like the following:

ID/Name	blob	url
name=0352c27a4dd-0	AMIfv97Of9vA4kUZp5trunoUTzjS0DqmC71-q6i2RG9mTpWkIVBd__Eh-kTLIlLyS-QUKwTqyow-Ua3Mb u8RvkV0I3kGV_ITUdohH9uwlCql3ROSI0XfnX_VGGi4Kyz-6ZjrvATjuZtkLeAe6soNDC-_CF3iq3kHbQ View blob	http://wami-recorder.appspot.com/audio?name=0352c27a4dd-0
name=0352c27a4dd-1	AMIfv95Imga1zBJ1-Kf2XJI9UW05quBMQyn2Lv7A2 0fs14CcgeNe3dukcckt_uzrZTZQtH3f_eBIRTkAYb-iFdpHARuUmm47XiaP1Qzj-y44_FnoLkLhCfz_k ZcVGPPmDAIg24kT0JHrNC7i7beA8-EZddehwlv5pA View blob	http://wami-recorder.appspot.com/audio?name=0352c27a4dd-1
name=0352c27a4dd-2	AMIfv97kRcNliyTrWs4YVcnE2hSUCOhRpT4Y9sewf 0ta1GAczeenM4XrBRpXMSywMB9ZlWUafAgX7tJ27 uxs9xi7TuYc3a9Ro42he6lsus5as-ISTOMW2QCFW dktLI33EWNiLnziflIJoGfi1xHRdB4Hc7LfmAL85w View blob	http://wami-recorder.appspot.com/audio?name=0352c27a4dd-2

There are a few additional considerations to keep in mind when using the GAE. First is that it has size limitations for a single query (a few minutes of speech recorded at 22,050 Hz.) Second, the programming environment does not allow the installation of native libraries, which might be useful for tasks such as audio resampling. Still, it suffices as a means to get an audio collection task up-and-running quickly.

3.4.3 Server Configuration Details

The preceding server descriptions suffice for the most basic uses of the WAMI recorder. In certain situations, however, additional configuration is necessary. In some cases, it may be necessary to play audio from locations other than where `Wami.swf` is hosted, requiring the use of a crossdomain policy file. Other times, it is necessary to ensure secure HTTP requests using HTTPS. Finally, there are certain applications that require audio streaming rather than the simple bulk-uploading we have presented so far. We describe how to achieve these advanced configurations where possible, and describe work-arounds to approximate their behavior when need be.

Crossdomain Policy File

Flash imposes a security measure called a crossdomain policy file that comes into play when the content being requested by Flash resides on a domain different from that of the Flash application itself. Suppose, for example that the SWF for the recorder is hosted in `http://website-1.com/recorder.swf`. Performing the action `Wami.startPlaying("http://website-2.com/audio.wav")` will then cause a runtime error in ActionScript unless you have the proper crossdomain policy file in place at `http://website-2.com/crossdomain.xml`. An example of what this file might look like, albeit one with very liberal permissions, is shown below:

```
<cross-domain-policy>
    <site-control permitted-cross-domain-policies="master-only"/>
    <allow-access-from domain="*" secure="false" />
    <allow-http-request-headers-from domain="*" headers="*"/>
</cross-domain-policy>
```

Note that manipulating `crossdomain.xml` implies that one needs to have a certain amount of control over the server. For example, in the preceding Google Apps Engine example, the following addition needs to be made to the `app.yaml` file:

```
- url: /crossdomain.xml
  mime_type: text/xml
  static_files: public/crossdomain.xml
  upload: public/crossdomain.xml
```

This can be problematic if one would prefer to make use of a third-party's service over which one has no server-side control. Suppose, for example, that a WAMI application needs to play audio from a TTS server setup by a third party to take requests of the form: `http://tts.com?words=hello%20world`. Without a crossdomain policy file in place, the application developer would not be able to play the audio through Flash. One work-around in this scenario is to proxy the audio through one's own servers to make it appear as though it is coming from a domain that has the proper policy file in place. Most web servers make proxies of this sort relatively easy to set up.

HTTPS

In some situations, it is imperative to use HTTPS to send requests rather than vanilla HTTP. This secure communication protocol combines HTTP with SSL to verify server identities and encrypt information going across the network. As an increasing number of web-related security concerns have become public in recent years, major websites have responded by supporting the `https://` scheme.

Similar to the Applets described in Section 3.2.2, website developers must acquire certificates from a certificate authority (e.g., VeriSign or Microsoft) to host an easily accessible secured site. All the popular browsers of today ship preconfigured to communicate with certain certificate authorities. Cheap or free certificates run the risk of not being fully supported by the popular browsers, however, causing the websites to retain the undesirable security pop-ups that a pricier certificate would resolve.

Including insecure content (anything communicated through an `http://` URL) in a secure site can also cause security pop-ups in some browsers. However, the precise behavior varies depending on the manner in which the content is embedded, the user's security settings, as well as the browser in question. For example, including `<iframe src="http://..." />` in a secure site will cause the following pop-up message in IE8:

Note that the default option is "Yes" to a question that asks if viewing only the secure content is OK. This is the opposite of most security dialog messages, even those for previous versions of Internet Explorer. The reasoning for the change is most likely that users often click "Yes" without thinking, which would have previously caused the browser to display insecure content. The error message shown above, however, will at least give the user pause, if only to try to figure out what actually happens when they click "Yes."

Streaming

Data transferred via HTTP are associated with a content-type that specifies the kind of data that will be found in either the request or the response body. For a WAV file, `audio/wav` or `audio/x-wav` are commonly used content types. For the AU container `audio/basic` is used. By default the WAMI recorder gathers up all the audio on the client side before shipping it over to the server via an HTTP POST with the `audio/x-wav` content type.

There are times, however, when waiting for the audio to finish recording is not ideal. Given that the audio is continuously buffering on the client side, for example, it is probably only reasonable to use the recorder for collecting files of a few megabytes in size. While a couple of minutes of data might be enough for many applications, there are certainly applications that require more. Unless silence can be detected and the microphone restarted automatically, collecting larger amounts of audio is made difficult with this technique. Moreover, some applications, such as server-side end-point detection or incremental speech recognition results, actually *must* receive the data before an utterance has ended. In these cases, only a streaming solution will suffice.

Oddly enough, HTTP 1.1 solves this problem by requiring HTTP 1.1 compliant applications to be able to receive a chunked transfer encoding. This method breaks the data down into a series of chunks, which are reassembled on the server side. Unfortunately, of the three plug-ins mentioned in previous sections, only Java supports the chunked transfer encoding. Flash does not provide a mechanism for performing POSTs in this fashion, and the only solution (without resorting to using a flash media player) is to implement a custom chunking protocol using multiple sequential HTTP POSTs. The individual POSTs, perhaps each containing a couple hundred milliseconds of audio, can be reassembled on the server side to perform a sort of psuedostreaming. Streaming a WAV file, however, is not really possible since the header information of a WAV expects a length, which would, of course, be unknown a priori. Since the WAV specification does not officially support data of unknown length, one could turn to an audio container that does, such as AU.

Since the streaming features are somewhat more involved to implement on both the server and client side of the audio collection equations, we leave them out of our default implementation in WAMI. Fortunately, for simple collection tasks like the one described Section 3.5, advanced streaming features are unnecessary.

3.5 Example: Speech Collection on Amazon Mechanical Turk

This section is a high-level overview of the steps necessary to set up an audio collection human intelligence task (HIT) on MTurk. More detailed, step-by-step instructions as well as the complete code for this tutorial can be found in the Google Code repository located here:

```
https://wami-gapp.googlecode.com
```

This example collection task will have workers read movie titles aloud, but generalizes easily to any prompted audio collection task. We begin by describing the necessary server-side setup for the task, making use of the GAE framework. We then discuss two ways of deploying the task. The first, and simplest, utilizes the MTurk web interface to deploy and manage the audio collection HIT. The second method of deployment is through MTurk's CLT by way of their `ExternalQuestion` HIT, which allows developers to embed a web page hosted outside of MTurk. The CLT themselves make it possible to script HIT management, making working with a large number of HITs easier.

There is a third, more advanced method, whereby one can deploy HITs programmatically using Java, C#, or a number of other languages. These SDKs not only offer fine-grained control over a HIT, they enable the development of human-in-the-loop applications that can ask questions of the outside world, such as "What does this person say in this audio clip?" or "How do you pronounce Caddyshack?" Using this method, an application could conceivably automatically collect its own training data without expert supervision. These SDKs, however, are beyond the scope of this tutorial. In Section 3.7.2, we describe nascent efforts in this arena.

3.5.1   Server Setup

The first step is to check out the code (perhaps to ~/wami-gapp) and run it on the local machine. To do so, however, one first needs to become familiar with GAE, since we will let this cloud service do the heavy lifting on the server side of our application. Signing up for an account is relatively straightforward.

With a new GAE account in hand, it's possible to use the web interface to create an application by giving it a unique application ID and a title. The application ID will be used in the URL, so it is preferable to choose something easy to type if need be. Choosing NAME as an ID reserves `NAME.appspot.com`; however, many IDs may already be taken. Once the application has been created, a dashboard is accessible, although no application has yet been deployed.

Before deploying an application, it is wise to run and test it locally. For this purpose, Google has provided an App Engine SDK for python. The GUI version of the SDK, available for both Mac OS X and Windows, is somewhat more user-friendly than the command-line scripts available for Linux. The following image depicts the GAE Launcher and the corresponding logging console:

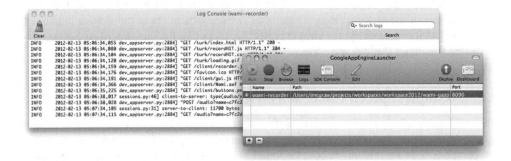

With the SDK installed, attaching the project is simply a matter of performing the *Add Existing Application* operation to attach ~/wami-gapp to the application launcher. A default port, such as 8080, will be provided such that running the app will make it available at local-host:8080. In particular, ~/wami-gapp/turk/index.html will now be available at http://localhost:8080/turk/index.html. This is the interface that will be deployed to MTurk for audio collection:

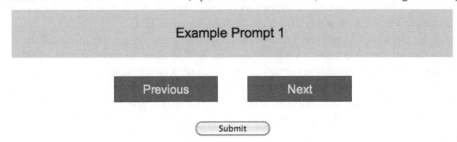

Even when everything is running smoothly, it is probably wise to check the logging console for anomalies. The sessions.py python code contains some examples of how a programmer can write messages to this console. If the logs look clean, however, there is just one small change that must be made before deployment. The top line of ~/wami-gapp/app.yaml, the file that configures the server, specifies the name of the app to be deployed. This must be changed to the unique ID of the app that was created through the GAE web interface. Failure to do so will result in an error message in the log along the lines of "this application does not exist."

A successful deployment will make the application available to the outside world at http://NAME.appspot.com/turk/index.html. Note that the special port is no longer part of the URL. We also provide an example of the recorder without the MTurk interface at http://NAME.appspot.com/client/index.html. With either site, it should now be possible to record an utterance and then see it appear in the *Blob Viewer* in the GAE web interface.

3.5.2 Deploying to Amazon Mechanical Turk

Now that audio collection has been tested end-to-end, we are ready to deploy a HIT on MTurk. There are a number of ways to access MTurk, but the web interface is arguably the simplest, so we use it here. We recommend starting with the MTurk *requester's sandbox*, which is a website identical to the real MTurk, but which will not cost money to experiment on. Of course, this means there are no workers willing to complete the tasks, but the sandbox is a great way to test out a HIT for the first time.

Once logged into the sandbox with an Amazon account, the first step is to navigate to the *Design* page, where HIT templates are created. Often it is easiest to start with a blank template. After choosing the blank template option, the requester is presented with a page to enter the properties of the HIT, such as a name, description and keywords. This is also where the payment is specified, as well as the number of assignments per HIT. Any particular worker will only be able to perform a given HIT one time, but if you specify a number of assignments per HIT greater than one, another worker will be allowed to come along and perform that HIT as well.

Perhaps the most important properties for a task, other than price, are the qualification restrictions placed on a HIT. With speech, in particular, the criteria by which workers are filtered has a large effect on the results. Restricting worker by location is one way to ensure that the language and accent roughly meet the criteria of the task, though it is by no means a guarantee. Qualification tests are another way to filter workers. With these, it may be possible to preapprove the voices that are allowed to perform a certain HIT.

Once the properties have been selected, the requester must design the layout of the HIT. Fortunately, Amazon has given us a large amount of flexibility. In particular, in addition to a WYSIWYG[1] editor, there is a button on the interface that allows one to edit the HTML directly: Edit HTML Source . Using the HTML editor we can replace everything in the text area with HTML and JavaScript of our choosing. In the case of our example audio collection HIT, we can replace all of the HTML code present in the HIT by default with just a few simple lines of code similar to those found in ~/wami-gapp/public/turk/index.html file:

```
<style type="text/css">@import url(recordHIT.css);</style>
<script type="text/javascript" src="recordHIT.js"></script>
<p><input type="hidden" id="wsession" /></p>
<script>
  var wsessionid = Wami.RecordHIT.create("${prompts}");
  document.getElementById('wsession').value = wsessionid;
</script>
<noscript>Please enable JavaScript to perform a HIT.</noscript>
```

Previewing the task should yield an interface similar to the one shown previously. It is possible, at this stage, to add a header describing the task, or perhaps a feedback text area to be submitted wth the HIT. Once the template is saved, we are ready to begin filling in the ${prompts} variable with prompts for the particular task at hand. In our case, we will insert the movie titles into our HIT using the list found in ~/wami-gapp/turk/movies.txt. To do this, we navigate to the *Publish* page of MTurk and select the audio collection template we have just saved. If all goes well, the interface will then prompt the requester to upload input data:

---

[1] What You See Is What You Get!

```
prompts
"The Green Mile<>Babe<>Waterworld<>Mad Max Beyond Thunderdome<>Caddyshack"
"The Thin Red Line<>Pitch Black<>Bridge to Terabithia<>Ed Wood<> Godspell"
...
```

Note that the prompts are separated by a delimeter: <>. This allows our JavaScript to bundle multiple prompts up into a single HIT. For large HITs, this bundling technique can save a significant amount of money on Amazon's commission. Although Amazon usually takes a 10% commission on a task, one that pays less than 5¢ is subject to Amazon's minimum commission charge of 0.5¢. Thus, tasks above 5¢ get more value-per-cent spent than those below.

Given the format described above, there are some special characters that cannot appear within a prompt. Fortunately, the prompts are inserted directly into the page as HTML. This means that HTML escape codes can be used to correctly display special characters. Moreover, HTML tags and styles can be used to insert any desired content in place of a prompt. For example, it would not be difficult to insert a YouTube video to prompt for a verbal response. In our simple example, we will stick with plain text.

After uploading the input data file, a preview of the interface should show the movie titles in place of the ${prompts} variable. The requester is then asked to confirm the cost (even the sandbox provides a sample cost summary), after which one can publish the HITs. Publishing HITs in the requester's sandbox will make them available in a corresponding worker's sandbox. From this interface it is possible to test the task in full:

Had we been deploying this HIT to the real MTurk interface, there is one very important step that we have not yet described. In many settings, research involving the use of human subjects requires the approval of an institutional review board (IRB). Typically, tasks in which workers do not remain anonymous are subject to additional scrutiny and one could reasonably consider a voice to be identifying information. If approved, IRBs often require that a consent form be supplied to the worker, either as part of the HIT or as a separate qualification step.

Make sure that the full submission process works from a few different browsers and always check to see that the results are being collected as expected. To do this, return to the requester's sandbox after submitting a few HITs. Find the page where requesters can *Manage* HITs, and get the results of the current task. They should look something like the following:

Flash	Platform	Session	Url 0	Url 1	Url 2	Url 3	Url 4
Flash Player 11.1.102	Mac: Chrome 16	dc1d4f9ba19	http://wami-recorder.appspot.com/audio?name=dc1d4f9ba19-0	http://wami-recorder.appspot.com/audio?name=dc1d4f9ba19-1	http://wami-recorder.appspot.com/audio?name=dc1d4f9ba19-2	http://wami-recorder.appspot.com/audio?name=dc1d4f9ba19-3	http://wami-recorder.appspot.com/audio?name=dc1d4f9ba19-4
Flash Player 11.1.102	Mac: Chrome 16	640021645ff	http://wami-recorder.appspot.com/audio?name=640021645ff-0	http://wami-recorder.appspot.com/audio?name=640021645ff-1	http://wami-recorder.appspot.com/audio?name=640021645ff-2	http://wami-recorder.appspot.com/audio?name=640021645ff-3	http://wami-recorder.appspot.com/audio?name=640021645ff-4
Flash Player 11.1.102	Mac: Chrome 16	3c5c749da4b	http://wami-recorder.appspot.com/audio?name=3c5c749da4b-0	http://wami-recorder.appspot.com/audio?name=3c5c749da4b-1	http://wami-recorder.appspot.com/audio?name=3c5c749da4b-2	http://wami-recorder.appspot.com/audio?name=3c5c749da4b-3	http://wami-recorder.appspot.com/audio?name=3c5c749da4b-4

Note that there is a URL for every piece of audio collected. To hear the audio, click on these URLs from a browser that can handle the `audio/x-wav` mime type. We have also saved some information about the user's system including their OS, browser, and Flash versions. This can be useful for debugging the HIT when behavior varies across system configuration.

3.5.3   The Command-Line Interface

We can perform the same HIT with MTurk's `ExternalQuestion` template using the CLT. Within the samples provided by the CLT, the `external_hit` sample is set up almost perfectly for the task at hand. We simply replace the URL in the `.question` file with our application's URL:

```
https://NAME.appspot.com/turk/index.html?prompts=${helper.
urlencode($prompts)}
```

The prompts file format is the same as in Section 3.5.2. The call to `helper.urlencode($prompts)` takes the prompts field of our input data, and escapes the special characters so that it can be inserted as a URL parameter. Our JavaScript then does the job of retrieving the parameter and parsing individual prompts. Note that with some browser/server configurations URLs with more than a few thousand characters cause errors. In these situations, it may be necessary to load data into the web page dynamically using AJAX. In many cases, however, embedding prompts in the URL should suffice.

MTurk HITs, even those deployed through the web interface, are really nothing more than HTML forms whose results are posted to Amazon's servers. Thus, we can create the same form, called `mturk_form` in our external hit. We can explicitly add HTML `input` elements to the form, or dynamically create them in JavaScript. The `hidden` input type is useful for saving result fields without the worker having to view them. At the very least, we must specify the assignment ID in this form. Finally, since we have full control over the submit button, we can validate results programmatically before submission.

The website used in the `ExternalQuestion` must be served via HTTPS. Recall that serving insecure content via HTTP will cause security alerts under some browser configurations. More importantly, the MTurk site recently began requiring the use of HTTPS. Fortunately, there is a free way to serve almost any content with a properly configured HTTPS certificate: one can piggyback off of the GAE certificates. We have, in fact, been doing this already for everything hosted through `https://NAME.appspot.com`. Although slightly more work, even content not hosted on GAE can be proxied through a GAE account, ensuring that the requests made from MTurk are secure.

3.6 Using the Platform Purely for Payment

In Section 3.5, we described a way to embed speech collection tasks directly into a single crowdsourcing platform. While this certainly makes for a smoother user experience, it is not strictly necessary to take advantage of paid crowdsourcing. Depending on the nature of the task and the choice of crowdsourcing platform (see Chapter 9 for alternatives to MTurk), it may not even be possible to integrate an audio collection task directly into the micropayment workflow. Fortunately, a work-around is not difficult. As previously described, some researchers have taken advantage of a paradigm whereby payment is processed through a crowdsourcing platform, but the worker is directed to an alternate means of interacting with a system (Gašić *et al.* 2011; Jurčíček *et al.* 2011). In essence, only a session identifier needs to be tracked to match a user interaction with a payable account on a crowdsourcing platform.

In this section, we explore the use of this paradigm to collect speech from smartphone users. Thus far, we have ignored the smartphone and focused on web-based approaches to streaming audio. Given the lack of microphone access on smartphone browsers there is little intersection between the mobile web and audio recording. Native applications must, therefore, be developed to transfer the audio to the server. We have found it beneficial to implement our own browser with audio streaming capabilities. With the help of WebKit, it was relatively straightforward to create a browser for both iPhone and Android devices that conforms to the WAMI audio recording protocol described in the previous sections.

Creating WAMI browsers was advantageous in a number of respects. First, we were able to wrap both the Flash and mobile application APIs so that they are exposed through JavaScript through identical function calls. This means that a WAMI-enabled site that works on a desktop browser is also compatible with a WAMI mobile browser. Second, unlike collecting data directly over the phone line, using the mobile phone's data connection gives us full control over the audio encoding used during collection. In the following experiment, we used the Free Lossless Audio Codec to encode 16 kHz audio data before posting it to a GAE server much like the one described in Section 3.4.2.

To experiment with the platform-for-payment approach, we created a read speech task identical to the desktop version described in Section 3.5.2, and seeded the prompts with short sentences and phrases. We still used MTurk for payment, but we asked users to download the WAMI Android application and install it on their phones to access the WAMI-enabled website used to record the audio. Just as on the desktop, the website asked workers to read a prompt. A small text box was added to the site, and we required the worker to enter a random code. We treated this code as a session ID and stored it both in the MTurk results and in our external application (via the GAE). Later, we were able to match utterances to individual MTurk HITs to process payment.

Over the course of a couple of days, we collected a batch of 10,000 prompts from Android users around the United States. Cumulatively, this amounted to 10 hours of audio data.

Fifty-eight individual workers performed the task. The following figure shows the distribution of audio collected across the 58 individual workers that performed the task. Although not always reported, typically most of the work in a crowdsourced task is completed by just a few individuals. As can be seen in the figure, a power-law distribution emerges when the individual workers are sorted by their contribution to the corpus. This may be of concern for certain applications and additional measures may be required to influence the distribution.

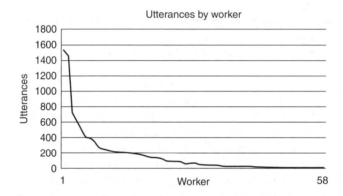

One of the benefits of having users download a native application is that we can collect metainformation to inform us of their environment relatively easily. The following chart depicts the fraction of smartphone users that were on their mobile phone network and those that were on WIFI. Moreover, we break down these numbers in terms of the version of the Android operating system is installed. It is clear from the figure that, to capture a relatively large cross section of users, a developer needs to ensure that their application is compatible, at least to some extent, with older versions of the operating system.

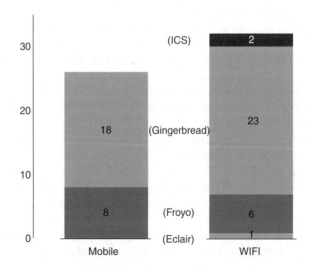

3.7   Advanced Methods of Crowdsourced Audio Collection

So far we have focused on the task of collecting read speech. Despite the apparent simplicity of the task, we have shown that it is not trivial to implement the required technology. Our hope is that providing these tutorials and case studies above will open a door to crowdsourced speech collection for a wider audience.

In this section, we take the next logical steps along two dimensions. First, we describe crowdsourced speech collection for spoken language interfaces. Using the same recording API described above, one can attach a speech recognizer or even embed a full-fledged spoken dialog system into a web page. Second, we describe how various speech-related crowdsourcing tasks might be stitched together with the help of a toolkit or an API. This technique allows the developer to insert *human computation* into any algorithm that cannot be performed fully automatically.

3.7.1   Collecting Dialog Interactions

As described in Section 3.1, the collection of in-domain speech data is crucial to the development of spoken language systems. While collecting read speech can be a helpful start in a new domain, ideally, the speech collected would be from actual user interactions with a prototype system. Here, we describe some recent work that suggests that crowdsourcing, and in particular MTurk, has the potential to assist with this aspect of system development.

We first summarize our own data collection effort *FlightBrowser* (McGraw *et al.* 2010). *FlightBrowser* is a web-based spoken dialog system that was derived from our ATIS originally developed under the DARPA Communicator project (Walker *et al.* 2001). Its design is based on a mixed-initiative model for dialog interaction. The system prompts for relevant missing information at each point in the dialog, but there are no constraints on what the user can say next. Using the WAMI Toolkit, we gave the system a multimodal web interface, which has been deployed to mobile phones and subsequently to MTurk.

The HIT we designed required workers to follow a scenario containing an itinerary and use the system to book the relevant flights. We deployed two tasks to MTurk at two different price points. The first was a 20¢-task that lasted 4 days and resulted in 876 dialog sessions. The second was a 10¢-task that ran for 6 days and resulted in an additional 237 sessions. In total, we collected 1113 dialog sessions containing over 10,651 utterances for about $200. We then transcribed this data, again using MTurk, and evaluated our dialog system along the dimensions of word error rate and task completion.

A similar evaluation was subsequently carried out by Rayner *et al.* (2011) to evaluate a computer aided language learning application. The interface, depicted in Figure 3.3, relied on an embedded Flash application to transmit speech from the browser to a remote server for recognition. The researchers deployed seven HITs to MTurk over seven consecutive days, recruiting a total of 26 subjects to their crash course in Japanese. They collected 129 sessions with a total of 9110 utterances for a cost of $170.

For some, porting an existing dialog system to the web may not be a realistic endeavor. For others, the web may not afford the desired interaction paradigm. Still, MTurk may be a valuable tool for subject recruitment. Jurčíček *et al.* (2011) use a telephone framework to deploy a restaurant information system, then use MTurk just to recruit subjects and collect feedback. Paying 20¢ per call, they were able to recruit 140 users to make 923 calls in a matter of days.

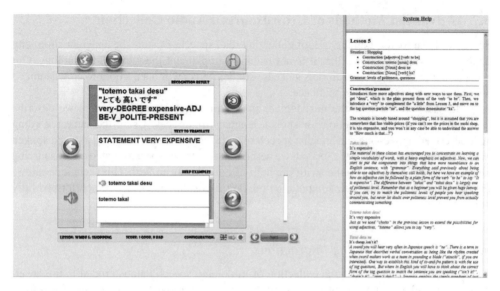

Figure 3.3   A spoken language system for computer-aided language learning. (From Rayner *et al.* 2011.) Reproduced by permission of Manny Rayner.

When using MTurk for speech data collection there are a number of considerations that must be made. Rough controls on the collection of accented speech can be performed through location restrictions. More precise controls might be achieved through the use of a secondary verification task, in which the speech data are sent back out to other MTurk workers for analysis. Moreover, microphone quality and acoustic environments can be highly variable. We have observed that forcing workers to hear their own audio in an extra "playback" step fixes many of these issues, but again, a secondary verification task might also suffice. With these extra precautions in place, we believe that MTurk has the potential to be a cheap source of high-quality speech data.

3.7.2   Human Computation

Many research communities are going beyond basic crowdsourcing and using a paradigm referred to as *human computation* (Quinn and Bederson 2011). In the context of crowdsourcing, a process is given the label human computation when it is entirely machine directed, and the human component is treated much like a function call in ordinary programming.

To give a concrete example of how the speech community might use this technique, imagine creating a compiler of language models (LMs). Rather than operating on text, however, this LM compiler would take as input a corpus of untranscribed speech. A secondary input might be a budget, restricting the compiler's spending. Given these inputs, the compiler could automatically crowdsource the transcription to MTurk, compute statistics over the transcripts, and output a LM that falls within the prescribed budget.

Many interesting research questions emerge from using human computation for speech-related tasks. One might wonder, for instance, how best to optimize the quality of the LM in the preceding example given the cost constraints. If cost were not an issue, another question to consider might be whether we could elicit *new* in-domain LM data from workers, perhaps by asking them to rephrase an utterance that they just transcribed. These questions would guide our research toward the field of *active learning*, in which a system actively participates in the creation or selection of its own training data.

To get to the state where these questions become pertinent, however, requires a significant amount of engineering and design work. Fortunately, researchers of human-computer interaction have led the way with an assortment of projects geared toward managing *human computation* tasks. *CrowdWeaver* provides a graphical user interface for constructing complex crowd workflows (Kittur *et al.* 2012). *CrowdForge* follows a map-reduce paradigm to split tasks and aggregate their results (Kittur *et al.* 2011). *TurKit* is a toolkit for writing iterative tasks using straightforward JavaScript snippets. Yet another line of research treats the crowd of humans much like a database. These efforts enable programs to access the crowd with declarative SQL-like languages (Franklin *et al.* 2011; Marcus *et al.* 2011; Parameswaran *et al.* 2012).

Whether by using one of the frameworks mentioned above or by accessing the MTurk API directly, wrapping nonexpert workers into an algorithm represents a fundamental shift in our ability to automate tasks for spoken language systems. Few experiments have been performed in the speech community using such techniques. In McGraw *et al.* (2011), we wrapped a LM compiler, similar to the one described above, into a deployed spoken language system. In this way, the system was able to incrementally improve on-the-fly. This work merely hints at the possibilities of what we call *organic* spoken language systems that grow and adapt based on input from the crowd.

3.8 Summary

This chapter has provided an overview of crowdsourcing speech data collection. We have given a brief history of speech collection and outlined the difficulty of these endeavors from both a technical and organizational standpoint. Even today, we have shown that the technology available for speech collection is limited, particularly on mobile devices and over the web. Despite the difficulties, the costs of data collection, both in time and money, can be substantially reduced using crowdsourcing techniques.

Since the crowd is most easily accessed via the web, we have developed a set of tools to accompany this chapter and provided tutorials that walk the reader through web-based speech collection. These tools can be used in any web page, but we demonstrate their deployment to MTurk, since it is arguably the most popular crowdsourcing platform today.

Finally, we have described the ways in which we can move beyond the typical tasks deployed through the MTurk web interface. Other crowdsourcing communities are already pushing the boundaries in terms of task complexity. Our hope is that, by combining our current research with these new approaches, we can turn this useful data collection technique into a platform for entirely new avenues of exploration at the intersection of human computation and spoken language systems research.

3.9 Acknowledgments

Thanks go to Stephanie Seneff, who provided helpful comments on early versions of this chapter, Scott Cyphers, who helped to shape the latest iteration of WAMI, and Edgar Salazar, who managed the collection of the mobile phone data. Finally, Alex Gruenstein was instrumental in getting early versions of WAMI out to the crowd.

References

Aust H, Oerder M, Seide F and Steinbiss V (1994) Experience with the philips automatic train timetable information system. *Interactive Voice Technology for Telecommunications Applications, 1994, Second IEEE Workshop on*, pp. 67–72.

Bernstein J, Taussig K and Godfrey J (1994) Macrophone: an american english telephone speech corpus for the polyphone project. *Acoustics, Speech, and Signal Processing, IEEE International Conference on* **1**, 81–84.

Canavan A, Graff D, Kimball O, Miller D and Walker K (1997). CALLHOME American English Speech. Linguistic Data Consortium, Philadelphia.

Cieri C, Corson L, Graff D and Walker K (2007) Resources for new research directions in speaker recognition: the mixer 3, 4 and 5 corpora. *Interspeech*, pp. 950–953. ISCA.

Dahl DA, Bates M, Brown M, Fisher W, Hunicke-Smith K, Pallett D, Pao C, Rudnicky A and Shriberg E (1994) Expanding the scope of the atis task: the atis-3 corpus. *Proceedings of the Workshop on Human Language Technology (HLT '94)*, pp. 43–48. Association for Computational Linguistics, Stroudsburg, PA.

David CC, Miller D and Walker K (2004) The fisher corpus: a resource for the next generations of speech-to-text. *Proceedings 4th International Conference on Language Resources and Evaluation*, pp. 69–71.

Di Fabbrizio G, Okken T and Wilpon JG (2009) A speech mashup framework for multimodal mobile services. *Proceedings of the 2009 International Conference on Multimodal Interfaces (ICMI-MLMI '09)*, pp. 71–78. ACM, New York.

Draxler C (2006) Exploring the unknown—collecting 1000 speakers over the Internet for the Ph@ttSessionz database of adolescent speakers. *Proceedings of Interspeech*, pp. 173–176, Pittsburgh, PA.

Fisher W, Doddington G and Marshall GK (1986) The DARPA speech recognition research database: Specification and status. *Proceedings of the DARPA Speech Recognition Workshop*, pp. 93–100.

Franklin MJ, Kossmann D, Kraska T, Ramesh S and Xin R (2011) Crowddb: answering queries with crowdsourcing. *Proceedings of the 2011 International Conference on Management of Data (SIGMOD '11)*, pp. 61–72. ACM, New York.

Freitas J, Calado A, Braga D, Silva P and Dias MS (2010) Crowd-sourcing platform for large-scale speech data collection. *VI Jornadas en Tecnologa del Habla and II Iberian SLTech Workshop FALA 2010*, pp. 183–186.

Gašić M, Jurčíček F, Thomson B, Yu K and Young S (2011) On-line policy optimisation of spoken dialogue systems via live interaction with human subjects. *IEEE Workshop on Automatic Speech Recognition and Understanding (ASRU), Hawaii*.

Godfrey J, Holliman E and McDaniel J (1992) Switchboard: telephone speech corpus for research and development. *Acoustics, Speech, and Signal Processing, IEEE International Conference on* **1**, 517–520.

Gorin A, Riccardi G and Wright J (1997) How may I help you? *Speech Communication* **23**(12), 113–127.

Gruenstein A and Seneff S (2007) Releasing a multimodal dialogue system into the wild: user support mechanisms. *Proceedings of the 8th SIGdial Workshop on Discourse and Dialogue, Antwerp, Belgium*.

Gruenstein A, McGraw I and Badr I (2008) The wami toolkit for developing, deploying, and evaluating web-accessible multimodal interfaces. *Proceedings of ICMI*.

Hemphill CT, Godfrey JJ and Doddington GR (1990) The ATIS spoken language systems pilot corpus. *Proceedings of the DARPA Speech and Natural Language Workshop*, pp. 96–101. Morgan Kaufmann.

Hirschman L (1992) Multi-site data collection for a spoken language corpus. *Proceedings of the workshop on Speech and Natural Language (HLT '91)*, pp. 7–14. Association for Computational Linguistics, Stroudsburg, PA.

Hurley E, Polifroni J and Glass JR (1996) Telephone data collection using the world wide web. *In the 4th International Conference on Spoken Language Processing, Philadelphia, PA*.

Jurčíček F, Keizer S, Gašić M, Mairesse F, Thomson B, Yu K and Young S (2011) Real user evaluation of spoken dialogue systems using Amazon Mechanical Turk. *Interspeech*, Florence, Italy.

Kittur A, Khamkar S, André P and Kraut R (2012) Crowdweaver: visually managing complex crowd work. *Proceedings of the ACM 2012 Conference on Computer Supported Cooperative Work (CSCW '12)*, pp. 1033–1036. ACM, New York.

Kittur A, Smus B, Khamkar S and Kraut RE (2011) Crowdforge: crowdsourcing complex work. *Proceedings of the 24th Annual ACM Symposium on User Interface Software and Technology (UIST '11)*, pp. 43–52. ACM, New York.

Kudo I, Nakama T, Watanabe T and Kameyama R (1996) Data collection of japanese dialects and its influence into speech recognition. *The 4th International Conference on Spoken Language Processing*, Philadelphia, PA. ISCA.

Lane I, Waibel A, Eck M and Rottmann K (2010) Tools for collecting speech corpora via mechanical-turk. *Proceedings of the NAACL HLT 2010 Workshop on Creating Speech and Language Data with Amazon's Mechanical Turk (CSLDAMT'10)*, pp. 184–187. Association for Computational Linguistics, Stroudsburg, PA.

Ledlie J, Odero B, Minkov E, Kiss I and Polifroni J (2009) Crowd translator: On building localized speech recognizers through micropayments. *Third Annual Workshop on Networked Systems for Developing Regions (NSDR)*. ACM, Big Sky, MT.

Marcus A, Wu E, Karger D, Madden S and Miller R (2011) Human-powered sorts and joins. *Proceedings of the VLDB Endow* **5**(1), 13–24.

McGraw I, Glass JR and Seneff S (2011) Growing a spoken language interface on Amazon Mechanical Turk. *Interspeech*, pp. 3057–3060.

McGraw I, Lee C-y, Hetherington L, Seneff S and Glass JR (2010) Collecting voices from the cloud. *The International Conference on Language Resources and Evaluation* (LREC).

Parameswaran A, Park H, Garcia-Molina H, Polyzotis N and Widom J (2012) Deco: declarative crowdsourcing. *Technical Report*, Stanford University, CA.

Peckham J (1993) A new generation of spoken dialogue systems: results and lessons from the SUNDIAL project. *Proceedings of Eurospeech 93*, pp. 33–40, Berlin.

Pitrelli J, Fong C, Wong S, Spitz J and Leung H (1995) Phonebook: a phonetically-rich isolated-word telephone-speech database. *Acoustics, Speech, and Signal Processing, IEEE International Conference on* **1**, 101–104.

Quinn AJ and Bederson BB (2011) Human computation: a survey and taxonomy of a growing field. *Proceedings of the 2011 Annual Conference on Human Factors in Computing Systems (CHI '11)*, pp. 1403–1412. ACM, New York.

Raux A, Bohus D, Langner B, Black AW and Eskenazi M (2006) Doing research on a deployed spoken dialogue system: one year of lets go! experience. *Proceedings of Interspeech*, pp. 65–68.

Rayner M, Frank I, Chua C, Tsourakis N and Bouillon P (2011) For a fistful of dollars: using crowd-sourcing to evaluate a spoken language call application. *SLaTE*, pp. 117–120.

Schultz T, Black AW, Badaskar S, Hornyak M and Kominek J (2007) Spice: web-based tools for rapid language adaptation. *The Proceedings of Interspeech*.

Seneff S, Hurley E, Lau R, Pao C, Schmid P and Zue V (1998) Galaxy-ii: A reference architecture for conversational system development. *Proceedings of the ICSLP*, pp. 931–934.

Walker MA, Passonneau R and Boland JE (2001) Quantitative and qualitative evaluation of darpa communicator spoken dialogue systems. *Proceedings of the 39th Annual Meeting on Association for Computational Linguistics (ACL '01)*, pp. 515–522. Association for Computational Linguistics, Morristown, NJ.

Wang H (1997) Mat—a project to collect mandarin speech data through networks in taiwan. *International Journal Computational Linguistics Chinese Language Process* **12**(1), 73–89.

Wheatley B and Picone J (1991) Voice across America: toward robust speaker-independent speech recognition for telecommunications applications. *Digital Signal Processing* **1**(2), 45–63.

Zue V, Seneff S and Glass J (1990) Speech database development at mit: timit and beyond. *Speech Communication* **9**(4), 351–356.

Zue V, Seneff S, Glass J, Polifroni J, Pao C, Hazen T and Hetherington L (2000) Juplter: a telephone-based conversational interface for weather information. *Speech and Audio Processing, IEEE Transactions on* **8**(1), 85–96.

4

Crowdsourcing for Speech Transcription

Gabriel Parent

Carnegie Mellon University, USA

4.1 Introduction

This chapter has been written for people with little or no previous experience with crowd-sourcing to guide them through the design, creation, and publishing of tasks for labeling and transcribing speech utterances. Because no two readers will have exactly the same require-ments for their annotation, this chapter addresses topics that apply to a large range of tasks. Based on the literature and on personal experience, some guidelines will be provided: how to achieve good-quality transcriptions, how to interact with the workers, how much to pay, and so on. The reader will be guided through the steps that lead to crowdsourced transcriptions: audio preprocessing, task design, submitting the open call and postprocessing for quality control.

4.1.1 Terminology

The terminology used in crowdsourcing publications widely varies. It does not help that the boundaries between crowdsourcing, human computation, and collective intelligence have not yet been clearly defined. For more information on this issue, refer to "Managing the crowd: toward a taxonomy of crowdsourcing processes" by Geiger *et al.* (2011) and "Human computation: a survey and taxonomy of a growing field" by Quinn and Bederson (2011). This chapter assumes that it is acceptable to refer to large-scale information extraction through search query analysis as "crowdsourcing," since the opinion of a large number of people is aggregated. Using audio CAPTCHA to transcribe speech (Schlaikjer 2007) is another good example of crowdsourcing. The present chapter mostly addresses a specific genre of

Crowdsourcing for Speech Processing: Applications to Data Collection, Transcription and Assessment, First Edition.
Edited by Maxine Eskénazi, Gina-Anne Levow, Helen Meng, Gabriel Parent and David Suendermann.
© 2013 John Wiley & Sons, Ltd. Published 2013 by John Wiley & Sons, Ltd.

crowdsourcing: **labor market crowdsourcing**. However, the preprocessing and quality control sections certainly apply to other crowdsourcing approaches.

This chapter uses the terminology set forward in the introduction of this book. Here is a list of some of the most frequently used terms:

Worker: Someone completing a task.

Requester: Someone publishing a task on a crowdsourcing platform.

Task designer: The person designing the task to be crowdsourced. This person can also have the role of requester.

Crowdsourcing platform: A platform to connect the requester with the workers, which often takes care of certain administrative tasks (payment, assignments, etc.). Examples, House-made platform, Amazon Mechanical Turk (MTurk), Crowd-Flower, JANA, oDesk.

Open call: The act of making a list of tasks publicly available to the workers.

Judgment: An answer/response from a worker for a given task. For example, a label or a transcription.

Worker pool: A set of workers that are available to a requester through a crowdsourcing platform.

One of the objectives of this book is to present information in a technology/platform-independent way. This should serve the reader better by allowing them to transfer relevant guidelines and knowledge to whichever platform best-fits their needs, whether it is a commercial platform or a homemade platform. However, because MTurk has had such a large influence on this field since its launch in 2005, many of the other platforms launched after that have adopted similar concepts and models. For this reason, it is common when we hear about "crowdsourcing platforms" to assume that they follow the MTurk model: requesters post a task for a certain price, then it becomes visible to workers who can complete the task in exchange for monetary compensation. For simplicity, this labor market crowdsourcing model is the one used in this chapter. However, there are many other possible models! The workers could be deciding how much they want to be paid to complete a task (e.g., fiveerr.com). Instead of being an anonymous relationship, where requesters' and workers' real names are not shared between parties, it could be fully identified. Perhaps, instead of having an authoritative relationship from the requester to the workers (as in MTurk where requesters can reject work), the relationship could be the other way around. If the requester is not careful in the way he or she publishes a task and pays the workers, the community as a whole could block that requester. If you are interested in learning more about crowdsourcing platforms, make sure to read Chapter 9.

4.2 Transcribing Speech

Speech transcription is defined as the written representation of speech, and can also refer to the act of transcribing speech (e.g., a worker is performing *speech transcription*). Because audio signals are continuous, and the written form is discrete, transcribing speech is inherently a hard task. A functional decomposition of speech transcription presented in (Roy and Roy 2009) is

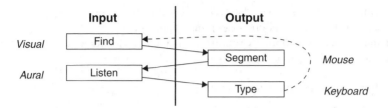

Figure 4.1 Functional decomposition of speech transcription (Roy and Roy 2009). Reproduced by permission of Brandon Roy.

shown in Figure 4.1. In the first step, FIND, the transcriber identifies speech in a stream of audio, which is SEGMENTed into a manageable chunk of audio. The transcriber then LISTENs to the segment and TYPEs it down. This decomposition into four basic tasks is interesting in that it allows the task designer to better understand how each one can be optimized. For example, certain transcription software will make the FIND step easier by providing a visual representation of the signal, such as the spectrogram, so identifying speech is easier. Also, certain processing techniques can be used to complete the FIND and SEGMENT step for the transcriber, thus making their life easier.

During the TYPE step, different types of symbols can be used to represent speech. The most common form is orthographic transcription, where the labels used by the transcribers are words. Note that the task designer can allow the usage of pseudowords, to designate, for example, lip smacks or background noise. Phonetic transcription is a more detailed type of transcription, where the transcriber maps phones to written form. The written form used by the transcriber is defined beforehand and is typically a phonetic alphabet, such as the International Phonetic Alphabet. Again, the transcribers have access to pseudosymbols that reflect certain nonlinguistic characteristics of the signal. Note that if an even more detailed representation of the speech is needed, a special case of phonetic transcription has to be used: analphabetic transcription. One such example is the system presented in (Pike 1943), where the transcriber uses a series of symbols to identify different components of speech instead of using an alphabet. Each type of transcription has a different learning curve for the transcriber. Because of the complexity of phonetic transcription, most attempts at crowdsourcing speech transcription focus on orthographic transcription. This form of speech transcription is intuitive to most nonexperts.

There are several other levels of transcription. For example, discourse analysis might require the annotation of interaction features (overlapping speech, turn-taking, grounding, etc.) and paralinguistic features (accent, stress, intonations, etc.). While there has been some attempts at crowdsourcing these annotations (Akasaka 2009), this will not be the focus of this chapter. More information on that topic can be found in *Transcribing Talk and Interaction* (Jenks 2011).

4.2.1   The Need for Speech Transcription

There are many reasons why a requester would want to transcribe speech. Feeding data to learning algorithms is probably one of the main ones at present. For example, training an automatic speech recognition (ASR) system requires training data in the form of phonetic transcriptions. The transcription can also be used to construct a language model. For example, when developing a spoken dialog system (SDS) for a new domain, developers often start by using a generic language model, possibly interpolated with a language model from a similar domain. Then, when enough speech data for that domain has been gathered from the interaction of the users with the SDS, the speech can be transcribed and used to train a domain-specific language model. Finally, there are many cases where the consumers of the transcribed speech will not be learning algorithms, but humans. Typical examples are movie subtitles and earnings call transcripts.

Transcription Systems

Speech transcription tasks are not all alike: different uses imply different transcription schemes. For example, if the transcription is used to build acoustic models, specific labels can be used to identify lip smacks and other noises. If the objective is to build a language model, then noise labels would not be necessary, and thus probably not annotated. A **transcription system** (Senia 1997) is used to formalize what should be transcribed, and how it should be accomplished. This is especially important when transcribing interactions and paralinguistic features. For example, such a system might define how to transcribe speech overlap, how to identify and represent turn-taking or how to annotate stress and intonation. While simple orthographic transcription does not require as much formalization, defining a clear transcription system is still important in order to obtain consistent data. If the task is not well defined, transcribers will use their own formalism (e.g., using the "%" character in place of noise, see Section 4.4.4) that will make aggregation more complex. Once you have defined a system, you need to communicate it in a simple way to the workers. Section 4.4.3 on instruction design will provide more guidelines on communicating the chosen transcription system to the workers.

4.2.2   Quantifying Speech Transcription

In order to understand some of the discussion in the following sections, the reader should be familiar with ways of quantifying transcription quality, such as word error rate (WER), as well as other metrics related to speech transcription. While these metrics are not specific to transcription accomplished through crowdsourcing, they are used in the literature for example to measure how efficient an interface is (using the RT metric, explained below), or to quantify the effect of wage on quality (using the WER metric, explained below).

Measuring Quality

The most common measures of quality of speech transcription assumes an authoritative reference transcript (also referred to as "gold standard (GS)"); see Section 4.3.3 for a discussion on the creation of such resources. Using this GS, the quality of transcription can be quantified by finding the *distance* between the transcription and its ground truth. There are two dimensions to be considered when doing so: the string metrics used to evaluate, and the unit at which the transcription will be evaluated (e.g., character, phone, word, and phrase).

A classic string metric used, edit distance, also known as Levenshtein distance, attempts to capture the minimum number of symbol edits to do in order to get from the reference string to the target string. An edit is one of the following: substitute (change one symbol for another), insert (insert a symbol in the reference), delete (delete a symbol from the reference). Let us take the example of the utterance *this pizza is cold* and compute its character edit-distance to one of the worker transcriptions: *this pita is cold*. In this case, the edit distance is 2. By replacing one *z* with *t* and removing the other *z*, we obtain the target transcription. Since there are multiple edit paths between a reference and a target transcription, a dynamic alignment of the two transcriptions has to be done to correctly identify the smallest distance. Note that this distance is sometimes normalized to obtain an error rate: simply take the distance obtained and divide by the number of symbols in the reference string. In our case, we have 18 characters in the reference string, thus giving an error rate of 1/9.

There are many more string metrics available (see also hamming distance and cosine similarity), but for the purpose of speech transcription, edit-distance is the one most commonly used. However, the unit for which the distance is computed is not always the character. For the utterance *the man was not there though*, both of the following have an edit distance of 4:

- **Ref:** *the man was not there though.*
- **T#1:** *the man was not there too.*
- **T#2:** *the can is not here though.*

For many speech applications, the **T#2** is worse than **T#1** because errors span multiple words. So, for example, if we are using the transcriptions to train an ASR system, the second transcription might insert noise in more phone models than the first one. For this reason, the quality of a transcription is often evaluated at other units. The three most common units of evaluations are character, word, and phone. The character corresponds to the example presented above: the string is composed of every character in the transcription. When considering the quality at the word level, you now look at every word as a distinct symbol. In our example, the reference is now made up of six symbols: *the, man, was, not, there, though*. Now, when we compare with the first transcription above, *the, man, was, not, there, too*, we see that only one symbol has to be substituted to obtain the target string. With the phone, the distance is computed using the phone representation of that string. So in our example, the reference would be *DH AH . M AE N . W AH Z . N AA T . DH EH R . DH OW*. Table 4.1 shows the error rates (word, character, and phone error rates (PERs)) of two candidate transcriptions.

The level to use depends on what you are trying to measure. For example, if you are assessing the quality of the phone models of an ASR system, looking at the number of errors in the phone representation of the output, also called PER, is preferable. Since crowdsourced transcription is generally orthographic transcription, the most commonly used metric in the literature is the WER, which corresponds to the edit-distance at the word level, normalized with the number of words in the reference string:

$$\text{WER} = \frac{S + I + D}{N},$$

where S is the number of substitutions, I is the number of insertions, D is the number of deletions, and N is the total number of words in the reference string.

Table 4.1 Example error rates of two transcripts.

		Word error rate (WER)
Ref:	[the, man, was, not, there, though]	
T#1:	[the, man, was, not, there, too]	1/6
T#2:	[the, can, is, not, here, though]	3/6
		Character error rate (CER)
Ref:	the man was not there though	
T#1:	the man was not there too	4/28
T#2:	the can is not here though	4/28
		Phone error rate (PER)
Ref:	DH AH . M AE N . W AH Z . N AA T . DH EH R . DH OW	
T#1:	DH AH . M AE N . W AH Z . N AA T . DH EH R . T UW	2/16
T#2:	DH AH . K AE N . IH Z . N AA T . HH IY R . DH OW	5/16

A related measure, called word accuracy, is sometimes used to describe the performance of a system:

$$\text{Accuracy} = 1 - \text{WER} = \frac{N - S - I - D}{N}.$$

Note that WER can be larger than 1, for example, if the transcription evaluated has an arbitrarily large number of inserts. WER is also used in other domains such as machine translation. For reference, the National Institute of Standards and Technology (NIST) reports that the average disagreement between two experts in the Rich Transcription evaluation series (NIST 2011) is between 2% and 4% WER.

Section 4.6.2 presents several ways to estimate the quality of a transcription without the use of ground truth. The majority of these approaches harness interworker variance (e.g., workers A, B, and C agreed, but worker D gave a completely different answer) or intraworker variance (e.g., worker A provided transcription X for a given utterance, and transcription Y for the same utterance 10 minutes later). A different approach that does not rely on the redundancy of the transcriptions is to use forced alignment to compare a worker transcription with the original audio clip. The forced alignment score (log likelihood) can then be used as a proxy for reliability of the transcription. This approach is described in more detail in Audhkhasi *et al.* (2011b).

Measuring Transcription Rate

To quantify how long it takes to complete a given transcription task, a typical measure is to take the ratio of time to transcribe a given utterance, over the duration of the utterance. For example, if, on average, it takes 10 minutes to transcribe a 1-minute utterance, that ratio is 10. It is usually written as 10xRT, to reflect the fact that it takes 10 times **R**eal **T**ime to transcribe the audio. The lower that ratio is, the less resources have to be allocated to transcribe a given dataset.

A major factor is the transcription system used. If you ask the workers to complete phonetic transcription along with interactional and paralinguistic annotations, it will take them longer to complete than if they were only doing the orthographic transcription. Try to keep it simple, asking for no more annotation than what you really need. Section 4.3.1 addresses how proper audio segmenting can improve the workers' transcription rate by reducing the number of times they have to replay an utterance, while Section 4.4.2 covers user interface optimization that can increase the workers' efficiency. If you are building a speech transcription interface on your own server (see Section 4.4.2), you have the option of measuring the transcription rate by starting a timer when the worker starts transcribing, stopping it when they submit their transcription and logging that information. This information can prove valuable if you want to perform A/B testing on different interfaces to identify the most efficient one.

4.2.3   Brief History

Before recording and playback devices were available, shorthand had to be used to transcribe speech. The process of writing in shorthand, called stenography, involves using symbols to represent syllables, words, or phrases. Although it can capture speech in real time, the process involves a *post hoc* decoding process where the shorthand is converted to the desired written form. Several shorthand systems have been developed throughout history, with some of the earliest evidence of such a system coming from Ancient Greece, around the mid-fourth century BC. A shorthand system for American English was developed by Thomas Lloyd in the late eighteenth century, which he used to transcribe the first meeting of the Federal Congress (Tinling 1961). The advent of audio recording and playback made it much easier to transcribe speech: once the speech is recorded, the transcriber can listen to it as many times as needed.

Recently, several projects have involved transcribing a large sized speech corpus. In 1992, DARPA funded the creation of the Switchboard corpus (Godfrey *et al.* 1992), which contains about 2500 conversations from over 500 speakers for a total of 240 hours of recorded speech and about 3 million words of text. While this is an interesting example of crowdsourcing speech *acquisition* through the telephony system, the transcription was carried out by professional transcribers and was expensive. Another example is the Fisher corpus (Cieri *et al.* 2004), created within the context of the DARPA EARS program. The speech acquisition protocol involved people receiving phone calls, and being asked to speak about a given topic. A total of about 2000 hours (16,000 conversations) were gathered. Given the scale of the annotation task, a special transcription system was developed, called Quick Transcription Specification (QTr). No special effort was required from the transcriber, such as providing punctuation, capitalization or indications of noise or mispronunciation. Cieri *et al.* report that transcribers using QTr achieve a transcription rate of $6 \times RT$.

In 2012, YouTube reported (YouTube 2012) that 1 hour of video was uploaded to their website **every second**. That is about the equivalent of one Fisher corpus every 33 minutes. In order to train ASR systems that can handle the diversity of such speech corpus, larger and larger transcribed speech corpus need to be created. While this was traditionally accomplished by professional transcribers, the scale of the task involves a cost that most small and medium-sized entities (e.g., research laboratories and small companies) cannot afford. This is where crowdsourcing comes into play. While the concept of crowdsourcing has been around for a while, the launch of MTurk in 2005 has had a significant impact on its development by allowing any tech-savvy person to have access to a large pool of workers. The platform has been used by large companies since its inception, but it took some time before the academic world showed interest in its potential. One of the first influential academic papers on the use of crowdsourcing was presented at EMNLP 2008 by Snow *et al.* (2008). It showed that crowdsourcing could provide good-quality data at very low cost for major natural language tasks: affect recognition, word similarity annotation, textual entailment annotation, temporal event ordering, and word sense disambiguation. The first academic paper to describe the use of a crowdsourcing platform for speech transcription came in 2009 (Gruenstein *et al.* 2009). CastingWords (castingwords.com) had been using MTurk for this purpose earlier. Since then, several studies have tested crowdsourcing of speech transcription in multiple settings, for multiple languages. It has been shown repeatedly to be fast, cheap, reliable, flexible, scalable, and to provide good-quality transcription. However, crowdsourcing has drawbacks that you should be aware of before deciding to use it.

4.2.4 Is Crowdsourcing Well Suited to My Needs?

Crowdsourcing is not the answer to every problem. Even if the cost of transcribing speech with crowdsourcing is low, there are still investments to be made. These include learning about the platforms as well as setting up and testing the task. Before you allocate resources to crowdsourcing, you should make sure that it is right for you. Here are some questions that can help you determine if crowdsourcing is the right choice:

- **Do I have the technical resources to do this correctly?** A lot of the work needed is performed by the platforms that you use. However, there are still tasks which require technical knowledge on your side. The steps involved in this process are described later in this chapter, and should be straightforward to anyone with an engineering, programming, or web development background. More information about the different platforms available and their respective technical constraints is presented in Chapter 9.
- **Am I dealing with sensitive information?** Most crowdsourcing platforms are open to anyone, essentially meaning that audio clips you submit with your task are disclosed publicly. Before publishing the data, make sure that you have correct ownership and usage rights to it and that you are permitted to make that data public.
- **Is there enough data to justify the setup time?** If you have only 1 hour of speech and you are not planning to process any more in the future, it might take you more time to learn, set up, and carry out quality control than the time it would take you to transcribe the speech yourself.
- **Are there enough native speakers of that language?** The target language for most crowdsourcing platforms at present is English. The interfaces are often defined in English. However, to the extent that you can define the instructions and provide the input data in any language, you can crowdsource speech transcription for the language you want. Audhkhasi *et al.* (2011a) used MTurk to transcribe Mexican Spanish, Novotney and Callison-Burch (2010) obtained transcriptions of Korean, Hindi, and Tamil speech, and Gelas *et al.* (2012) crowdsourced transcription of Swahili and Amharic recordings. As a matter of fact, a 2010 survey

(Ipeirotis 2010) showed that about 47% of the workers on MTurk were from the United States, 34% from India and 19% from other countries. While these demographics certainly have changed since then, and vary depending on the platform used, it does indicate that some workers are native speakers of languages other than English.

- **Is orthographic transcription sufficient?** As previously mentioned, phonetic transcription is much harder to carry out than orthographic transcription, and thus, requires more training. For this reason, while the literature indicates that crowdsourced orthographic transcription is of good quality (assuming that quality control is applied), the same does not hold for phonetic transcription. If the application for which you need the transcription requires phonetic transcription, you can use a pronouncing dictionary, such as the CMU pronouncing dictionary for English (http://www.speech.cs.cmu.edu/cgi-bin/cmudict), to convert the orthographic transcription to a string of phonemes. This will not take into account regional and other variants that may have been pronounced. Besides dictionaries, grapheme-to-phoneme converters may be of help to obtain a phonetic transcription out of an orthographic transcription (Reichel and Schiel 2005).
- **Does the speech contain specialized vocabulary?** The more common the vocabulary used in the audio, the more likely the nonexperts are to understand the utterance and to properly transcribe it. For example, crowdsourcing the transcription of university-level lectures on quantum physics might not give the best results.

The rest of this chapter focuses on instructions and guidelines for how to create an interface with crowdsourcing platforms for transcribing speech. To avoid obsolescence, there is no absolute reference to specific application programming interface (APIs) or tools, since they may change at any time. Instead, there are pointers to where that information may be found. The following is presented in the most platform-independent way possible; however, in order to add clarity, two platforms (CrowdFlower and MTurk) will be used to show concrete example tasks.

4.3   Preparing the Data

In the decomposition of speech transcription presented in Roy and Roy (2009), the transcriber executes four distinct tasks: FIND, SEGMENT, LISTEN, and TYPE. In most cases of crowdsourced transcription in the literature, the first two steps are completed for the workers before submitting the open call. In this way, the same segment can be presented to multiple workers to allow for aggregation. It also gives them the opportunity to focus on the harder tasks of listening and typing words. Segmenting, along with other steps that can be taken before submitting the speech for transcription, is covered in this section.

4.3.1   Preparing the Audio Clips

Segmenting the Signal

It is advantageous if the entire utterance to be transcribed can be contained in the working memory of a worker. Several factors have been shown to have an impact on the memory recall of a speech utterance, two of which are speech rate (the slower the better), and the type of utterance (the more structured the better) (Stine and Wingfield 1987). Individual differences, such as age, also have an effect on working memory capacity. While you cannot directly control for these factors, you can easily control the length of the utterance given to the workers. A good length is one that allows the worker to transcribe the utterance without having to replay the clip, while giving enough context for them to properly disambiguate words if needed. Varying lengths have been used in the literature: around 2–3 seconds (Parent and Eskénazi 2010; Audhkhasi *et al.* 2011), 5 seconds (Novotney and Callison-Burch 2010), 10 seconds (Liem *et al.* 2011; Marge *et al.* 2010b), and 30–60 seconds (Evanini *et al.* 2010). The large majority of the papers surveyed segment the speech into utterances shorter than 10 seconds. *mp3splt*, distributed under GNU GPL 2, will split audio into multiple small pieces. It also has the capability of splitting at silence, which is preferred since the important context is more likely to be present between pauses. More advanced segmenting algorithms for speech transcription are presented in Roy and Roy (2009).

Compressing the Signal

In an architecture where the audio clips are stored remotely from the worker interface, compressing the data will ensure that workers do not have to wait before starting to listen to the utterance. For speech transcription, the most important feature of a codec is that it sufficiently captures the frequencies carrying the voice signal. The typical hearing range is 20 Hz to 20 KHz. A typical male adult has a fundamental frequency (i.e., the frequency of the lowest frequency sinusoidal in the signal) between 85 and 180 Hz and the range is 165–255 Hz for woman (Titze 1994). A good codec will capture the fundamental frequency, and keep as many harmonics as possible. The bandwidth for a single human voice transmission channel is 4000 KHz. Note that humans can still understand speech when the fundamental frequency is not directly present in the signal since it can be reconstituted through its harmonics. For example, the plain old telephony system (POTS) uses a frequency band of around 300–3400 Hz, thus dropping most people's fundamental frequency and the first harmonic. The first major lossy compression to use this perceptual slackness was the MPEG-1 (known as .mp3 files). This encoding scheme goes even further, and harnesses another limitation of sound perception: auditory masking. This phenomenon occurs when a sound is masked by the presence of another sound. While they are commonly encoded at 128 Kbps rate, the algorithm can be parametrized to change the quality/compression rate trade-off.

The major drawback of MPEG-1 is the patent situation. The technology implemented in all MP3 encoders is covered by patents owned by the Fraunhofer society. Consequently, if you are going to use MP3, make sure to use an encoder that pays royalties to avoid any legal issues. Visit mp3licensing.com for more information. Another option is to use a patent-free technology, such as the Vorbis codec and its companion .ogg container. There are several open-source encoders available on the web, and many Flash and HTML5 audio players are compatible with .ogg files.

4.3.2   Preprocessing the Data with a Speech Recognizer

The first reason why you might want to use ASR on your input is to detect clips that are not worth sending through the transcription process. For example, if a clip is empty, too noisy, or in another language, it might be better to filter it out and not send it to workers for transcription (unless you need this type of utterance for some reason, e.g., noise models). Most ASR systems will provide a confidence score with their output, which usually corresponds to the likelihood of the output text being the true written representation of the speech. Using an empirical threshold on that score, clips that are not worth transcribing can be filtered out.

The other use case for ASR is to provide its output to the worker along with the audio clip. This has been used in a task where workers were asked to label the ASR output as *correct* or *incorrect* (Parent and Eskénazi 2010). The advantage of this approach is that the worker does not have to type words for utterances that were correctly recognized by the ASR, thus reducing the cost of the task. The higher the performance of the ASR, the fewer clips that have to be transcribed. In the study above, 54% of the utterances were correctly recognized by the ASR, leaving much less to be transcribed in another transcription task. Another possibility is to allow the workers to edit the ASR output (McGraw *et al.* 2011). The idea behind this approach is similar: if the ASR transcription is of good quality, the worker will be providing transcriptions much faster, perhaps only changing one or two words of the ASR output from time to time.

4.3.3   Creating a Gold-Standard Dataset

The use of a GS dataset has been widespread among the early adopters of crowdsourcing. In the case of speech transcription, the GS contains expert transcriptions for a subset of the data to be annotated. Reusing an existing transcribed speech corpus, such as the Wall Street Journal speech corpus (Paul and Baker 1992), is a possibility if no in-domain expert transcription resources are available. However, it is much better to build your own GS from speech from the same domain for which you will be crowdsourcing transcription. The most important use case for GS is quality control: by injecting certain utterances of the GS in the utterances to be annotated by the workers by observing the match between the worker and the GS, it is possible to measure whether the worker is doing well or not (see Section 4.6 on quality control). CrowdFlower (crowdflower.com) will provide instant feedback (Figure 4.2) to the worker according to a GS provided by the task designer.

The transcription system used for the creation of the GS should be the same as the one the worker will be using. It is also good if the creation of the GS uses the same interface

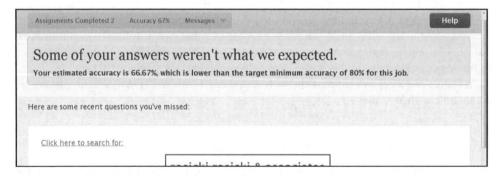

Figure 4.2   Instant feedback in a CrowdFlower task. Reproduced by permission of CrowdFlower.

that the workers will be using, to avoid possible interface bias. Asking multiple experts to transcribe the same GS utterances allows you to compute expert–expert agreement. Having multiple GS transcriptions also gives you the option to only inject the "unanimous" subset of the GS, the subset where every expert agreed on the label. For example, you have a GS comprised 200 utterances and 3 different experts label it, they agree on the same label 150 times. Instead of presenting all 200 utterances in the data to the worker, you provide the 150 unanimous instances. The estimation of the quality of the worker will be more precise since the GS should contain less hard-to-transcribe utterances. However, note that if you do this, you will not be able to carry out a comparison of crowd–expert versus expert–expert agreement, since expert–expert agreement on the 150 utterances subset is perfect by definition.

4.4   Setting Up the Task

There are several options available to you as you decide which platform to use. In the last few years, many companies have developed web-based crowdsourcing platforms. Largely influenced by MTurk, they encompass the concept of workers and requesters, providing a way for requesters to specify what the task is and handling the payment system that makes this labor market possible. A large number of startups have built specialized services on top of those platforms. For example, SpeakerText (speakertext.com) will charge a fixed cost per minute to transcribe audio clips. Other start ups have focused on other domains: text translation, image labeling, and so on. If you do not have the technical resources to build the interface with a crowdsourcing platform, and perform your own quality control, these all-in-one solutions may be the way to go. However, the cost is likely to be higher than what you would pay if you did not have an intermediary between you and the crowdsourcing platform.

In general, there are two approaches to setting up a task on a crowdsourcing platform. The first is to use a platform-rendered task, where you provide certain information like the task title and instructions, along with input data, and the platform takes care of building the HTML for you. The other is to design the whole task on your own, and host the task on your own server. In that case, you are merely using the crowdsourcing platform to bring traffic to your server.

4.4.1   Creating Your Task with the Platform Template Editor

The most user-friendly way to post a task is to use a template provided by a crowdsourcing platform. While most platforms offer templates, they differ in how you can customize them. For example, with CrowdFlower you can specify a title for your task, instructions, and various input forms. A simple way to achieve speech transcription through these rendered tasks is to host the audio clips on some external servers, such as Amazon Simple Storage Service (S3), and have the worker click on the link to listen to the audio. If their browser is configured properly, the audio clips should then be streamed from S3, and played within their browser. Typically, you tell the platform what data to use to render the tasks by uploading a file. An example input file for a speech transcription task could be the following:

```
File: input.csv
audio_id
```

Add Questions
Single Line Text
Label
Your transcription
Optional instructions
Default value
Required?
Select a validator:
(optional)
Multi Line Text
Checkbox
Title
Phone calls transcription
Instructions
Your task will be to listen to a short audio recording of spoken words, then write the corresponding words in the text box below. The audio recordings consist of real life recordings of customers calling a bus information system to get the bus schedule. The speaking style is [...]
Click on the following link to listen to an audio clip. Write down the
Your transcription (required)

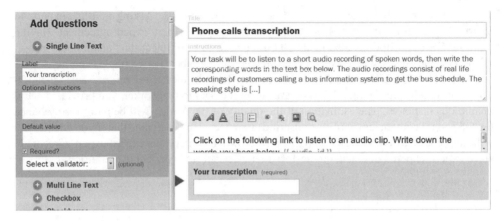

Figure 4.3 An example task designed with the CrowdFlower interface. Reproduced by permission of CrowdFlower.

```
https://s3.amazonaws.com/crowdsourcing/clip1.ogg
https://s3.amazonaws.com/crowdsourcing/clip2.ogg
https://s3.amazonaws.com/crowdsourcing/clip3.ogg
```

Figure 4.3 shows an example speech transcription task designed using CrowdFlower template editor. Once the task designer defines all the required elements (e.g., the title and the instructions) as well as the `input` elements (e.g., `textbox` and `radio` buttons), CrowdFlower renders tasks for as many lines were in the input file (i.e., 3 in this example). Figure 4.4

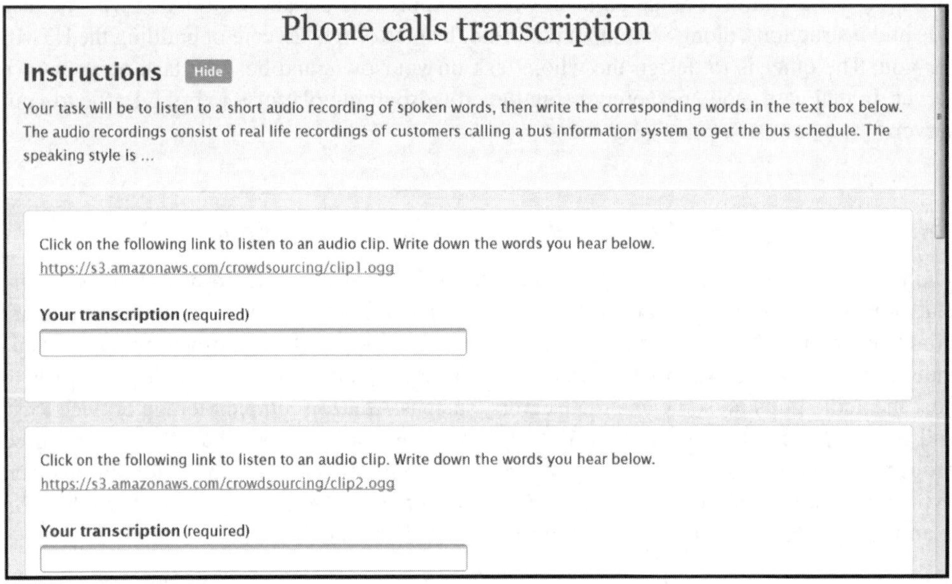

Phone calls transcription
Instructions
Hide
Your task will be to listen to a short audio recording of spoken words, then write the corresponding words in the text box below. The audio recordings consist of real life recordings of customers calling a bus information system to get the bus schedule. The speaking style is ...
Click on the following link to listen to an audio clip. Write down the words you hear below.
https://s3.amazonaws.com/crowdsourcing/clip1.ogg
Your transcription (required)
Click on the following link to listen to an audio clip. Write down the words you hear below.
https://s3.amazonaws.com/crowdsourcing/clip2.ogg
Your transcription (required)

Figure 4.4 The example task once rendered by CrowdFlower.

shows the task produced by CrowdFlower for this example. Visit crowdflower.com for more information on what the platform can do.

An HTML editor is offered with most platforms, so you can build more complex tasks. This approach has several drawbacks: very limited or no scripting capabilities (e.g., no JavaScript), no way to record client-side events (e.g., no way to record time between clicks), and no way to provide customized real-time feedback to the user. The next example provides an alternative approach, where the worker directly interacts with your own server, thus giving you maximum flexibility. Also, in the previous example, the task designer/requester interacts with the platform through a web interface, which does not scale well if you need to publish a task daily because it requires human intervention. The next example uses an API to communicate with the crowdsourcing platform.

4.4.2   Creating Your Task on Your Own Server

One of the crowdsourcing platforms offering the most flexibility is MTurk. There is a special category of task called *external HIT*. HIT stands for "human intelligence task" and external refers to the fact that the HIT lives on an external server, controlled by you and not by Amazon. The actual task interface that you design, build, and host on your server is displayed to the workers through an `iframe` on the MTurk website. Figure 4.5 shows the interaction between MTurk and your server during the lifespan of an external HIT.

The first event is task creation, during which the requester uses the command-line interface (CLI) or the MTurk API to create task assignments. In Figure 4.5, this request originates from

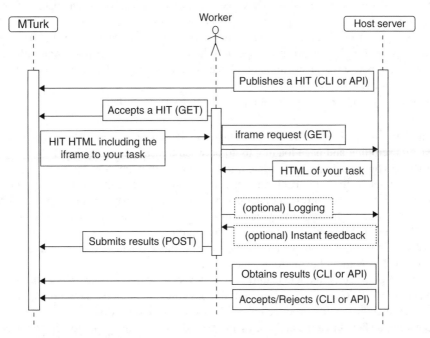

Figure 4.5   Interaction between MTurk and the host server during the lifespan of an external HIT. Reproduced by permission of CrowdFlower.

the host server, but that is not a requirement. When a worker browses MTurk's website and clicks on your task, they obtain an HTML page with an `iframe` to your task. The next step in the interaction is the one that requires the most work on your part: your server has to send the HTML of the task so that it can be displayed in the `iframe`. While the worker is completing your task, you can optionally log certain types of information (e.g., clicks, duration, and answers) and provide instant feedback. Finally, once the worker has completed your task and submits the form, some information, including assignment ID and worker ID, is sent back to the MTurk server so that the payment information can be updated. From that point on, you have the option to download the results from Amazon's server and accept/reject individual results. An advantage of this approach is that you have the choice of using the technology and web server you are the most familiar with. If you are not familiar with web development, there are a lot of good books on the topic, and several free online tutorials, one of the most popular being w3schools.com. This is only a brief overview of the possibilities MTurk offers. Read the Developer Guide, the API Reference, and the CLI Reference (MTurk 2012) for more details.

The Body of the Task

The benefit of hosting the task on your own server is that you have 100% control over what you want to display to the worker. Along with instructions and a feedback text box (covered later), one of the main components is the body of the task, where the worker can listen to audio clips, and transcribe them in an input box. For playing audio, there are multiple free solutions available, some Flash-based while others use the native mechanisms offered by HTML5 browsers. Flash-based players will not be available to workers who have not installed the necessary plug-ins. On the other hand, the support for audio playback in HTML5 depends on the browser used by the workers. See w3schools.com (2012) for up-to-date information on browser support. JW Player (longtailvideo.com) offers both a Flash and an HTML5 video player, free of use.

There are several tweaks that can augment the transcriber's efficiency:

- Fit multiple audio clips and transcription boxes in the same task body. This way, the workers have less scrolling and page loading to do between transcriptions. Budget for paying more per HIT if you are asking for multiple transcriptions within the same task.
- Add a handler on the transcription input text box so that when the worker clicks on it, the audio player plays the corresponding clip (possible depending on the audio player API).
- Set the tab order right so that they can quickly jump from one transcription text box to the next. Use the HTML *tabindex* attribute for that purpose.
- Keyboard shortcuts can be created to allow the workers to pause, replay, and fast forward the audio clips without having to use their mouse.
- Possibly, provide autocomplete features for the text box, so that when workers start transcribing certain words (e.g., name of streets), the possible completions show up automatically. Be aware that this could introduce bias in the worker's transcription.
- Depending on how large the audio files are, consider buffering the audio clips on page load so that they are immediately available to the workers when they click on play.
- Make sure to mention these optimizations in your instructions, so the workers can use them.

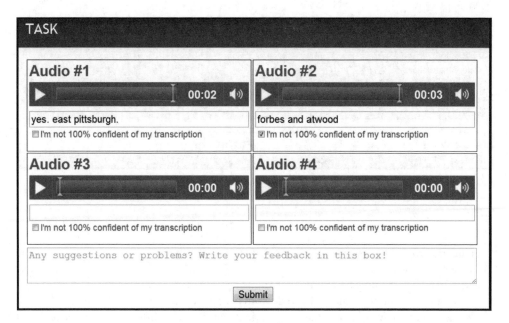

Figure 4.6   A possible interface for speech transcription.

Finally, do not forget to test your web interface on the major browsers (yes, that includes older versions of Internet Explorer). Nothing will annoy the worker more than an interface which does not display or work properly.

A possible interface is shown in Figure 4.6.

4.4.3   Instruction Design

No matter which crowdsourcing platform is used, the task designer must communicate what is expected from the workers. This step is crucial, since a misunderstanding on the worker's part will decrease the overall quality of the transcriptions obtained. Here are a couple of guidelines:

- **Start by clearly stating what the task is:** While "transcribing" and "transcription" are easily understandable by most people, it is wise to use words that are unambiguous to as many people as possible. A possible alternative could be "You will write down in words what you hear in the audio clips."
- **Clearly explain your transcription system:** Unlike formal written form, common speech is continuous and noisier. False starts, repetitions, interjections, and poor syntax are very common. Because most nonexperts are used to writing formally, it is important for the worker to understand that transcription is different. To achieve this, here is a list of common points that you may want to emphasize:
  - **Capitalization:** Tell them if proper casing matters for you.
  - **Punctuation:** Should they use punctuation to express pauses and intonations?
  - **Abbreviations:** In most cases, you do not want them to abbreviate words.

Figure 4.7   Collapsible instructions.

  ○ **Contractions**: In most cases, you want them to use whatever best corresponds to the speech they hear.
  ○ **Numbers**: Should they be written with digits or letters?
  ○ **Profanity**: Should they annotate swear words or replace them with a placeholder?
  ○ **Noise**: You may want them to use a generic placeholder, or different labels for different noises.
  ○ **Partial words**: What if the speaker said: "I want the bl-red one?"
• **Cover edge cases:** What should happen if the audio snippet is empty, or if there is nothing but noise? What if it looks like no label applies to this utterance? What if the audio clip is in another language? Fleshing out these edge cases will also help you understand your dataset better.
• **Provide many, many, many examples**: This cannot be emphasised enough: do not expect the workers to know what the task is if you do not fully explain it. The number of workers with previous experience in speech transcription and labeling is limited. The results you will obtain are directly influenced by how well you explain your task. Experience shows that one of the best ways to describe how to complete a task is to provide numerous examples. If the examples are starting to take too much real-estate on the screen, make sure to use a collapsible instructions/examples section (Figure 4.7).
• **Provide a list of uncommon words that workers may encounter**: If your speech corpus is domain-specific, and contains words that workers are not likely to know how to spell, you might save them a lot of time looking up words if you provide them with a list of such words. For example, if your data contains a lot of street and district names, you could compile a list and make it available to the workers.
• **If there is remuneration with the task, set the expectation**: Workers are often concerned that the time spent on certain tasks will be wasted if the requester rejects their work. For your task to be attractive to the worker, you need to clearly define the cases that would cause you to reject their work. A fictive example could be "We will audit 10% of the task you submit. If more than 50% of your audited answers are wrong, we will reject all of your work." This approach also has the advantage of weeding out workers who were not planning to provide correct answers. To reassure them further, you can specify that in the case where there is a bug on your side of the interface, you will still pay them for their work.

Tutorial and Qualification

It is possible to pair instructions and examples with a proper tutorial, where the workers are walked through several examples, ending with a pretest questionnaire where their understanding of the task is assessed. MTurk offers this built-in feature that allows you to block access to workers who have not passed the pretest, referred to as qualification in MTurk terminology. If this feature is not built-in to the platform you are using, you can implement it yourself, given that the task is residing on your server.

4.4.4   Know the Workers

Do not forget the market nature of most crowdsourcing platforms: when a worker is browsing tasks and is looking at the one you have published, if they think their time will not be well spent working on the task, they will work for someone else. If you are interested in learning more about what workers are concerned about, try visiting workers' discussion boards. Currently, two active ones are *turkernation.com* and *mturkforum.com*.

Another good idea is to offer the workers the opportunity to give you feedback on each task they complete. This can be implemented by adding a text box titled "Please, leave feedback" at the end of each task. Following are real comments obtained for a speech transcription task and a transcription validation task (Parent and Eskénazi 2010). Hopefully, the following will show how useful workers' feedback can be.

Workers Comments

Some of the comments will help you identify bugs and technical problems with your task. Consider logging worker ID along with information of their browser (which can be obtained through JavaScript). It might be useful for investigating issues:

- "audio #2, 3, 4, 5, 6, 7, 8, 9, 10 don't play. just showing at the bottom of the browser connecting to reap.cs.cmu.edu but not playing the audio file"
- "Always hide instructions does not appear to work on Safari. it is getting annoying. :("

Other workers leave comments on the design of the task, and how it could be improved:

- "Seeing the text before hearing it sets up a listener bias such that what might have seemed ambiguous seems instead to be what has been suggested."
- "Good thing i do not have to add what i hear phonetically."
- "it would be nice if the audio went through some preprocessing to normalize the volume levels."
- "It would be great if you introduced a hotkey for replaying an audio."

One of the main comments left by workers concerned profanity in the audio clips. At that time, we had not given any particular instructions for these words, and some workers even came up with their own guidelines. We later included specific instructions to solve this issue:

- "I guess you want us to go ahead and transcribe profanity. It wasn't addressed specifically in the instructions. see audio #9"
- "Used % in place of profanities."
- "sorry for the obscenity, but that's what the guy said."

Asking for feedback helped us quickly realize that it was very helpful to the workers to have a list of common streets in our corpus:

- "My biggest difficulty with these HITs is not being familiar with the place and street names that will be used. If there is a way to include a list of these in the HIT instructions that would be very helpful. I resolved as many as I could by doing Google searches for what I thought I was hearing."
- "please tell me what city these street names refer to, or give me a list to double check against."
- "It would be very helpful for the instructions to list most of the streets the buses run on."

It is also just good to hear that some of your workers are enjoying the task. According to a survey by Ipeirotis (2010), 258 of the 808 MTurk workers surveyed (32%) participate because the tasks are fun, so try to make your task fun!

- "Thanks for giving me a chance to do this. I'm originally from Pittsburgh and its great to hear the accent again!"
- ":D Fun. Thank you."
- "Love these hits. Great how tabbing to the next file makes it play!!!"

Workers E-mail

Some crowdsourcing platforms will provide a way for the workers to send e-mail to the requesters. For some of the tasks, we published in the past (Parent and Eskénazi 2010), there have been as many as 30 e-mails per day from workers worried that they had not been or were not going to be paid, or wanting to say that something was wrong with the task. Some examples, so you can see what to expect:

- **Workers whose work has not yet been approved**:

  Hi. I would like to do some of your "Compare recorded speech with its transcription" hits, but my work from 2 days ago still has not been approved. Could you let me know if or when you'll be approving these? Thank you.

- **Workers thankful for payment**:

  Thank you very much for accepting my work and the payment. Definitely I will go through the work carefully to get maximum quality work.

- **Workers wanting to do your task, but not sure they qualify for it**:

  Dear sir, I am interested in your hit, please do allow me to do the hit! thanking you

- **Workers wanting clarifications on the instructions**:

  I've been enjoying working on this project. Hearing some people get annoyed with the automated system has been so funny. I read the instructions and submitted lots of hits. The only problem is with the no answer... it was clear and correct so a lot of them I said understandable. I went back and reread the instructions and it says they should be not understandable. Just to be clear, it should be the latter. Is that correct?

- **Workers complaining about their work being rejected:**

  Thank you for rejecting what I did on that I had everyone of them right, there is no questioning that.

4.4.5   Game Interface

The interface described in this section is task oriented: workers have clear instructions, and a list of tasks to complete. An alternative is to build a game-like interface (a.k.a game with a purpose or GWAP (von Ahn 2009)) where players transcribe speech utterances in the context of a game. One of the advantages of this approach is the added incentive: workers are not working, they are playing. Such interfaces can be used in conjunction with crowdsourcing platforms, for example, the game is posted on MTurk and a little monetary compensation is given to play the game. However, they are more frequently found as stand-alone games without monetary incentive. Some of these games are published through their own domain, such as www.gwap.com, and some are published through gaming platforms, such as the Facebook games.

In his masters thesis, Rio Akasaka presented an example implementation (Akasaka 2009) of a Facebook game where players transcribe speech (Figure 4.8). Over the course of 3 weeks, more than 2500 transcriptions were obtained from this game. Liem (2011) presents another transcription game and discusses the aspect of enjoyability, including having a multiplayer mode, providing feedback on performance and imposing a timing constraint. Another option is to show all recognition hypotheses to the players, and ask them to click on the correct one (Luz *et al.* 2010). There is less effort required to play this game because the players do not have to use their keyboard. A possible extra incentive for playing a transcription game is to gain knowledge of a language. Gruenstein *et al.* (2009) presents an educational game that can serve the purpose of transcribing speech.

4.5   Submitting the Open Call

This section addresses considerations when submitting an open call through a crowdsourcing platform. Most of this information only applies to models similar to the MTurk one, where the task designer needs to specify certain parameters such as the number of workers who will be completing the task, and how much the workers will be paid. If, for example, you are building

Figure 4.8 Speech transcription with a game interface (Akasaka 2009). Reproduced by permission of Ryo Akasaka.

a GWAP through Facebook, then you might want to skip this section, and jump to Section 4.6 on quality control.

More precisely, this section addresses the issues of deciding how much to pay workers to complete your transcription and deciding how many distinct workers should be transcribing each utterance. It also presents some general guidelines about submitting an open call.

4.5.1   Payment

One of the parameters the requester has to set when submitting an open call is the wage the workers will be paid. Some platforms (CrowdFlower is an example) can suggest a wage based on an estimation of how long it will take workers to complete your task. If you are using a platform that does not offer this feature, here is some information that might be useful in determining payment.

Intuition dictates that the more you pay a worker, the more motivation they have to complete your task, which in turn will increase throughput and quality. While there is clear evidence that paying more does increase the rate at which workers complete your task, studies suggest that it does not have the expected effect on the quality. Novotney and Callison-Burch (2010) and Marge *et al.* (2010a) submitted transcription tasks at different wages and observed no difference in the quality of the transcriptions. The latter study notes that a task that took 13 hours to complete with a wage of $0.05 took 62 hours to complete when the wage was set to $0.005. Studies that crowdsourced several hours of transcription reported payments of $0.01 and $0.005 per 5 seconds utterances transcribed, respectively (Novotney and Callison-Burch 2010; Parent and Eskénazi 2010).

While these numbers can give you an idea of how much you might have to pay, the reality of labor markets is that the interest workers will have in your task will depend on what other tasks are being offered to them at that time. There is no absolute answer to how much you should pay; the practical solution is to start with a reasonable wage, and adjust the price according to the throughput you observe and the feedback you receive from the workers. This trial and error method has been used in several studies (Novotney and Callison-Burch 2010; Parent and Eskénazi 2010; Williams *et al.* 2011). Equally important to setting a price that will bring workers to your task is setting a price that is fair to the workers. See Chapter 11 for more information on ethical aspects of crowdsourcing.

Another important reason to set a good price is to maintain a good reputation within the workers' circles, and to ensure that they come back to your tasks when you publish more. In a project that collected more than 1,000,000 judgments from over 1000 workers through MTurk (Parent and Eskénazi 2010), after the first batch of transcriptions were completed (+100,000 judgments), the best workers were identified (see Section 4.6.3) and awarded bonuses between $5 and $10 depending on their performance. This strategy, along with paying the workers quickly, and personally answering each question asked by the workers proved to be effective for worker retention. The next batch of transcriptions published were completed quickly, and, overall, workers transcribed 3900 utterances per hour, which translates to an average of about 5 hours of speech translated per hour.

4.5.2   Number of Distinct Judgments

One of the key features of crowdsourcing is that it allows a requester to obtain labels from multiple workers for the same target utterance. Under the right conditions, aggregating those multiple labels will provide a more reliable final label. This phenomenon applies to speech transcription tasks as well and can be explained by the individual perceptual differences between the workers. Workers inherently perceive audio clips differently, while some utterances might be hard to understand for one worker, and easy for the others, the reverse situation might apply with a different utterance. Whether these differences are explained by differences in background, accent, or expertise, the result is that as a whole the crowd is able to provide transcription approaching expert quality (Marge *et al.* 2010a; Novotney and Callison-Burch 2010; Parent and Eskénazi 2010). So far, there has not been evidence that adding more people to the crowd can degrade the results, unless the added workers are biased toward poor work.

Another parameter that the task designer has to set is how many distinct workers will provide judgment for a given task. This parameter has a large impact on the quality of the final aggregated dataset, and, with most crowdsourcing platforms, will cause the cost of the annotation to grow linearly. The right setting will depend on the quality control used and on the nature of the task.

For speech transcription tasks, some studies asked for three judgments per utterance (Novotney and Callison-Burch 2010), while others asked for five (Akasaka 2009; Parent and Eskénazi 2010). Evanini *et al.* (2010) showed an improvement of 0.4% WER when performing string merging on five utterances instead of three. If you want to precisely measure the effect of the number of judgments, a simple way to do this is:

(a) Create a GS dataset (see Section 4.3.3).
(b) Submit the task for that dataset, asking for a large number of judgments from the crowd (say 15 per utterance). Note that this should not be too expensive since the size of the GS dataset should be small.

1. For $i=1$ to the total number of judgments:
   (i) Keep the first i judgments.
   (ii) Aggregate the selected judgments (see Section 4.6.4).
   (iii) Calculate the agreement between the aggregated solution and the GS.

This approach has been used in Snow *et al.* (2008) for several natural language tasks and in Parent and Eskénazi (2010) for a speech labeling task. The pattern observed is that the precision gained by going from one worker to three and five workers is considerable, but after that the precision seems to converge and there is not much of gain when aggregating more judgments.

Quality versus Quantity

You assign the same utterance to multiple workers in order to increase the transcription quality for that utterance. Section 4.6 on quality control will discuss various ways to use that redundancy to obtain better quality annotation. However, there are applications where the money spent to improve quality through redundancy would be better spent annotating more data. This "more data is better data" argument has often been advanced. In the case of speech transcription, if the end goal is to present the transcription to a reader (for example, as a caption), then the quality of the transcription is what really matters. In this case, redundancy of annotations is perfectly justified. However, if the transcriptions are to be used to train a speech recognition system, Novotney and Callison-Burch (2010) shows that more data is better than better data. In that study, two speech transcription corpora were crowdsourced: one with 20 hours of speech transcribed by three different workers (best transcription selected) and another with 60 hours of speech. The ASR trained on the former corpus outperformed the ASR trained on the latter corpus by 3.3% WER (40.9% WER and 37.6%, respectively). Note that this trade-off is not an issue if the budget is large enough, or if the size of the total dataset is small (e.g., you really only have a total of 100 utterances you could transcribe).

Dry Run

When the size of the data to be labeled is large, it is judicious to first submit a small subset (100 items) of the data before publishing the whole dataset. This way, if a bug is present in your interface, or something is wrong with the input data, you will not end up having to resubmit everything. In case you realize that something is wrong after the tasks have been submitted, most crowdsourcing platforms provide a mechanism that allows you to pause a task. You could also implement a similar safeguard mechanism in your own code, if you are hosting the task on your own server.

Apart from bugs and issue identification, a dry run can be used to empirically determine the optimal number of workers you need for your task and the optimal wage. That dry run can also be used to obtain feedback on the task from the crowd: a "Leave feedback" text box is recommended at all times in order for the crowd to leave comments about the task.

Table 4.2 Transcriptions from workers, and how they are affected by quality control mechanisms.

	Worker A	Worker B	Worker C (bot)	Worker D
Original	Forbes and hatwood	Forbes at atwood	Hello there	Forbes and the woods
Normalized	Forbes and atwood	Forbes at atwood	Hello there	Forbes and the woods
Filtered	Forbes and atwood	Forbes at atwood		Forbes and the woods
Agreggated	Forbes and atwood			

4.6   Quality Control

Quality control (QC) is typically implemented by a set of mechanisms that eliminate noise, filter bad judgments, and improve the quality of the final dataset. It is arguably the most important aspect of the crowdsourcing process. If your current design does not incorporate any form of quality control, you will most definitely end up with suboptimal transcriptions. There have been surveys on how to accomplish proper quality control when crowdsourcing different types of tasks, see Quinn and Bederson (2011) for an exhaustive list of the strategies used so far. This section focuses on quality control for the specific task of speech transcription. For that purpose, QC mechanisms can be divided in three categories : **normalization**, **filters**, and **aggregation**. The effect of each category can be seen in Table 4.2.

Normalization is an optional step where every transcription is cleaned up and possibly autocorrected (e.g., punctuation removed and "Hatwood" autocorrected to "Atwood"). The role of a filter is to remove noise from the input data, whether that noise is caused by bots, bad workers, ill-defined tasks, or bad input data (e.g., remove the outlier transcription "hello there"). The role of the aggregation step is to improve the quality of the data that made it through the filters by merging multiple signals together. Most of the QC mechanisms found in the literature can be classified in one of these three categories. Note that some designers end up using only filters with no aggregation mechanism, or only aggregation and no filter. The next subsections address normalization, filters, and aggregation. The last subsection is reserved for a discussion on using multiple passes, where the output of a pool of workers is provided to another pool of workers for validation, rating, or correction.

4.6.1   Normalization

Normalizing the transcriptions obtained is optional, but can help reduce the number of distinct transcriptions, and thus make some of the subsequent processing more effective. Let us take the example of a worker who provided the following transcription: "At what time?" Another worker, transcribed it in the same way, but without capitalization as "at what time?" If no normalization is done, the WER between the two strings will be 1/3 (the first word, "At," can be replaced by "at"), while it should really be 0 since they are both equally good transcriptions.

The type of normalization that can be applied depends on what you need the transcriptions for. Removing punctuations such as commas and semicolons can help if you do not need the punctuation, for example, if you are building a n-gram language model. However, if you are collecting transcriptions to present directly to humans (e.g., earnings call transcripts), then retaining the punctuation will help improve readability. Several experiments on crowdsourcing speech transcription have used some form of normalization (Parent and Eskénazi 2010; Williams *et al.* 2011).

Here is a list of some normalization strategies you might want to consider:

- Changing casing ("At what time?" to "at what time?").
- Removing punctuation ("Yes, goodbye!" to "yes goodbye").
- Expanding abbreviations ("forbes ave." to "forbes avenue").
- Expanding contractions ("I'm there" to "I am there").
- Numbers ("20" to "twenty").
- Removing diacritics ("J'ai mangé une poutine" to "J'ai mange une poutine").

You should be aware that noise can be introduced during normalization. For example, in the case of contraction expansion, if the speaker effectively said "I'm" (EY M) and not "I am" (EY AE M), then the normalization is introducing a phone that is not present in the signal.

All these strategies can be implemented with a mix of regular expressions and external resources, such as number-to-word mappings and English contraction lists. While you can find abbreviation lists on the web, you might want to complement them with abbreviations relevant to the domain of the speech that is being transcribed.

Another strategy that can be used to reduce the number of distinct transcriptions is to perform spellchecking and autocorrection on the transcriptions, for example, correcting "first avenu" to "first avenue." GNU Aspell is a good starting point if you are looking for an open-source spellchecker. If you believe a full spellchecker might be overkill for your task, you can identify common errors workers make and only autocorrect those. For example, in Parent and Eskénazi (2010), workers had a lot of difficulty in transcribing "Duquesne University," which was transcribed as "ducane," "duquene," and so forth. A list of common misspellings was created for "duquesne" (and other words such as "bigelow" and "squirrel"), and was used to automatically correct transcriptions containing these errors.

4.6.2   Unsupervised Filters

There are three general types of workers that we want to filter out: bots, workers who are trying to game the system, and poor workers. Bots are specialized web crawlers that automatically interact with crowdsourcing platforms and respond to open calls in the hope of obtaining compensation. A typical bot will obtain the HTML definition of the task, parse it in order to understand what the required HTML input elements are, and will submit answers either randomly (unbiased error) or consistently, which is more damaging because it is more likely to bias your final dataset. It is hard to evaluate how widespread the problem is, but bots are definitely being used for this purpose. In working with MTurk, the first evidence that bots were completing tasks (Parent and Eskénazi 2010) was that some of the workers completing the tasks were leaving strange feedback in the "Leave feedback" text box. These were obviously not real feedback comments rather, they were copy/pasted from the task instructions or snippets from web corpus. Another example: in a large-scale project, the postanalysis of how much of the work had been performed by the top 5% workers revealed that the top contributor had submitted more than 6000 judgments, each time providing the same answer, and the following comment in the feedback box: "Sorry task unclear." All submissions by this worker were rejected, and the worker did not complain about the rejections, thus corroborating the hypothesis that the worker was a bot.

CAPTCHA and Qualification

The naive solution for detecting a bot is to insert a CAPTCHA once in a while in your task. Some platforms will do that for you. The main drawback of this approach is that you are slowing down legitimate workers, thus making your task less attractive for them. Also, reverse Turing tests such as CAPTCHA are in theory crackable with... crowdsourcing! There have been rumors of pornography websites asking their users to solve CAPTCHA in exchange for access to images, also known as a pornography attack (CAPTCHA 2012). While the official CAPTCHA website asserts that no evidence of such attacks have been detected, and that in any case, they would have little impact because of the limited throughput such methods offer, it remains a possibility.

Another solution is to require some kind of prerequirement to complete the task. It could be passing some pretest (as described in Section 4.4.3), or having a history of being a good worker. For example, MTurk provides a way to filter workers by the percentage of their answers that have been accepted in the past. Even with these two filters, there is the possibility that some bots work will make it through to submission, and introduce noise in the judgments you obtain. At that point, bots and system gamers submissions can be treated in the same way. A system gamer is a human who can solve the CAPTCHA test and prerequirement, but will not genuinely answer the bulk of the task. Their goal is to go as fast as possible to maximize profit, while minimizing the odds of being detected and rejected.

Time on Task

One effective filter for these two categories of workers uses the time required for task completion and is based on the intuition that if a worker can complete a task faster than it takes to read the questions, the worker cannot be doing genuinely valid work. This is a sufficient condition to reject the worker's submission, but not it is not necessary. Sophisticated bots can be programmed to simulate spending time on the task. Still, it is worth implementing this filter because it is fairly simple to do. To configure this filter, work on your own task several times (or have someone else do this to lessen bias), and record how long it takes to complete the task with accuracy. This average time can then be used to flag suspicious workers. If some of the workers are answering too quickly, no matter if this worker is a bot or a gamer, you can reject their work (and/or contact the worker). Note that some of the honest workers could be using automated scripts on top of your task to legitimately make the task faster to complete. Such an example would be using Greasemonkey to automatically modify the layout of your task (padding, margin, font size, etc.) to reduce the amount of scrolling required to complete the task.

Intraworker Variance

Another filter is based on intraworker variance: if, over the course of a transcription session, the worker transcribes the **same utterance in five different ways**, it is likely that they are not paying much attention, or are simply answering randomly. To detect this, you need to have control over when and how the tasks are displayed to the workers. This might not possible using a template-based platform, but it becomes possible once you host the task on your own server, such as with an MTurk external HIT. When assigning a task to a worker, it is possible to use a heuristic similar to the following: If the worker is not new AND if the number of tasks completed by that worker is a multiple of X, THEN pick an utterance that was transcribed by that worker some time ago and assign it to her again. Then, either at runtime or after you have collected all of the tasks, for every worker and each utterance for which you have multiple transcriptions, compare how much variance there is in their answers using an appropriate string metric, such as WER. Some variance is tolerable, since the worker's own definition of the task evolves over time. However, if the variance is too high, something is most likely wrong with that worker. You then need to decide what threshold you want to use, and what strategy to adopt when a worker does not pass this filter: reject the work, manually investigate or contact the worker.

Interworker Variance

The intraworker variance filter will not be of much help when the worker is a bot that consistently provides the same transcription. Efficient filters in this case are based on inter-worker variance. Intuitively, given four workers transcribing the same utterance, if three of their transcriptions are very similar and one is a complete outlier (such as the transcript "hello there" from worker C in Table 4.2), then it is likely that the worker who provided the outlier gave you a worthless response. Note that all filters in this subsection, including the one based on interworker variance, are unsupervised: there is no need for a human to intervene (except to set a threshold). Contrary to the intraworker variance filters, you do not need to ask the same worker multiple answers for the same utterance, but you do need to ask multiple workers to transcribe the same utterance, which can become costly. Also, it is harder to filter workers in real time, since you need to obtain transcriptions from multiple workers before the filter becomes reliable. This is best implemented as a postprocessing step. For speech transcription, a simple implementation could be as follows:

1. For all utterances, obtain judgments from X workers.
2. After you have collected data for every X workers, compute the WER between their transcription, and every other worker's transcription for utterances that worker completed. If that worker transcribed Y utterances, you should have $Y*(X-1)$ WER values.
3. Compute the average of these values.
4. Filter the outliers, either relatively (say, the 5% workers with worst average WER), or with a certain threshold found empirically.

There are several modifications to this simple heuristic such as using some other string metrics, averaging at the utterance level instead of at the worker level, or measuring the distance between a worker transcription and the aggregation of all worker transcriptions (Audhkhasi *et al.* 2011a). The key idea remains: use the transcriptions from multiple workers for a single utterance to identify workers who should be filtered out.

4.6.3   Supervised Filters

The interworker-based filters have one obvious flaw: if there are more poor workers then good workers, you might end up filtering out the good workers. Also, it requires creating multiple assignments for each utterance, which can become costly. An alternative is to use a human to do part of the quality control. For example, have the task designer (i.e., you) evaluate the quality of all (or a subset of) the transcriptions provided by the workers. This approach does not require obtaining multiple labels for each utterance. If you do not like the idea of having to manually validate each transcription provided by the workers, you can use the crowd to do that validation! Add an extra step to your crowdsourced transcription pipeline, where a different set of workers is asked to evaluate the transcription provided by the first workers (see Section 4.6.5 on using multiple passes). If some workers are consistently being flagged as providing bad transcriptions, then it is a sign that they should be filtered out.

A similar idea uses a GS: instead of having a human validate transcription after the worker has submitted the work, have the human create a GS, and inject some of the GS utterances before submitting the open call. Section 4.3.3 presents other use cases for GSs, as well as general guidelines for their creation. Once GS have been injected into the tasks completed by the workers, we have a simple way to evaluate the quality of the worker by comparing the answers provided by the workers to the GS. Depending on the type of task that is being completed, a metric is used to evaluate the quality of the workers answers. Examples that have been used in the literature include the Pearson correlation for grading tasks, Fleishman correlation for ranking task, and Cohen's kappa for a labeling task. For speech transcription tasks, a GS has been used in several studies, including studies by Gelas *et al.* (2011), Marge *et al.* (2010a), and Lee and Glass (2011). In all cases, WER was used to evaluate the quality of a given transcription compared with the GS. One should keep in mind that since not every worker's answer is known and present in the GS, the metrics obtained are only estimates of the true value and they all have a different confidence interval. A good balance of GS utterances and unlabeled utterances should be achieved in order to ensure that the estimation is accurate without greatly increasing the cost of the task. It would be possible to present 9 GS utterances for every 10 utterances, but then 90% of the cost would go to obtaining judgments for known utterances. Parent and Eskénazi (2010) used a ratio of 10% (1 GS utterance for every 10 utterances). Note that the quality of the estimator is also a factor of the number of tasks completed by the worker: if a worker only completed 10 utterances, then the estimation will be based on only 1 GS utterance. This problem can be solved by showing more GS utterances at first, and reducing the number of such utterances when the precision of the estimator is high enough. CrowdFlower appears to employ such a strategy.

Filters can be designed based on the difference between a worker's answer to a GS utterance and the GS true value. For example, imagine the following scenario. Workers A, B, and C work on one of your transcription tasks. They each transcribe around 500 utterances, thus allowing you to calculate the accuracy of the workers on 50 GS utterances (you injected 10% GS). Worker A has an average WER of 1, B displays a WER of 0.5, while worker C has an average WER of 0.1. This means that on average, worker C will only produce an error every ten words. Then you have to choose how you will use that information to filter out the workers. One possibility is to filter out workers with a WER average above a certain level, and pass the rest to be aggregated. In the example, you would reject the transcriptions from worker A, and keep the ones from workers B and C to be aggregated (see Section 4.6.4).

Note on Filtering Workers

Filtering workers, along with all the instances they provided, can be drastic in some cases. What if the worker just had a moment of inattention while they were transcribing that GS utterance? An alternative is to filter at the transcription level: identify potentially bad transcriptions, and either do not include them in your final dataset, or resubmit them for re-transcription. Most of the techniques presented so far can be applied at the transcription level instead of at the worker level. For example, instead of filtering out outlier workers, a filter can be designed to remove outlier transcriptions.

4.6.4   Aggregation Techniques

Aggregating multiple transcriptions together is conceptually similar to the problem of merging the output of multiple ASR: **given multiple candidate transcriptions of a given utterance, what is the best possible transcription?** Note that the best possible transcription does not have to be exactly one of the candidate transcription, it can be a combination of them.

Naive Aggregation

A naive approach is to keep the most frequent transcription out of all candidate transcriptions. This works best when normalization has been applied to the candidate transcriptions. However, the longer the utterance, the less likely this approach is to succeed. This approach is used in the literature as a baseline for other aggregation techniques. Another trivial aggregation approach is to keep the candidate transcription that was provided by the most reliable worker, as defined by some score (e.g., a score based on the intra/interworker variance or based on the agreement with the GS).

ROVER

ROVER—Recognition Output Voting Error Reduction—is the most widely used form of aggregation for crowdsourced transcriptions. Among others, Marge *et al.* (2010a), Novotney and Callison-Burch (2010), and Williams *et al.* (2011) described the use of ROVER for quality control. This approach described in Fiscus (1997) first performs iterations of dynamic programming alignments between pairs of transcriptions, which produces a word transition network. Then, for each correspondence set, a certain voting function is used to obtain the composite transcription. It is in that voting function that worker reliability scores can be integrated. For example, words from a worker whose agreement with the GS is very high can be set to weigh more than words from other workers. In all the studies cited above, ROVER outperformed the other naive aggregation approaches.

The NIST Scoring Toolkit (available at www.itl.nist.gov/iad/mig/tools/) contains a ROVER tool that provides an implementation of Fiscus' algorithm.

Lattice

Evanini *et al.* (2010) describes another way of aggregating multiple transcriptions: "a word lattice is formed from the individual transcriptions by iteratively adding transcriptions into the lattice to optimize the match between the transcription and the lattice." The lattice must contain an end-of-utterance symbol in order for this approach to work. Once the lattice has been built, the optimal path is selected to be the composite transcription. Optimality can be defined in several ways, for example, by taking the path that has the largest weight. Again, the weight can be a combination of the number of transcriptions using a given edge and the reliability of the workers providing the transcription. This definition of optimality tends to favor a longer path: for example, imagine that one of the workers has provided a transcription with 1000 words, while the three other workers agreed on one with five words. By picking the path that has the largest weight, the longest (and probably wrong) transcription will be selected. To solve this problem, path weight can be normalized using the number of nodes that composes it.

Results

It is very difficult to compare studies one another because the dataset and the conditions under which experiments were run are not the same. Table 4.3 provides an overview of the results obtained from similar studies. All studies took place on MTurk involve a one-pass transcription (except for Parent and Eskénazi (2010) and Lee and Glass (2011)) and use an English dataset (except for Audhkhasi *et al.* (2011a)).

Of all the studies in Table 4.3, only Novotney and Callison-Burch (2010) mentions the average transcription rate of the workers: $11 \times RT$. Also, four out of seven studies used some kind of normalization on the workers' transcriptions. The transcription quality results (WER) are not directly comparable. For example, most utterances in Parent and Eskénazi (2010) were short and had already been correctly processed by an ASR, thus bringing the average WER down. In most of the studies in Table 4.3, the authors conclude that the quality of the transcriptions approaches the quality of expert transcription.

4.6.5   Quality Control Using Multiple Passes

Some studies have investigated using multiple passes of MTurk to reduce the load on the workers. For example, in Marge *et al.* (2010b), workers first transcribe an utterance without fillers. The latter are added to the transcripts in a second pass. Parent and Eskénazi (2010) also used two passes. In the first one, workers classified the ASR output as *correct/incorrect*

Table 4.3   Results obtained from crowdsourced speech transcription studies.

Paper	Filters	Aggregation	WER
Novotney and Callison-Burch (2010)	Interworker	ROVER, best worker	18%
Williams *et al.* (2011)	n/a	ROVER, most frequent	15.6%
Marge *et al.* (2010b)	n/a	ROVER	15.3%
Lee and Glass (2011)	Classifier	ROVER	10.20%
Evanini *et al.* (2010b)	n/a	ROVER, lattice	5.1%-22.1%
Parent and Eskénazi (2010)	GS, qualification	n/a	5.4%
Audhkhasi *et al.* (2011a)	GS, qualification	ROVER	2.3%

and as *understandable/nonunderstandable*. The second pass was used to transcribe only the utterance that had been classified as *incorrect* and *understandable*, thus reducing the overall cost of the task.

Another important reason to use multiple passes is for quality control. In a first pass, workers transcribe utterances, which are then submitted to other workers in a different task for validation, rating and/or correction. Lee and Glass (2011) implemented this strategy with two distinct tasks, *Short Transcription* and *Transcription Refinement*. The output of multiple instances of the first task were aggregated into larger transcripts (75 seconds) for another pool of workers to correct. A similar approach is used in Liem *et al.* (2011). As described in the paper, "each segment is sent through an iterative dual pathway structure that allows participants in either path to iteratively refine the transcriptions of others in their path while being rewarded based on transcriptions in the other path, eliminating the need to check transcripts in a separate process." For those interested in experimenting with iterative tasks, Greg Little developed an open-source Java/JavaScript API for running iterative tasks on MTurk (TurKit 2012).

At least two companies that provide a speech transcription service based on crowdsourcing use multiple passes for quality control. CastingWords segments long audio recordings into smaller overlapping ones, which are then submitted to MTurk for workers to transcribe. Other passes ask the workers to edit, rate, and correct the transcripts obtained with the first pass. In one of their early implementations, SpeakerText also implemented quality control with multiple passes. Audio recordings were first segmented in small chunks of around 10 seconds and transcribed by a first set of workers. This was followed by another pass where multiple workers verified the quality of the transcription. Several transcriptions were then concatenated, to create larger audio segments, and were presented to workers for a round of correction. The intuition behind concatenating multiple short transcriptions is that the workers need to have context in order to identify certain errors. Finally, a fourth pass handled the validation of the edits made during the third pass.

Matt Mireles, co-founder of SpeakerText, mentions two main drawbacks to using multiple passes (Matt Mireles, personal communication, July 2012). First, the total cost per minute of audio increases rapidly with the number of passes, especially if the first pass focuses on small chunks of audio. For every edit pass, another pass of validation is required. Second, since more workers are working on the same utterance *in parallel*, more of them have to get accustomed to the speaker's vocabulary and voice. This also drives the price up since you end up spending more for training the workers.

4.7   Conclusion

In the last 3 years, more than 15 papers on crowdsourced transcription were published. This chapter aggregates that information in one place. It also introduced several general concepts related to speech transcription such as what is included in a transcription system and how to quantify transcription quality. Section 4.3 gave two examples of tasks, one created with the CrowdFlower graphical editor and the other implemented as a MTurk external task. Several guidelines were suggested, including optimizations of the transcription interface and communication tips. Finally, quality control mechanisms (i.e., normalization, filters, aggregation, and multiple passes approaches) were presented, along with results obtained from various studies.

Crowdsourcing has several advantages over conventional transcription. It is easier to scale, cheaper and, given the proper quality control, the quality of the transcriptions obtained can approach expert quality. However, it is not well suited to every situation such as when the speech data is confidential, when there are not a lot of native speakers of the language or when the domain is extremely specialized (e.g., university-level physics lectures). On a final note, remember that the workers transcribing your data should be treated adequately. Their work should not be rejected wrongly, and they should be paid well and on time.

4.8 Acknowledgments

I thank Maxine Eskénazi, Chris Callison-Burch, and Christoph Draxler for their constructive feedback. Also, many thanks to Magali Lemahieu and Elizabeth Huff for their help proofreading this chapter, and to the MTurk workers who participated in my "Proofread a book chapter" task.

This work was supported by National Science Foundation grant IIS0914927. Any opinions, findings, and conclusions and recommendations expressed in this material are those of the author and do not necessarily reflect the views of the NSF.

References

Akasaka R (2009) Foreign Accented Speech Transcription and Accent Recognition Using a Game-based Approach. Masters Thesis, Swarthmore College, Swarthmore, PA.

von Ahn L (2006) Games with a purpose. *IEEE Computer* **39** (6), 92–94.

Amazon Mechanical Turk. http://aws.amazon.com/documentation/mturk/ (accessed 9 July 2012).

Audhkhasi K, Georgiou P and Narayanan S (2011) Accurate transcription of broadcast news speech using multiple noisy transcribers and unsupervised reliability metrics. *Proceedings of International Conference on Acoustics, Speech and Signal Processing (ICASSP 2011)*.

Audhkhasi K, Georgiou P and Narayanan S (2011) Reliability-weighted acoustic model adaptation using crowd sourced transcriptions. *Proceedings of Interspeech 2011*.

The Official CAPTCHA Site. http://www.captcha.net (accessed 8 July 2012).

Cieri C, Miller D, and Walker K (2004) The Fisher corpus: a resource for the next generations of speech-to-text. *Proceedings of the Fourth International Conference on Language Resources and Evaluation (LREC 2004)*.

Evanini K, Higgins D and Zechner K (2010) Using Amazon Mechanical Turk for transcription of non-native speech. *Proceeding of Conference of the North American Chapter of the Association of Computational Linguistics Workshop: Creating Speech and Language Data with Amazon's Mechanical Turk* (NAACL 2010).

Fiscus J (1997) A post-processing system to yield reduced error rates: recognizer output voting error reduction (ROVER). *Proceedings of IEEE Workshop on Automatic Speech Recognition and Understanding (ASRU-1997)*.

Geiger D, Seedorf S, Schulze T, Nickerson R and Schader M (2011) Managing the crowd: towards a taxonomy of crowdsourcing processes. *Proceedings of the Americas Conference on Information Systems (AMCIS 2011)*.

Gelas H, Abate ST, Besacier L and Pellegrino F (2011) Quality assessment of crowdsourcing transcriptions for African languages. *Proceedings of Interspeech 2011*.

Gelas H, Besacier L and Pellegrino F (2012) Developments of Swahili resources for an automatic speech recognition system. *Proceedings of the 3rd Workshop on Spoken Language Technologies for Under-resourced Languages (SLTU 2012)*.

Godfrey JJ, Holliman EC and McDaniel J (1992) SWITCHBOARD: telephone speech corpus for research and development. *Proceedings of the International Conference on Acoustics, Speech and Signal Processing (ICASSP 1992)*.

Goto M and Ogata J (2011) PodCastle: recent advances of a spoken document retrieval service improved by anonymous user contributions. *Proceedings of Interspeech 2011*.

Gruenstein A, McGraw I and Sutherland A (2009) A self-transcribing speech corpus: collecting continuous speech with an online educational game. *Proceedings of ISCA Workshop on Speech and Language Technology in Education (SLATE 2009)*.

Ipeirotis PG (2010) Demographics of Mechanical Turk. Working paper CeDER-10-01, New York University, Stern School of Business. http://hdl.handle.net/2451/29585 (accessed 1 October 2012)

Jenks CJ (2011) *Transcribing Talk and Interaction: Issues in the Representation of Communication Data*. John Benjamins Publishing Company.

Lee C and Glass J (2011) A transcription task for crowdsourcing with automatic quality control. *Proceedings of Interspeech 2011*.

Liem B (2011) Designing a Transcription Game. Undergraduate Thesis, Department of Applied Mathematics, Harvard College, MA.

Liem B, Zhang H and Chen Y (2011) An iterative dual pathway structure for speech-to-text transcription. *Proceedings of Association for the Advancement of Artificial Intelligence Workshop (AAAI 2011)*.

Luz S, Masoodian M and Rogers B (2010) Supporting collaborative transcription of recorded speech with a 3D game interface. *Knowledge-Based and Intelligent Information and Engineering Systems*, pp. 394–401. Springer, Berlin.

Marge M, Banerjee S and Rudnicky A (2010a) Using the Amazon Mechanical Turk for transcription of spoken language. *Proceedings of the International Conference on Acoustics, Speech and Signal Processing (ICASSP2010)*.

Marge M, Banerjee S and Rudnicky A (2010b) Using the Amazon Mechanical Turk to transcribe and annotate meeting speech for extractive summarization. *Proceedings of Conference of the North American Chapter of the Association of Computational Linguistics Workshop: Creating Speech and Language Data with Amazon's Mechanical Turk (NAACL 2010)*.

McGraw I, Glass J and Seneff S (2011) Growing a spoken language interface on Amazon Mechanical Turk. *Proceedings of Interspeech 2011*.

NIST—Rich Transcription Evaluation Project. http://www.itl.nist.gov/iad/mig/tests/rt/ (accessed 8 July 2012).

Novotney S and Callison-Burch C (2010) Cheap, fast and good enough: automatic speech recognition with non-expert transcription. *Proceedings of Conference of the North American Chapter of the Association of Computational Linguistics (NAACL 2010)*.

Parent G and Eskenazi M (2010) Toward better crowdsourced transcription: transcription of a year of the Let's Go Bus information system data. *Proceedings of IEEE Workshop on Spoken Language Technology (SLT 2010)*.

Paul DB and Baker JM (1992) The design for the Wall Street Journal-based CSR Corpus. *Proceedings of the DARPA SLS Workshop*.

Pike KL (1943) *Phonetics: A Critical Analysis of Phonetic Theory and a Technic for the Practical Description of Sounds*. University of Michigan Press, Ann Arbor, MI.

Quinn AJ and Bederson BB (2011) Human computation: a survey and taxonomy of a growing field. *Proceedings of Human Factors in Computing Systems (CHI 2011)*.

Reichel U and Schiel F (2005) Using morphology and phoneme history to improve grapheme-to-phoneme conversion. *Proceedings of Interspeech 2005*.

Roy BC and Roy D (2009) Fast transcription of unstructured audio recordings. *Proceedings of Interspeech 2009*.

Schlaikjer A (2007) A dual-use speech CAPTCHA: aiding visually impaired web users while providing transcriptions of audio streams. *Technical Report 07-014*, Language Technologies Institute, Carnegie Mellon University, Pittsburgh, PA.

Senia F and van Velden J (1997) Specification of orthographic transcription and lexicon conventions. *Technical Report SD1.3.2*, SpeechDat-II LE-4001.

Snow R, O'connor B, Jurafsky D and Ng AY (2008) Cheap and fast—but is it good? Evaluating non-expert annotations for natural language tasks. *Proceedings of Conference on Empirical Methods on Natural Language Processing (EMNLP 2008)*.

Stine EL and Wingfield A (1987) Process and strategy in memory for speech among younger and older adults. *Psychology and Aging* **2**(6), 272–279.

Suendermann D, Liscombe J and Pieraccini R (2010) How to drink from a fire hose: one person can annoscribe 693 thousand utterances in one month. *Proceedings of Special Interest Group on Discourse and Dialogue (SIGDIAL 2010)*.

Tinling M (1961) Thomas Lloyd's Reports of the First Federal Congress. *The William and Mary Quarterly* **18**(4), 519–545.

Titze IR (1994) *Principles of Voice Production*. Prentice-Hall.

TurKit Homepage. http://groups.csail.mit.edu/uid/turkit/ (accessed 9 July 2012).

w3schools—HTML5 audio. http://www.w3schools.com/html5/html5_audio.asp (accessed 7 July 2012).

Williams JD, Melamed ID, Alonso T, Hollister B and Wilpon J (2011) Crowd-sourcing for difficult transcription of speech. *Proceedings of IEEE Workshop on Automatic Speech Recognition and Understanding (ASRU 2011)*.

YouTube—Broadcast Yourself. Statistics. http://www.youtube.com/t/press_statistics (accessed 4 April 2012).

5

How to Control and Utilize Crowd-Collected Speech

Ian McGraw[1] and Joseph Polifroni[2]
[1]*Massachusetts Institute of Technology, USA*
[2]*Quanta Research, USA*

Armed with tools to record audio data from a browser, a researcher can seamlessly deploy speech collection interfaces to large crowds over the web. Although these tools make collection simple, caution must be exercised before deploying any speech interface to a sizable crowd. Researchers often do not make explicit in their publications the significant amount of trial and error that takes place with any new audio-based task. Typically, getting a corpus "right" (in other words, making sure the data meet minimum requirements and can be useful for research) is an iterative process (Draxler and Jänsch 2006; Dickie *et al.* 2009).

In this chapter, we describe several efforts at collecting audio data using MTurk, focusing on lessons learned. We address two overarching characteristics of speech resources that researchers may be particularly concerned with: quality and variety. The first experiments, initially presented in McGraw *et al.* (2010), cover the simple collection of read speech with Amazon Mechanical Turk (MTurk), using the mechanisms presented in Chapter 3. In these efforts, noisy utterances were filtered with a recognizer, as in Chapter 4. The collection of speech for a spoken dialog system using MTurk is then described in Section 5.1. Variety is a concern when training such systems, and this section describes a technique that makes use of crowdsourced scenarios to elicit a variety of spoken language.

The final set of experiments in this chapter describes an attempt to move beyond paid crowdsourcing and collect data from a large, preexisting user base. These experiments, first described in McGraw *et al.* (2009) and Gruenstein *et al.* (2009), deploy speech-enabled educational games to learners around the world. Given the wide deployment, the audio was collected under a large variety of linguistic and acoustic conditions. However, these experiments show

Crowdsourcing for Speech Processing: Applications to Data Collection, Transcription and Assessment, First Edition.
Edited by Maxine Eskénazi, Gina-Anne Levow, Helen Meng, Gabriel Parent and David Suendermann.
© 2013 John Wiley & Sons, Ltd. Published 2013 by John Wiley & Sons, Ltd.

that, despite this enormous variance, we cannot only collect speech data but also transcribe a significant portion of it for free.

In this final set of experiments, we place speech data collection in the context of games with a purpose (GWAP) (von Ahn 2006). In doing so, we examine the implications of placing a speech recognizer in the loop in the game scenario. Games provide a mechanism for *self-transcription*, a way of determining automatically the acceptability of a recognition hypothesis as a gold-standard transcription. To prove the reliability of these self-transcriptions, we report on experiments in which we retrain a speech recognizer using a variety of inputs.

While the purpose of this book is to describe speech data collection and not its use, the two are intertwined at the level of quality control. We want to insure that the speech we collect is useful, and one way to do that is to place it in the context of one of its ultimate uses, that is, making automatic speech recognition (ASR) systems better. While we end this chapter with a discussion of how various automated mechanisms for verification of transcription quality affected word error rate (WER), the purpose of the discussion is not the improvement of ASR results, but rather the improvement of speech collected from the crowd, as measured, along one dimension, by WER.

5.1   Read Speech

Although location-based services, and systems that interface with them, have been a traditional domain for spoken dialog systems (see Gruenstein *et al.* 2006), data collection for these domains has been traditionally accomplished in the laboratory or through *ad hoc* web interfaces. Large marketing campaigns to ensure heavy usage are rarely undertaken. Given the millions of possible addresses that a user might speak, it is discouraging that so little data has been collected. In this pilot experiment, a speech-enabled web interface was distributed using MTurk, and elicited a total of 103 hours of speech from 298 users. This simple task demonstrates the feasibility of large-scale speech data collection through MTurk.

5.1.1   Collection Procedure

The addresses in our reading task were taken from the TIGER 2000 database provided by the US Census Bureau. Each address is a triplet: (*road, city, state*). There are over six million such triplets in the TIGER database. To ensure coverage of the 273,305 unique words contained in these addresses, we chose a single address to correspond to each word. Our pilot experiment formed by 100,000 such triplets; MTurk workers were paid one US cent to read each prompt. Figure 5.1 shows an example human intelligence task (HIT). After the worker has recorded an address, they are required to listen to a playback of that utterance before moving on, to help mitigate problems with microphones or the acoustic environment.

Since we are employing anonymous, nonexpert workers, there is little incentive to produce high-quality utterances, and a worker may even try to game the system. We propose two distinct ways to validate worker data. The first is to have humans validate the data manually. Given the success of MTurk for transcription tasks in previous work, we could theoretically pay *other* workers to listen to the cloud-collected speech and determine whether the expected words were indeed spoken. At this stage, redundancy through voting could be used to verify the speech transcription. A less expensive approach, explored below, is to integrate a speech recognizer into the data-collection process itself.

Figure 5.1   A sample human intelligence task (HIT) for collecting spoken addresses.

Since the VoIP interface employed was used by our dialog systems, we were able to incorporate the recognizer in real time. Thus, we could block workers who did not satisfy our expectations immediately, by analyzing their input with a speech recognizer. For the pilot experiment, however, we decided not to block any workers. Running the recognizer in a second pass allows us to examine the raw data collected through MTurk and experiment with different methods of blocking unsuitable work, which might be deployed in future database collection efforts.

5.1.2   Corpus Overview

The reading tasks were posted to MTurk on a Wednesday afternoon. Within 77 hours, 298 workers had collectively read all 100,000 prompts, yielding a total of 103 hours of audio. Figure 5.2 depicts the average number of utterances collected per hour plotted according to the worker's *local* time of day. Workers tended to talk with our system during their afternoon; however, the varying time zones tend to smooth out the collection rate with respect to the load on our servers.

The majority of our data, 68.6%, was collected from workers within the United States. India, the second largest contributor to our corpus, represented 19.6% of our data. While some nonnative speakers produced high-quality utterances, others had nearly unintelligible accents.

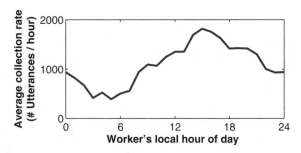

Figure 5.2   Data collection rate as a function of the worker's local time of day.

This, as well as the fact that the acoustic environment varied greatly from speaker to speaker, make the resulting MIT address corpus particularly challenging for speech recognition.

To determine the properties of our corpus without listening to all 103 hours of speech, two researchers independently sampled and annotated 10 utterances from each worker. Speakers were marked as male or female and native or nonnative. Anomalies in each utterance, such as unintelligible accents, mispronounced words, cutoff speech, and background noise, were marked as present or absent. We then extrapolate statistics for the overall corpus based on the number of utterances contributed by a given worker. From this, we have estimated that 74% of our data is cleanly read speech.

This result raises the question of how to effectively manage the quality of speech collected from the cloud. One immediately apparent solution is to run an automatic method that incorporates our speech recognizer into the validation process. In particular, we run the recognizer that we have built for the address domain over each utterance collected. We then assign a quality estimate, q, to each worker by computing the fraction of recognition hypotheses that contain the US *state* expected given the prompt. Figure 5.3 shows the recognizer-estimated quality of each worker, plotted against the number of utterances that worker contributed. Note that a single worker may actually be two or more different people using the same Amazon account.

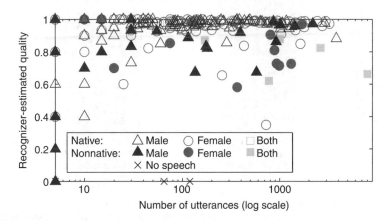

Figure 5.3   Individual workers plotted according to the recognizer-estimated quality of their work, and the number of utterances they contributed to our corpus.

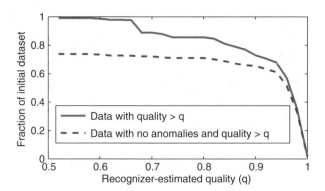

Figure 5.4 By filtering out users whose estimated quality does not meet a certain threshold, we can simulate the effect of using the recognizer to automatically block workers.

MTurk provides requesters with the ability to block workers who do not perform adequate work. Using our automatic method of quality estimation, we simulate the effects of blocking users according to the quality threshold q. It is clear from Figure 5.4 that, while the data collection rate might have slowed, requiring a high q effectively filters out workers who contribute anomalous utterances. Figure 5.5 depicts how the corpus properties change when we set $q = 0.95$. While not unexpected, it is nice to see that egregiously irregular utterances are effectively filtered out.

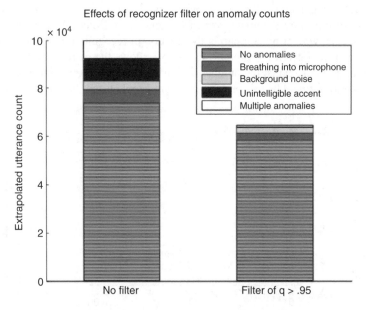

Figure 5.5 Breakdown of anomalies present in the corpus as a whole and the subcorpus where workers have a high-quality estimate, $q \geqslant 0.95$. The recognizer-filtered subcorpus still retains 65% of the original speech data. Though not explicitly shown here, we found that nonnative speakers were still able to contribute to this subcorpus: 5% of the filtered data with no anomalies came from nonnative speakers.

5.2 Multimodal Dialog Interactions

The experience with the address corpus inspired us to deploy a fully functional multimodal spoken dialog system to MTurk. Since the very same toolkit used in the prompted speech experiments acts as the web front-end to most of our spoken dialog systems, it is relatively straightforward to collect a corpus for our *FlightBrowser* dialog system. WAMI embeds an audio recorder into a web page to stream speech to MIT's servers for processing. Recognition results are then sent either to another server-side module for processing, or straight back to the client via Asynchronous Javascript and XML (AJAX).

We explored a range of price points for web tasks deployed to MTurk that ask the worker to book a flight according to a given scenario. The scenarios themselves were also generated by MTurk workers. Finally, once the data had been collected, we manually posted it back on MTurk for transcription and use the transcripts to evaluate the WER of the system. Had we automated these last few steps, we might have called this a human computation algorithm for spoken language system assessment. As is, we refer to this as crowdsourced spoken language system evaluation.

5.2.1 System Design

FlightBrowser was derived from the telephone-based Mercury system, originally developed under the DARPA Communicator project Seneff (2002). Mercury's design was based on a *mixed-initiative* model for dialog interaction. When flights are found, the system describes verbally the set of database tuples returned in a conversational manner. It prompts for relevant missing information at each point in the dialog, but there are no constraints on what the user can say next. Thus, the full space of the natural language understanding system is available at all times.

Using the WAMI Toolkit, we adapted Mercury to a multimodal web interface we call FlightBrowser. The dialog interaction was modified, mainly by reducing the system's output verbosity, to reflect the newly available visual itinerary and flight list display. A live database of flights is used for the system. About 600 major cities are supported worldwide, with a bias toward US cities. The interface was designed to fit the size constraints of a mobile phone, and multimodal support, such as clicking to sort or book flights, was added. Figure 5.6 shows FlightBrowser in a WAMI app built specifically for the iPhone.

5.2.2 Scenario Creation

When designing a user study, many spoken dialog system researchers struggle with the question of how to elicit interesting data from users without biasing the language that they use to produce it. Some have tried to present scenarios in tabular form, while others prefer to introduce extra language, hoping that the user will only pick up on the important details of a scenario rather than the language in which it is framed. Continuing the theme of crowdsourcing research tasks, we take an alternative approach.

To generate scenarios, we created an MTurk task that asked workers what they would expect from a flight reservation system. They were explicitly told that we were trying to build a conversational system that could handle certain queries about flights, and we provided them with a few example scenarios that our system can handle. Their job was then to construct a

Figure 5.6 *FlightBrowser* interface loaded in the iPhone's WAMI browser. The same interface can be accessed from a desktop using any modern web browser.

set of new scenarios, each starting with the word "You. . . " and continuing to describe "your" desired itinerary. We paid $0.03 per scenario, and within a few hours 72 distinct workers had given us 273 scenarios, examples of which are shown below.

Not all of these 273 scenarios were suitable for a user study. As shown in the examples in Box 5.1, some workers did not fully follow the directions. Other crowdsourced scenarios had dates that were in the past by the time we deployed our system. For the most part, however, the scenarios generated were far more creative and varied than anything we could have come up with ourselves in such a short amount of time. Although it was clear that some tasks would cause our system trouble, we did not explicitly exclude such scenarios from our study. For example, our system does not have *Disneyland* in its vocabulary, let alone a mapping from the landmark to the nearest airport. Ultimately, we chose 100 scenarios to form the basis of the data collection procedure described in Section 5.2.3.

We view the scenario collection procedure described above as a step toward constructing user studies that are relatively unbiased with respect to system language and capabilities. One could envision formalizing a framework for soliciting relevant scenarios for evaluating spoken dialog systems from nonexpert workers.

5.2.3   Data Collection

Although the interface shown in Figure 5.6 is optimized for a mobile device, the WAMI Toolkit allows us to access it from modern desktop browsers as well. The following paragraphs describe how we were able to collect over 1000 dialog sessions averaging less than $0.20 apiece in under 10 days of deployment on MTurk.

Box 5.1 Sample scenarios crowd collected for the *FlightBrowser* domain.

1. You need to find the cheapest flight from Maryland to Tampa, Florida. Find a cheap flight out of your choice of Philadelphia, Dulles or Baltimore airports.
2. You won a prize to visit Disneyland but have to provide your own airfare. You are going the week of Valentine's Day and you need two tickets from Seattle. You only have $500 to spend on tickets.
3. Destination: London, England
   Departs: January 15, 2010, anytime in the morning
   Returns: March 1, 2010 anytime after 3:00 p.m.
   Price: Anything under $1500. To make things interesting, I want as many layovers as possible!
4. You would like to take a vacation in Puerto Rico for 2 weeks. The departure and arrival dates must be on a Saturday.
5. You are a cartoonish crime boss in New York City, but Batman has caught on to you and you need to skip town for a while. You decide to head for Memphis, a city not normally known for costumed villainy. Set up a one-way flight with no layovers first thing tomorrow morning; cost is no object.

HIT Design

The design of a HIT is of paramount importance with respect to the quality of the data we collected using MTurk. Novice workers, unused to interacting with a spoken language interface, present a challenge to system development in general, and the MTurk workers are no exception. Fortunately, MTurk can be used as an opportunity to iteratively improve the interface, using worker interactions to guide design decisions.

To optimize the design of our system and the HIT, we deployed short-lived MTurk tasks and followed them up with improvements based on the interactions collected. Since the entire interaction is logged on our servers, we also have the ability to *replay* each session from start to finish, and can watch and listen to the sequence of dialog turns taking place in a browser. By replaying sessions from an early version of our interface, we discovered that many workers were not aware that they could click on a flight to view the details. This inspired the addition of the arrows on the left-hand side, to indicate the potential for drop-down details.

Although initially we had hoped to minimize the instructions on screen, we found that, without guidance, a number of MTurk workers just read the scenario aloud. Even after providing them with a short example of something they could say, a few workers were still confused, so we added an explicit note instructing them to avoid repeating the scenario verbatim. After a few iterations of redeploying and retuning the dialog and scenario user interfaces, we eventually converged on the HIT design shown in Figure 5.7.

In order to complete a HIT successfully, a worker was required to book at least one flight (although we did not check that it matched the scenario); otherwise they were asked to "give up." Whether the task was completed or not, the worker had the option of providing written feedback about their experience on each scenario before submitting.

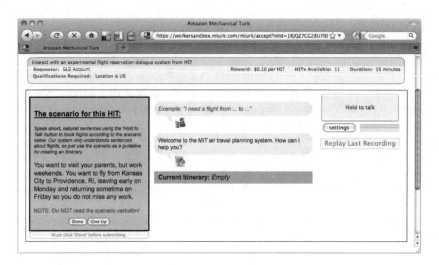

Figure 5.7   The screenshot depicts a sample HIT as shown to an MTurk worker.

Extended Deployment

With the design stage complete, we decided to leave our HIT on MTurk for an extended period of time to collect a large amount of data. Beginning on a Tuesday, we deployed FlightBrowser to MTurk and paid workers $0.20 for each scenario. We restricted the deployment to workers who had Amazon accounts in the United States. Each worker was limited to submitting sessions corresponding to the 100 scenarios described previously. In just under 4 days from our initial deployment, we had collected 876 dialogs from 63 distinct users, totaling 9372 audio files. An example dialog collected is shown in Box 5.2.

Curious about how price affected the rate of collection, we deployed the same task for $0.10 around a month later. This task was started on a Thursday and left running for 6 days. Though clearly there was less interest in the HIT, we were still able to collect 2595 utterances over 237 dialogs from 43 distinct workers. It should be noted that we made no special effort to exclude workers from the earlier task from participating in the $0.10 HIT a month later.

Figure 5.8 shows histograms for each price point of sessions collected from individual workers, as well as the number of tasks they marked "finished" and "give up." As shown in the plots, five workers participated in the $0.10 task despite being paid twice as much the month before. It is interesting to note that they were still willing to participate despite earning half as much. In fact, three of the top four contributors to the second round of data collection were repeat visitors. This is consistent with our general observation that many workers from the United States do these tasks for fun, and that the small sums of money involved are viewed as an added bonus.

In both deployments, a nontrivial number of audio files were recognized as noise or silence. This phenomenon has been observed previously when utterances come from more realistic sources Ai *et al.* (2007). Listening to these in context, it became apparent that some users required time to familiarize themselves with the recording software. We decided to ignore the 1316 files associated with empty recognition results, leaving 10,651 utterances for analysis. Table 5.1 summarizes statistics from both deployments.

Box 5.2 Sample MTurk-collected interaction. Parts of the system responses have been omitted for brevity. Also not shown are the updates to the visual display that show the user query results.

S: [...] How can I help you?

U: I need a flight from kansas city to providence rhode island rhode island

S: [...] What date will you be traveling?

U: Monday morning.

S: [...] One moment please.

S: [...] There are at least 27 flights departing between 5:30 a.m. and 12:39 a.m.

U: I will take the five thirty.

S: [...] What date will you be returning on?

U: [noise] Friday afternoon.

S: [...] One moment please.

S: [...] There are at least 20 flights departing between 12:23 p.m. and 6:45 p.m.

U: I will take the twelve twenty three.

S: Can you provide a departure or arrival time? There are at least 6 Delta flights from Providence to Kansas City on Friday February 5th.

U: Arrive at five oh four.

S: [...] How about a flight arriving at 5:04 p.m. connecting in Detroit? Would that work? [...]

U: Yes

S: Delta flight 6695 connecting to Delta flight 7377 in Detroit has been added to your itinerary. [...] Is there something else I can do for you?

U: No

S: Have a nice day.

5.2.4 Data Transcription

To transcribe the newly collected data, we once again turn to the MTurk cloud service. Previous work has explored the use of MTurk for transcription to generate highly accurate orthographies. In exploring this area further, we found that seeding the transcription interface with recognizer hypotheses enables an automatic detection method for "bad" transcripts.

Figure 5.9 depicts a flowchart of our transcription procedure. We deployed our entire corpus to MTurk in a $0.05 HIT, which asked workers to listen to utterances and correct recognizer hypotheses. Each HIT contains a bundle of 10 utterances for transcription. Once a set of candidate transcripts is complete, we automatically filter transcripts that are likely to be erroneous before moving on to a voting stage where transcripts are combined given the candidates they have accumulated so far. The process was iterated until 99.6% of our data were accepted by our voting scheme.

We use two filters to remove poor transcript candidates from the pool before voting. First, since the average number of words per HIT is around 45, the likelihood that *none* of them need to be corrected is relatively low. This allows us to detect lazy workers by comparing the submitted transcripts with the original hypotheses. We found that 76% of our nonexpert

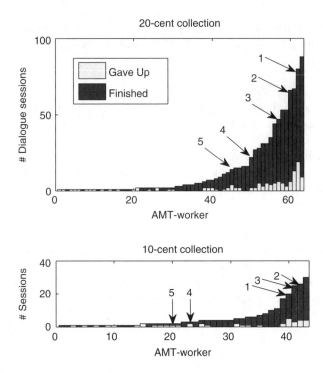

Figure 5.8 A breakdown of the data collection efforts by worker. For each price point, the workers are sorted in ascending order of the number of dialogs they contributed to the corpus. Numbers 1–5 identify the five workers who participated in both data collection efforts.

transcribers edited at least one word in over 90% of their hits. We assumed that the remaining workers were producing unreliable transcripts, and therefore discarded their transcripts from further consideration. Second, we assume that a transcript *needs* to be edited if more than two workers have made changes. In this case, we filter out transcripts that match the hypothesis, even if they came from otherwise diligent workers.

The question of how to obtain accurate transcripts from nonexpert workers has been addressed by Marge *et al.* (2010), who employ the ROVER voting scheme to combine transcripts. Indeed, a number of transcript combination techniques could be explored. In this work,

Table 5.1 Corpus statistics for $0.10 and $0.20 MTurk HITs.

	$0.20 HIT	$0.10 HIT
# Sessions	876	237
# Distinct workers	63	43
# Utterances	8232	2419
Avg. # utts. / session	9.5	10.2
% Sessions gave up	14.7	17.3

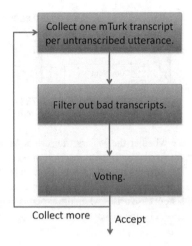

Figure 5.9 Flowchart detailing the transcription procedure. For a given utterance, one transcript is collected at a time. Based on the editing habits of each worker, bad transcripts are filtered out. Provided enough good transcripts remain, punctuation and capitalization is removed and voting then proceeds. If there is not yet a majority vote, another transcript is collected. If five good transcripts are collected without a majority vote, we begin to accept a plurality.

we take a simple majority vote, unless there is no agreement among five unfiltered transcripts, at which point we accept a plurality. We found that 95.2% of our data only needed three good transcriptions to pass a simple majority vote. Table 5.2 indicates the amount of data we were able to transcribe for a given number of good transcripts.

The total cost of the transcription HIT was $207.62 or roughly 2¢ per utterance. Fifty three audio files did not have an accepted transcript even after collecting 15 transcripts. We listened to this audio and discovered anomalies such as foreign language speech, singing, or garbled noise that caused MTurk workers to start guessing at the transcription.

In order to assess the quality of our MTurk transcribed utterances, we had two expert transcribers perform the same HIT for 1000 utterances randomly selected from the corpus. We compared the orthographies of our two experts and found sentence-level exact agreement to be 93.1%. The MTurk transcripts had 93.2% agreement with the first expert and 93.1% agreement with the second, indicating that our MTurk-derived transcripts were of very high quality.

Figure 5.10 shows a detailed breakdown of agreement, depicting the consistency of the MTurk transcripts with those of our experts. For example, of all the data edited by at least

Table 5.2 The fraction of the data that required only G good transcripts for a majority (or plurality for $G > 5$).

# Good transcripts required (G)					
% Corpus transcribed (T)					
G	2	3	5	6	7+
T	84.4	95.2	96.3	98.4	99.6

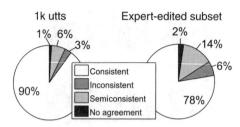

Figure 5.10   These charts indicate whether the MTurk transcripts were consistent, semiconsistent, or inconsistent with the two expert transcribers. Since transcription was seeded with recognition results, not all transcripts required edits. The agreement breakdown is shown for the full 1000 utterances and for the subset that was actually edited by the experts. The consistent case is when the experts and MTurk transcripts all agree. The semiconsistent case arises when the experts disagreed, and the MTurk transcript matched one of the expert transcripts. The inconsistent case occurs when the MTurk transcript does not agree with either expert.

one expert, only 6% of the MTurk transcripts were inconsistent with an expert-agreed-upon transcript. Where the experts disagree, MTurk labels often match one of the two, indicating that the inconsistencies in MTurk transcripts are often reasonable. For example, "I want a flight to" and "I want to fly to" was a common disagreement.

Lastly, we also asked workers to annotate each utterance with the speaker's gender. Again, taking a simple vote allows us to determine that a majority of our corpus (69.6%) consists of male speech.

5.2.5   Data Analysis

Using the MTurk transcribed utterances, we can deduce that the WER of the MTurk-collected speech was 18.1%. We note, however, that, due to the fact that the dialog system task was not compulsory, this error rate may be artificially low, since workers who found FlightBrowser frustrating were free to abandon the job. Figure 5.11 shows the WER for each worker plotted against the number of sessions they contributed. It's clear that workers who experienced high error rates rarely contributed more than a few sessions, likely due to frustration. To provide a fairer estimate of system performance across users, we take the average WER over all the *speakers* in our corpus and revise our estimate of WER to 24.4%.

Figure 5.11 also highlights an interesting phenomenon with respect to system usability. It appears, that workers were willing to interact with the system so long as their WER was less than 30%, while workers who experienced higher WERs were not likely to contribute more than a few sessions. We imagine this threshold may also be a function of price, but did not explore the matter further in this study.

Upon replaying a number of sessions, we were quite happy with the types of interactions collected. Some dialog sessions exposed weaknesses in our system that we intend to correct in future development. The workers were given the opportunity to provide feedback, and many gave us valuable comments, compliments, and criticisms, a few of which are shown in Box 5.3.

To analyze the linguistic properties of the corpus quantitatively, we compared the recognition hypotheses contained in the worker interactions with those of our internal database of developer utterances. From March 2009 to March 2010, FlightBrowser has been under active

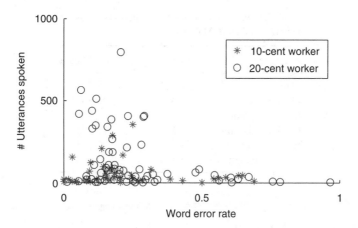

Figure 5.11   The number of sessions contributed by each worker is plotted against the WER experienced by that worker. It is clear that a high WER discourages users from participating despite the financial incentives. In this domain, a WER of less than 30% appears to be required before the task is deemed to be worthwhile.

development by five members of our laboratory. Every utterance spoken to FlightBrowser during this time window has been logged in a database. User studies have been conducted in the laboratory and demos have been presented to interested parties. The largest segment of this audio, however, comes from developers, who speak to the system for development and debugging. In total, 9023 utterances were recorded, and these comprise our internal database. We summarize a number of statistics common to the internal and the 20¢ MTurk-collected corpora in Table 5.3.

While the average recognizer hypothesis is longer and the number of words is greater in our internal data, the overall language in the cloud-collected corpus appears to be more complex, as illuminated by distinct n-gram counts. This is striking because our data collection was domain-limited to 100 scenarios, while the developers were unrestricted in what they could say. These results suggest that, because system experts know how to speak with the system, they can communicate in longer phrases; however, they do not formulate new queries as creatively or with as much variety as MTurk workers.

Box 5.3 Feedback on the *FlightBrowser* MTurk HIT.

1. There was no real way to go back when it misunderstood the day I wanted to return. It should have a go back function or command.
2. Fine with cities but really needs to get dates down better.
3. The system just cannot understand me saying "Tulsa."
4. Was very happy to be able to say 2 weeks later and not have to give a return date. System was not able to search for lowest fare during a 2-week window.
5. I think the HIT would be better if we had a more specific date to use instead of making them up. Thank you, your HITs are very interesting.

Table 5.3 Comparison between internal corpus and one collected from MTurk.

	Internal	MTurk collected
# Utts.	9,023	8,232
# Hyp tokens	49,917	36,390
# Unique words	740	758
# Unique bigrams	4,157	4,171
# Unique trigrams	6,870	7,165
Avg. Utt. length	4.8	4.4

5.3   Games for Speech Collection

In the experiments above, we relied largely on MTurk to provide the crowd in our human computation tasks, via micropayments. While these payments are typically quite low, we were interested in using the GWAP paradigm (von Ahn 2006) as a mechanism for transcribing speech that is both educational for the user and free for us. To this end, the two games we deployed were designed for speech data collection, but with a twist. As we will describe, the games were instrumented such that clues as to the reliability of the collected speech were derivable from the interaction itself. As a metric for reliability, we choose WER, in particular WER as measured after using the collected data to retrain a speech recognizer. As we've noted, the purpose of our experiments was to collect speech that would prove useful across a range of Human Language Technologies; WER reflects just one of those technologies, but it is one that we had easy access to. It is hoped that results along those lines are indicative of overall quality of speech collected.

The first game we devised for this purpose, called *Voice Race*, collects isolated words or short phrases. The second, called *Voice Scatter*, elicits continuous speech in the form of longer sentences. Both of these games were hosted on a popular educational website for a little less than a month, during which time data were logged and gathered. We then ran a suite of experiments to show that some of these data could be accurately transcribed simply by looking at the game context in which they were collected. Using these data, we were able to adapt the acoustic model to yield improvements otherwise unattainable in an unsupervised fashion. This protocol does not fall under our definition of human computation, since we perform these experiments in an off-line fashion; however, we believe this represents a significant step toward what *could* become a fully organic spoken language interface, as described in Chapter 3.

The experiments described below point to a broader characteristic of spoken language interfaces, that is, many spoken language systems come with a crowd built-in. By definition, spoken language interfaces, when outfitted with an Internet connection, have a suite of nonexpert individuals who can provide valuable data. Since they are unpaid, however, we are restricted to collecting the data that are a by-product of normal interaction with the system. To clarify this point, we examine the voice-dialing work of Li *et al.* (2007). Here, data are collected from mobile phone users who speak names from their contact lists to dial a call. A confirmation is then given, to which the user must reply "yes" or "no." If the caller says "yes" and the call goes through, Li *et al.* make the assumption that the spoken name was recognized correctly. While

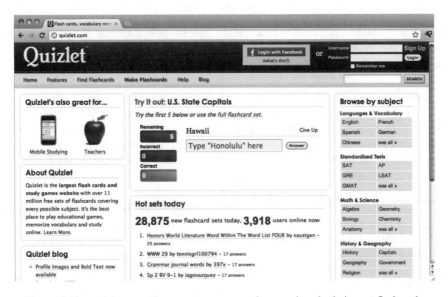

Figure 5.12   quizlet.comallows users to create, share, and study their own flashcards.

there is some noise added in the form of false positives, they find this confirmation question is a useful means of collecting labeled training data.

The work below uses a similar approach to determine when a recognition result is correct. We do so, however, without the aid of an additional confirmation step, by using domain-specific context to validate the semantics of a recognition result. While particularly useful in the education domain, where words are known, we contend that such techniques might be applied to other spoken language interfaces to obtain similarly *self-transcribed* data.

5.4   Quizlet

The speech-enabled games described here were designed as a fun way to review flashcards to help memorize vocabulary words, scientific terms, mathematical concepts, and so on. They can be played by anyone with a web browser and a microphone. *Voice Race* and *Voice Scatter* were built and deployed to quizlet.com[1] with the help of its founder. The Quizlet website, shown in Figure 5.12, is a place where users can create, share, and study virtual flashcards. Like real flashcards, each virtual card has two sides: typically one is used for a *term* (a word or short phrase) and the other for its *definition*. At the time we deployed our games to Quizlet, the site had over 420,000 registered users, who had contributed over 875,000 flashcard sets, comprised more than 24 million individual flashcards. As of this writing, there are closer to 6 million users studying over 9 million sets with more than 300 million study terms.

Voice Race and *Voice Scatter* are based on two text-based games that had already existed on Quizlet for some time. The study material for each game is loaded from a single set of flashcards chosen by the user. The ASR component was simple to incorporate into these

---

[1]http://www.quizlet.com.

games with the help of the publicly available WAMI Javascript API described in Section 3.3.1. When a game session is initialized, a simple context-free grammar is constructed using the terms in the chosen set of flashcards. The audio is streamed to a server, where speech recognition is performed and sends the results back to the Javascript event handlers in the browser.

On the server side, The SUMMIT speech recognizer is used with a dictionary containing 145,773 words, and an automatic L2S module to generate pronunciations for those words that are found to be missing at runtime. A small context-free grammar built over the terms and definitions in the chosen set serves as the language model (LM) for the recognition component of each game. While each instance of these games is a small-vocabulary recognition task over a particular set of flashcards, in aggregate the utterances we collected cover a large vocabulary, and should be useful in a variety of speech tasks.

In the spirit of other GWAPs, our games were designed to elicit data as a by-product of game-play. Previous to this work, the only GWAP for a speech-related task that we are aware of is *People Watcher* (Paek *et al.* 2007). *People Watcher* elicits alternative phrasings of proper nouns, which are used to improve recognition accuracy in a directory assistance application. The use of an educational website to transcribe data was explored in Cai *et al.* (2008), in which a task intended to help students learn a foreign language was deployed via a prototype website, and used by 24 students to label 10 sentences. Neither of these applications elicit new speech data, or are applied on a large scale.

Our games, on the other hand, collect audio recordings, as well as the context in which they occur. With the help of Quizlet, we find it easy to recruit a large number of willing subjects, which gives rise to a diversity of ages, genders, fluency, accents, noise conditions, microphones, and so forth. Furthermore, the games we design allow for the automatic transcription of a significant subset of the collected data, and for cheap transcription of the vast majority of the remainder. This means that an arbitrary amount of transcribed utterances may be collected over time at no, or slowly increasing, cost.

Our games are different from typical GWAPs in a few key respects. First, these are single-player games. Whereas typical GWAPs rely on the agreement of two humans to obtain labels, *Voice Race* instead uses the artificial intelligence of an automatic speech recognizer. Contextual constraints and narrow-domain recognition tasks are paired to bootstrap the collection of transcribed corpora, which cover a larger vocabulary and a variety of noise conditions. Second, GWAPs label existing data, whereas *Voice Race* both elicits new speech data, and automatically transcribes much of it. Thus, while *Voice Race* cannot label arbitrary speech data, it can *continuously* provide new, transcribed speech without any supervision. Third, unlike GWAPs, which offer only diversion, these educational games directly benefit their players by helping them to learn. Finally, users have control over their game content, which is not randomized as in many GWAPs.

In Sections 5.5 and 5.6, we describe the two games we deployed for speech data collection. For each, we describe the game itself, as well as providing an overview of the speech collected using the game. Of particular interest is the use of the structure of the game itself to self-transcribe the data. Our goal was to see how effectively we could exploit aspects of the game context to enable us to identify a trusted set of ASR hypotheses that could serve as gold-standard transcriptions, with no human intervention. In order to demonstrate the effectiveness of these hypotheses, we describe the results of using self-transcribed data to bring down WER, our proxy for overall data quality.

5.5 Voice Race

In this section, we describe *Voice Race*, a speech-enabled educational game that we deployed to Quizlet over a 22-day period to elicit over 55,000 utterances representing 18.7 hours of speech. *Voice Race* was designed such that the transcripts for a significant subset of utterances can be automatically inferred using the contextual constraints of the game. Game context can also be used to simplify transcription to a multiple choice task, which can be performed by nonexperts. We found that one-third of the speech collected with *Voice Race* could be automatically transcribed with over 98% accuracy, and that an additional 49% could be labeled cheaply by MTurk workers. We demonstrate the utility of the self-labeled speech in an acoustic model adaptation task, which resulted in a reduction in the *Voice Race* utterance error rate.

In *Voice Race*, shown in Figure 5.13, definitions from a set of flashcards move across the screen from left to right. Players must say its matching term before a definition flies off the screen. Each such "hit" earns points and makes the definition disappear. If a definition is never hit, then the player is shown the correct answer and prompted to repeat it aloud. As the game progresses, the definitions move more quickly and appear more frequently.

Because each utterance is collected in the course of playing the game, a combination of recognition *N* best lists and game context can be used to automatically infer the transcripts for a significant subset of the utterances. Intuitively, when the top recognition hypothesis is known to be a correct answer, this is a strong indication that it is accurate. Using such constraints, 34% of the collected utterances were *automatically* transcribed with near perfect accuracy. For the remaining utterances, game context can also be used to simplify the task of human transcription to one of choosing among several alternative transcripts on a short list. Such a simple task is easy to complete with no training, so we used MTurk for transcription. To the best of our knowledge, when this work was performed there was no precedent in academia

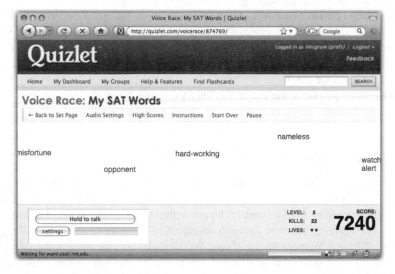

Figure 5.13 The *Voice Race* game with vocabulary flashcards. As a definition moves from left to right, the player must say the corresponding vocabulary word before it flies off the screen. Each such "hit" earns points and makes the definition disappear.

for using MTurk in this fashion. We found that the transcripts produced by MTurk workers agreed well with those of two experts.

5.5.1 Self-Transcribed Data

Although it has been shown that transcriptions need not be perfect to be useful (Novotney and Callison-Burch 2010), we were interested in seeing if aspects of the game itself could help us identify automatically ASR hypotheses that are close to transcriptions obtained through human effort. We begin by precisely defining the way in which *Voice Race* is able to *self-transcribe* a portion of its data. Recall that each utterance occurs in a context where the correct answer (or answers) is known. This information, when combined with the recognition results, can be used to automatically infer the transcript for certain utterances, and greatly limit the set of likely transcripts for the rest. The subsets of interest are as follows:

Hit: In *Voice Race*, a "hit" occurs when the top speech recognition hypothesis contains the correct term associated with a definition visible on the screen. Players typically aim for the right-most definition, so such "hits" are likely to be the most reliable indicators of an accurate recognition hypothesis. Suppose, for example, that a student is learning state capitals. At any given time, only a few state names are shown on the screen. The probability of a *misrecognition* resulting in a state capital that has a corresponding state onscreen is low. Even when terms are phonetically similar, such as the words "Austin" and "Boston," they are unlikely to co-occur in the game context, making them effectively distinguishable.

Miss: A "miss" occurs when the user has spoken, but a hit has not been detected. There is no way of knowing if a miss is due to a human error or a speech recognition error. However, when misses are due to recognition errors, the ground-truth transcript for the user's utterance is likely to be one of the correct answers. As such, when considered in aggregate, misses may be useful for automatically identifying difficult terms to recognize.

Prompted hit/miss: The taxonomy above applies to most *Voice Race* utterances. *Voice Race* also provides an additional category of labeled data: when a definition flies off the screen without being "hit," players are shown the correct answer and prompted to read it aloud. As such, when players are cooperative, the transcript of their utterances should be known in advance. Moreover, we run these utterances through the same small-vocabulary recognizer used for the game to notify the player of whether or not he or she was understood. These utterances can therefore again be classified as "hits" or as "misses."

5.5.2 Simplified Crowdsourced Transcription

The contextual game constraints identified in Section 5.5.1, and in particular the "hits," are useful for automatically transcribing a significant portion of the data. In addition, the same constraints may be used to greatly decrease the *human* effort required to transcribe the remainder. The correct transcript for each utterance is likely to be either one of the correct answers, one of the utterances on the *N best list, or both*. This means that the task of human transcription for most utterances can be reduced to one of choosing a transcript from a short list of choices drawn from these two sources. Given that it requires no expertise or knowledge of the task domain to listen to a short audio clip and choose a transcript from a list, we designed a transcription task that could tap the large pool of MTurk workers.

Box 5.4 The construction of the multiple-choice MTurk task selects candidate transcriptions in the order listed above.

1. The prompted term, if the user was asked to repeat aloud
2. The top two distinct terms in the recognition N best list
3. The terms associated with the two right-most definitions
4. Any remaining terms on the N best list
5. Random terms from the flashcard set

We designed the MTurk transcription task such that workers listen to a *Voice Race* utterance and then choose from one of four likely transcripts. They can also choose "None of these" or "Not Speech." The likely transcripts were drawn in order from the sources found in Box 5.4, until four unique candidate transcripts were obtained to create the six answer multiple choice transcription question. An example MTurk task is shown in Figure 5.14.

After transcribers select a transcript, they can optionally label two additional attributes. *Cutoff* indicates that the speech was cutoff—this happens occasionally because players release the space bar, which they must hold while speaking, before they finish. Future iterations of the game will likely correct for this by recording slightly past the release of the key. Transcribers may also select *Almost* if the utterance was understandable, but contained hesitations, extra syllables, mispronunciations, and so on.

5.5.3 Data Analysis

Voice Race was made available on Quizlet.com for a 22-day trial period. No announcements or advertisements were made. The two games were simply added to the list of activities

Figure 5.14   An MTurk task for transcribing *Voice Race* utterances.

Table 5.4   Properties of *Voice Race* data collected over 22 days.

Games played	4,184	Mean words per utt.	1.54
Utterances	55,152	Total distinct phrases	26,542
Total hours of audio	18.7	Mean category size	53.6

available to study each (English) flashcard set. Nonetheless, as Table 5.4 shows, a total of 55,152 utterances were collected, containing 18.7 hours of speech.

5.5.4   Human Transcription

Ten thousand utterances representing 173 minutes of audio were drawn from 778 *Voice Race* sessions and then submitted for transcription to MTurk. Within 16 hours, each utterance had been labeled by five different MTurk workers using the simplified transcription task discussed in Section 5.5.3, at a cost of $275.

Table 5.5 shows agreement statistics for the workers. A majority agreed on one of the transcript choices for 97% of the utterances, agreeing on "None of these" only 13% of the time. Thus, the simple forced choice among four likely candidates (and "no speech') yielded transcripts for 84% of the utterances.

To judge the accuracy of the produced labels, two experts each labeled 1000 utterances randomly drawn from the set of 10,000. The interface these experts used to transcribe the data was identical to the MTurk worker's. Their transcript choices showed a high level of agreement, with a Cohen's kappa score of 0.89. Each of their label sets agreed well with the majority labels produced by the MTurk workers, as measured by kappa scores of 0.85 and 0.83.

Using the MTurk majority labels as a reference transcription, the utterance-level recognition accuracy on the set of 10,000 *Voice Race* utterances was found to be 53.2%. While accuracy is low, it's important to note that the task is a very difficult one. The two experts noted while transcribing that (1) the vast majority of the utterances seemed to be from teenagers, (2) there was often significant background noise from televisions, music, or classrooms full of talking students, and (3) many microphones produced muffled or clipped audio. While these problems lead to imperfect speech recognition accuracy, they also lead to a richer, more interesting corpus. Moreover, usage levels suggest that accuracy was high enough for many successful games. In Section 5.5.5, we show that, despite relatively poor recognition performance overall, it is nonetheless possible to use game context to automatically obtain *near-perfect* transcriptions on a significant subset of the data.

Table 5.5   Agreement obtained for transcripts and attributes of 10,000 utterances, each labeled by five MTurk workers.

Five-way agreement	69.2%	Majority "None of these"	12.9%
Four-way agreement	18.0%	Majority "cutoff"	12.1%
Three-way agreement	9.8%	Majority "almost"	7.2%

Table 5.6 Percentage of 10,000 MTurk-labeled utterances and recognition accuracy grouped by game context. Hits are further broken down in terms of the position of the item on the screen at the time the hit occurred. Statistics for the four right-most positions are shown.

Game context:	miss	hit	prompted-miss	prompted-hit
Correct (%):	13.9	86.4	12.7	97.5
Total data (%):	43.7	43.8	8.9	3.6

Hit context:	4-hit	3-hit	2-hit	1-hit
Correct (%):	41.3	69.4	81.7	**98.5**
Hit data (%):	1.8	3.4	9.0	**69.4**

5.5.5 Automatic Transcription

As previously described, because each utterance occurs in the course of playing *Voice Race*, we hypothesized that it should be possible to identify a subset of the data for which transcripts can be inferred automatically with high accuracy. We evaluate this hypothesis using as reference the transcripts agreed upon by a majority of MTurk workers. We explore the utility of *hits*, *misses*, *prompted-hits*, and *prompted-misses*. Table 5.6 (top half) shows the amount of speech data collected in each category out of the 10,000 MTurk-labeled utterances.

More than 4000 of the 10,000 utterances were *hits*, and the recognition accuracy on this data is 86.4%. In addition, *prompted-hits* yield an accuracy of 97.5%, meaning that they yield nearly perfectly transcribed data. Unfortunately, they represent less than 5% of the data.

Using game-context to filter data for accurately labeled utterances can be taken further in the case of a *hit*. Students are most likely to aim for the right-most label on the *Voice Race* screen. It stands to reason then, that *hits* of definitions that are not the right-most one are more likely to be due to a misrecognition. We call a hit that occurred while the item was in the nth position on the screen (from right to left) an *n-hit*. Recognition accuracies for $n = 1, \ldots, 4$ are presented in Table 5.6 (bottom half).

It is exciting to note that *1-hit* constitute 30.4% of the total MTurk-labeled data, and are recognized with 98.5% accuracy. Of all 55,152 utterances collected, 18,699—representing 5.8 hours of audio—are self-labeled in this fashion.

5.5.6 Self-Supervised Acoustic Model Adaptation

One common use of transcribed speech data is to perform acoustic model adaptation. While typically this requires human transcription, we explore using the automatically transcribed utterances to adapt the telephone acoustic models used by *Voice Race* in a fully automatic fashion. We performed MAP adaptation, using 1-hits and prompted hits. The 10,000 utterances transcribed by MTurk served as our test set, while the remaining 45,152 utterances without human labels were used for adaptation. Using these updated acoustic models to decode the test set resulted in a decrease in utterance error rate from 46.8% to 41.2%. To show that this *self-supervised* adaptation algorithm outperforms typical unsupervised approaches, we use the confidence module of our recognizer, Hazen *et al.* (2002), to extract high-quality utterances for a similarly sized training set. The utterance error rate for a decoder based on models trained from these utterances is 43.9%.

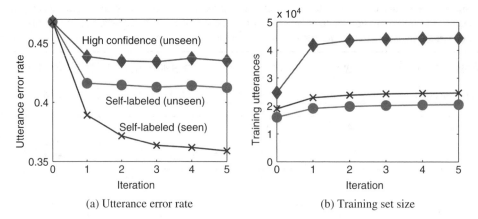

(a) Utterance error rate   (b) Training set size

Figure 5.15 Iterative acoustic model adaptation, trained using: (1) iteratively calculated high-confidence utterances, excluding the 10,000 MTurk-transcribed test set (i.e., the test data are unseen), (2) an iteratively calculated set of self-labeled utterances (unseen), and (3) an iteratively calculated set of self-labeled utterances, including test utterances (seen).

This self-supervised adaptation algorithm can also be run iteratively. Since we have saved the game context, we can actually recompute new hits and misses using the updated acoustic model. In other words, we can *re*recognize a set of utterances and recompute the "hits," yielding a larger pool of self-labeled training data, which is then used in the following iteration. In theory, the new hits should still yield highly accurate transcripts. To test this hypothesis we iteratively trained and relabeled all 55,152 utterances until convergence. We found that 44.8% of the data could be *self*-labeled while maintaining an accuracy of 96.8%—computed on the subset corresponding to the MTurk-transcribed data.

With these promising results, we turn to analyzing the effects of the iterative approach on acoustic model adaptation. Note that, despite the self-supervised nature of our approach, training on all 55,152 utterances simulates having already *seen* the test data. We therefore perform separate experiments on the 45,152 utterance training set to simulate the case when the test data are *unseen*. We again use a confidence-based baseline approach in which, at each iteration, utterances that received high recognition confidence are selected for the new adaptation set. As Figure 5.15 shows, however, the confidence-based approach is not amenable to this iterative adaptation procedure. By the second iteration, this approach uses over twice the data of the self-supervised method. However, the self-supervised method retains a 2.2% absolute improvement in error rate over the iteratively calculated high-confidence utterances through multiple iterations.

Figure 5.15 also shows the results of iteratively selecting from *all* 55,152 utterances (without using the MTurk labels) treating the 10,000 utterance test set as *seen* data. Iteratively selecting high-confidence training utterances from all the data achieves error rates similar to those found when selecting self-labeled utterances from the original 45,152 utterances, and is omitted from the graph for clarity. Iteratively selecting *self-labeled* utterances from all of the data, however, improves performance significantly, even across iterations. The *iterative* gains are likely due to the fact that the self-adaptation set now includes utterances gathered from the same session, meaning that the speaker, acoustic environment, and vocabulary are the same. This hints at the potential for games like *Voice Race* to improve in a personalized, fully automatic, online fashion. The elicited utterances, however, typically contain only a single word or short phrase. To explore collecting labeled continuous speech, we devised *Voice Scatter*, which elicits significantly longer utterances.

5.6   Voice Scatter

Voice Scatter again relies on the resources of Quizlet to provide a speech-enabled educational game to a large user-base. A by-product of its use is the collection and orthographic transcription of a significant amount of *continuous* speech. In our experiments, the game was made available on Quizlet for a 22-day period and resulted in the collection of 30,938 utterances, constituting 27.63 hours of speech, over a 22-day period. Each individual game uses only eight flashcards, and speech recognition was again performed using a narrow-domain CFG grammar. Despite the limited domains of each individual game, an estimated 1193 speakers played the game with 1275 distinct flashcard sets, so recognition hypotheses in the corpus cover 21,758 distinct words.

A screenshot of *Voice Scatter* is shown in Figure 5.16. Players first choose (or create) a set of flashcards to study. Then, up to eight terms and definitions are "scattered" randomly across the screen. Using a microphone and a web browser, players speak short commands to connect each term to its definition; for example, "match cell to a membrane bound structure that is the basic unit of life." Players hold the space bar, or click an on-screen hold-to-talk button, while speaking. When a term is correctly paired with its definition (a "hit"), they come together in a fiery explosion, and then disappear from the screen, as shown in Figure 5.16. When they are incorrectly paired (a "miss"), they collide and then bounce off of each other. A timer counts

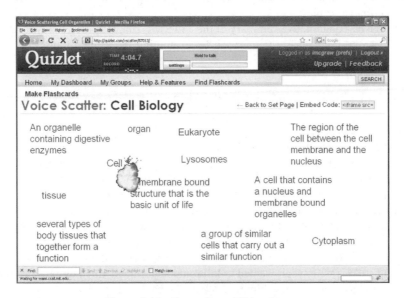

Figure 5.16   Screenshot of *Voice Scatter*.

Table 5.7 Properties of *Voice Scatter* data collected over 22 days.

Games played	4,267	Distinct words recognized	21,758
Utterances	30,938	Total number of "hits"	10,355
Hours of audio	27.63	Recognized words per "hit"	8.327
Distinct speakers	1,193[a]	Distinct flashcard sets	1,275

[a]Distinct speakers estimated as one speaker per IP address.

upward at the top of the screen, encouraging (though not requiring) players to set a speed record for the flashcard set.

Again, speech recognition was incorporated using the WAMI Javascript API and the SUMMIT speech recognizer. The following simple context-free grammar is used as the speech recognizer's LM:

```
[match] <TERM> [to|with|and|equals] <DEF>
[match] <DEF>  [to|with|and|equals] <TERM>
```

where the brackets indicate optionality, and TERM and DEF are any of the terms or definitions on the screen as the game begins.

5.6.1 Corpus Overview

Voice Scatter elicits utterances containing spontaneous continuous speech; however, because terms and definitions are visible on the screen, utterances—especially long ones—sometimes have the feel of being read aloud. While there is no specific requirement that players read the terms and definitions verbatim, there is a strong incentive to do so to avoid speech recognition errors. In addition, some (but certainly not all) players speak quickly because of the timer displayed during game play.

Table 5.7 gives a quantitative summary of the collected data. However, the type and variety of the data can be immediately understood by examining the sample transcripts shown in

---

Box 5.5 Example transcripts drawn from the corpus.

match aimless to drifting
match robust to strong and vigorous
local area network lan
match silk road with an ancient trade route between china and europe
anything that makes an organism different from others variation
match malaise to a physical discomfort as a mild sickness or depression
match newtons first law of motion to an object at rest tends to stay at rest and a moving
    object tends to keep moving in a straight line until it is affected by a force
match what does friar lawrence point out to get romeo to see that life isnt so bad juliet is
    alive and still his wife tybalt wanted to kill romeo but romeo killed him instead the
    prince could have condemned him to death but he banished him instead

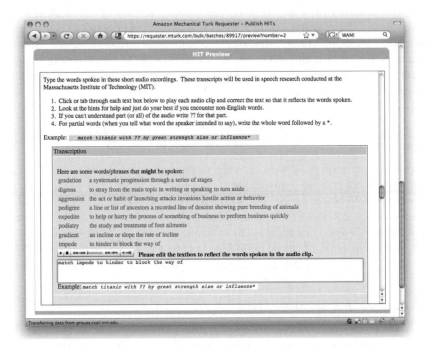

Figure 5.17   An MTurk task for transcribing *Voice Scatter* utterances.

Box 5.5. As is shown, even though each individual *Voice Scatter* game is restricted to a small vocabulary, in aggregate there is a large and varied vocabulary. Moreover, by examining a random sample of utterances, we noted that almost all speakers appeared to be teenagers, and that utterances were recorded both in quiet and noisy environments. Noise typically came from televisions, music, computer noise, and people talking in the background. Finally, since players are trying to master unfamiliar material, some words are mispronounced. We observed one player, for example, who consistently mispronounced vocabulary words like "proliferate," "unanimity," and "steadfast."

5.6.2   Crowdsourced Transcription

We used MTurk to orthographically transcribe 10,000 *Voice Scatter* utterances drawn from 463 random users (as determined by IP address), which totaled 11.57 hours of speech. The MTurk task is shown in Figure 5.17. Workers were given 10 utterances per page to transcribe. A text box for transcription was initialized with the speech recognizer's top hypothesis, and workers were asked to edit it to reflect the words actually spoken. To guide the transcriber, each utterance was accompanied by a list of terms and definitions from the game associated with that utterance. Each utterance was transcribed by three different workers, yielding 30,000 transcripts created by 130 workers for a total cost of $330.

Since we have three transcripts for each utterance, we must combine them somehow to form a gold-standard *MTurk transcript*. We chose the majority transcript if there was exact agreement by at least two of the workers, and selected a transcript at random if all three workers disagreed. There was majority agreement on 86.7% of utterances.

To assess the reliability of transcripts obtained in this manner, two experts each performed the same transcription task on a 1000-utterance subset of the MTurk-transcribed data. Inter-transcriber "word disagreement rate" (WDR) was computed, given N transcripts from two transcribers A and B, as follows:

$$\text{WDR} = \left(\frac{\sum_{i=1}^{N} \text{Sub}_i + \text{Del}_i + \text{Ins}_i}{\sum_{i=1}^{N} \frac{1}{2}(\text{Length}_{i,A} + \text{Length}_{i,B})} \right).$$

WDR is simply a symmetric version of WER, as the denominator is the sum of the average length of each pair of compared transcripts.

The interexpert WDR was 4.69%. The WDRs between the MTurk transcripts and each of the two experts were 5.55% and 5.67%. Thus, it seems reasonable to treat the MTurk transcripts as a near-expert reference orthography. In addition, the average WDR produced by pairing the three sets of transcripts produced by MTurk workers was 12.3%, indicating that obtaining multiple transcripts of each utterance is helpful when using MTurk to procure a reference.

5.6.3   Filtering for Accurate Hypotheses

Because *Voice Scatter* players often read terms and definitions verbatim, a significant portion of the utterances ought to be recognized with no, or very few, errors. We explored the usefulness of three sources of information in identifying this subset of utterances, with the goal of selecting a subset of the data that can be automatically transcribed with human-like accuracy. First, we consider the utility of speech recognition confidence scores, which provide a measure of uncertainty based on acoustic and lexical features. Second, we look at information from the game context associated with each utterance. Much like in *Voice Race*, speech recognition hypotheses that produce "hits" are unlikely to occur by chance. In this case, however, a "hit" occurs when a term is correctly matched to its definition. Third, we explore the importance of using a small vocabulary, strict grammar during recognition by comparing our results to those produced by a trigram trained on all flashcards appearing in the corpus.

Figure 5.18 explores the usefulness of each of these factors in identifying high-quality subsets of the data. The curves shown are produced from three experiments performed on the 10,000 utterance MTurk-transcribed development set. First, we ordered the set of hypotheses logged from game play based on their confidence scores, as produced by the module described in Hazen *et al.* (2002). We then drew utterances from the set in order from high to low confidence, and measured their cumulative WDR to produce the curve indicated with squares. Second, we performed the same experiment, using only the 4574 utterances that were identified as "hits" according to their recognition hypotheses. This produced the curve of triangles. Third, to explore the effect of vocabulary and LM size, we trained a trigram on all flashcard terms and definitions that appeared in the corpus. Using this n gram as the LM, we rerecognized each utterance to produce a new hypothesis and confidence score. We then drew hypotheses from these results in order of confidence score, to create the curve of circles. Finally, the dotted line shows the average WDR between the MTurk transcripts and each expert on the 1000 utterance expert-transcribed subset. It represents an expectation of human transcription agreement on the set.

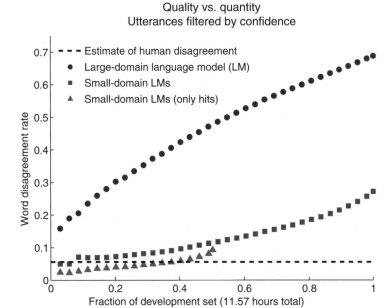

Figure 5.18 Cumulative word disagreement rate (WDR) for recognition hypotheses produced using either a large-domain trigram or many small-domain grammars on the 10,000 utterance MTurk-transcribed set. Cumulative subsets are created by incrementally adding hypotheses ordered by confidence score. An estimate of human WDR, calculated using the 1000 utterance expert-transcribed subset, is shown for comparison.

First and foremost, it is clear from Figure 5.18 that the small-domain nature of our recognition tasks is essential. The n-gram LM had an overall WDR of 68.8% when compared to the MTurk transcripts on all 10,000 utterances, whereas the narrow-domain LMs achieved a WDR of 27.2%. Moreover, using only confidence scores, it is possible to select a subset containing 15% of the original data with a near-human WDR of 7.0%.

Finally, by considering only "hits," it is possible to select a subset containing 39% of the data at a human-like WDR of 5.6% by discarding just 78 minutes of low-confidence "hits." Indeed, ignoring confidence scores altogether, and simply choosing all "hits," yields 50.2% of the data at a WDR of 9.3%. It is worth noting, however, that on these filtered subsets, human transcripts are still likely to be better. For example, the average WDR between experts and the MTurk transcripts on the 511 expert-transcribed "hits" was only 3.67%.

5.6.4   Self-Supervised Acoustic Model Adaptation

The self-transcribed *Voice Scatter* subcorpora became another resource to use in the common task of acoustic model adaptation. We adapt the original acoustic model, used by both *Voice Scatter* and *Voice Race*. To show that these self-transcriptions are useful across domains, we explore how the quantity and quality of orthographically transcribed *Voice Scatter* data influences the effectiveness of the adapted acoustic model on the *Voice Race* recognition task.

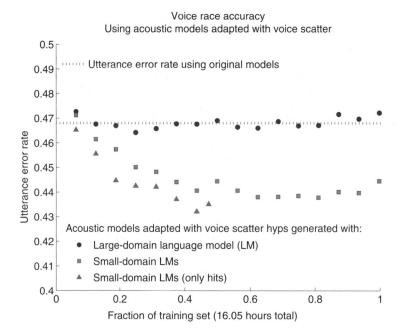

Figure 5.19   *Voice Race* utterance error rate using an acoustic model trained with incrementally more self-transcribed *Voice Scatter* utterances (sorted by confidence). The self-transcripts are generated using the original acoustic model via: a "large-domain" *n* gram, the small-domain grammars used in the online system, and the "hits" found in hypotheses generated from these small-domain grammars.

We drew self-transcribed utterances from the 16.05 hours of data that were *not* transcribed by MTurk workers, so that we can analyze the usefulness of these transcribed data as a development set. Utterances and their self-"transcripts" were accumulated in 1-hour increments using each of the three filtering methods described above. After each new hour of data was added to the set, acoustic model MAP adaptation was performed using forced alignments of the self-transcripts. Each adapted acoustic model was then used by the speech recognizer to produce hypotheses for 10,000 MTurk-labeled utterances collected from *Voice Race*.

Figure 5.19 shows the utterance error rate found on the the MTurk-labeled *Voice Race* data using successively larger sets of *Voice Scatter* utterances filtered via the three methods for adaptation. First, it is clear that using errorful hypotheses produced by the *n*-gram LM does not result in an improvement in utterance error rate, regardless of the amount of training data used. Second, using high-confidence hypotheses of utterances recognized with a small-domain LM achieves significant gains, and appears to reach a local minimum when between 60% and 80% of the *Voice Scatter* data are used. Third, when only "hits" are used, error rates fall faster, and achieve a better local minimum, even though less than half of the total data are available.

Finally, by comparing Figures 5.18 and 5.19, we can see that the manually transcribed utterances serve as a useful development set, both to select a filtering method and set a confidence threshold at which to consider data self-transcribed. According to the development set, selecting the high-confidence "hit" data that comprises roughly 39% of the total corpus should yield a human-like WDR. Choosing a training set from utterances based on this operating point would achieve an utterance error rate in Voice Race quite close to the best local minimum shown in Figure 5.19. Moreover, in the absence of a development set, a 7.8% relative reduction in utterance error rate would have been attained simply by using all of the "hit" data. With *Voice Scatter* and *Voice Race* we have shown the feasibility of large-scale data collection without the help of MTurk. While MTurk certainly offers a significant amount of control, where feasible, unpaid crowdsourcing can be an economic alternative, which can yield data useful in a number of ways. While here we have only explored acoustic model adaptation, we imagine this same data might be used for pronunciation modeling, among other speech recognition tasks.

5.7 Summary

In this chapter, we have examined a number of ways to collect both spontaneous and read speech using MTurk. We described our first efforts at collecting read speech, along with experiments in collecting realistic, user-generated scenarios leading to spontaneous speech collection. In both cases, traditional transcription methods were subsequently employed using MTurk to complement the speech data. We then turned our attention to games as a mechanism for speech data collection, also with MTurk support. The games were designed to allow us to obtain self-transcribed data, that is, data whose gold-standard transcription could be inferred from the context provided by the MTurk task. To verify these transcriptions, we described experiments in which we used the data to retrain an ASR engine and reduced WER.

The results of all of these experiments show the usefulness of the MTurk paradigm for speech. There is overhead associated with collecting speech via MTurk, but once the infrastructure is in place, it is relatively easy to collect large quantities of speech quickly. By adding in a gaming aspect to the data collection, we were able to exploit a symbiotic relationship between speech recognizers, which perform better with limited vocabularies, and gamers, who like to provide correct input. Thus, an answer elicited by a user in context is typically reliable when it is both correct and correctly recognized.

We feel this paradigm has promise in any game-like interaction where context can effectively restrict recognition choices. For example, data collection on mobile devices could exploit GPS coordinates to constrain the recognition of location-based entities in a geocaching-type game. Language teaching games could use a combination of geolocation and multiple choice questions to create "live" games that learners could play as they wander around town. Our experiments with retraining show that it is possible to bypass the added, and sometimes costly, step of transcribing every utterance before using it to improve the underlying technology.

5.8 Acknowledgments

The crowdsourced transcription for the flight domain was managed by Chia-ying Lee. The educational game experiments were conducted with Alex Gruenstein and would not have been possible without the contributions of Andrew Sutherland, founder of quizlet.com. This work was funded by Quanta Computer Inc.

References

Ai H, Raux A, Bohus D, Eskenazi M and Litman D (2007) Comparing spoken dialog corpora collected with recruited subjects versus real users. *Proceedings of SIGdial*.

Cai J, Feldmar J, Laprie Y and Haton JP (2008) Transcribing southern min speech corpora with a web-based language learning system. *Proceedings of International Workshop on Spoken Language Technologies for Under-Resourced Languages (SLTU)*.

Dickie C, Schaeffler F, Draxler C and Jänsch K (2009) Speech recordings via the internet: An overview of the VOYS project in Scotland. *Proceedings of Interspeech*, pp. 1807–1810, Brighton.

Draxler C and Jänsch K (2006) Speech recordings in public schools in Germany—the perfect show case for web-based recordings and annotation. *Proceedings of LREC*, pp. 2112–2115, Genova.

Gruenstein A, McGraw I and Sutherland A (2009) A self-transcribing speech corpus: Collecting continuous speech with an online educational game. *Proceedings of Speech and Language Technology in Education Workshop (SLaTE)*.

Gruenstein A, Seneff S and Wang C (2006) Scalable and portable web-based multimodal dialogue interaction with geographical databases. *Proceedings of Interspeech*.

Hazen TJ, Seneff S and Polifroni J (2002) Recognition confidence scoring and its use in speech understanding systems. *Computer Speech and Language* **16**, 49–67.

Li X, Gunawardana A and Acero A (2007) Adapting grapheme-to-phoneme conversion for name recognition. *Proceedings of ASRU*.

Marge M, Banerjee S and Rudnicky A (2010) Using the Amazon Mechanical Turk for transcription of spoken language. *Proceedings of ICASSP*.

McGraw I, Gruenstein A and Sutherland A (2009) A self-labeling speech corpus: collecting spoken words with an online educational game. *Proceedings of Interspeech*.

McGraw I, Lee C-y, Hetherington L, Seneff S and Glass JR (2010) Collecting voices from the cloud in *Proceedings of LREC*.

Novotney S and Callison-Burch C (2010) Cheap, fast and good enough: automatic speech recognition with non-expert transcription. *Proceedings of HLT NAACL*.

Paek T, Ju YC and Meek C (2007) People watcher: A game for eliciting human-transcribed data for automated directory assistance. *Proceedings of Interspeech*.

Seneff S (2002) Response planning and generation in the MERCURY flight reservation system. *Computer Speech and Language* **16**, 283–312.

von Ahn L (2006) Games with a purpose. *IEEE Computer* **39**(6), 92–94.

6

Crowdsourcing in Speech Perception

Martin Cooke[1,2], Jon Barker[3], and Maria Luisa Garcia Lecumberri[2]

[1] *Ikerbasque, Spain*
[2] *University of the Basque Country, Spain*
[3] *University of Sheffield, UK*

6.1   Introduction

Our understanding of human speech perception is still at a primitive stage, and the best theoretical or computational models lack the kind of detail required to predict listeners' responses to spoken stimuli. It is natural, therefore, for researchers to seek novel methods to gain insights into one of the most complex aspects of human behaviour. Web-based experiments offer the prospect of detailed response distributions gleaned from large listener samples, and thereby provide a means to ask new types of questions. Instead of instructing listeners to classify speech stimuli into one of a small number of categories chosen by the experimenter, large sample experiments allow the luxury of meaningful analysis of what is effectively an open set of responses. This freedom from experimenter bias is more likely to lead to unexpected outcomes than a traditional formal test that, of necessity, usually involves far fewer participants. Web-based experimentation involving auditory and linguistic judgements for speech stimuli is in its infancy, but early efforts over the last decade have produced some useful data. Some of these early crowdsourcing experiences are related in Section 6.2.

However, the promise of web-based speech perception experiments must be tempered by the realisation that the combination of audio, linguistic judgement and the web is not a natural one. Notwithstanding browser and other portability issues covered elsewhere in this volume, it is relatively straightforward to guarantee a consistent presentation of textual elements to web-based participants, but the same cannot be said currently for audio stimuli, and speech signals in particular. Similarly, while it may be possible using pre-tests to assess the linguistic ability

Crowdsourcing for Speech Processing: Applications to Data Collection, Transcription and Assessment, First Edition.
Edited by Maxine Eskénazi, Gina-Anne Levow, Helen Meng, Gabriel Parent and David Suendermann.
© 2013 John Wiley & Sons, Ltd. Published 2013 by John Wiley & Sons, Ltd.

of a web user whose native language differs from that of the target material in a text-based web experiment, it is far more difficult to do so for auditory stimuli. Here, performance alone is not a reliable indicator of nativeness, since it can be confounded with hearing impairment or equipment problems. Section 6.3 examines these issues in depth.

Nevertheless, we will argue that with careful design and post-processing, useful speech perception data can be collected from web respondents. Technological advances are making it easier to ensure that stimuli reach a listener's ears in a pristine state and that the listener's audio pathway is known. New methodological techniques permit objective confirmation of respondent-provided data. Ingenious task selection can lead to the collection of useful data even if absolute levels of performance fall short of those obtainable in the laboratory.

In the latter part of this chapter, we present a comprehensive case study that illustrates one approach that seems particularly well suited to web-based experimentation in its current evolutionary state, namely, the *crowd-as-filter* model. This technique uses crowdsourcing solely as a screening process prior to the selection of exemplars that are pursued further in formal tests. As we will see in this application, tokens that have the potential to say something valuable about speech perception are rare, and the great benefit of crowdsourcing is to increase the rate at which interesting tokens are discovered.

6.2 Previous Use of Crowdsourcing in Speech and Hearing

As early as 10 years ago psychologists were realising the potential of the Internet as an alternative to laboratory-based experimentation. In an early comparison of web-based and laboratory-based experimentation, Reips (2000) identifies a list of 18 advantages of the former. These include a range of obvious factors such as the availability of a large number of subjects and the ability to reach out to demographically and culturally diverse populations, as well as cost savings in laboratory space, equipment and subject payments. However, Reips argues that there are also subtler advantages that may be no less important. For example, participants of Internet-based experiments are 'highly' voluntary, meaning that there may be less motivation to produce deceptive responses. Likewise, results may have high external validity and generalise to a larger number of settings (e.g. Laugwitz 2001), and findings are likely to be more applicable to the general population (Horswill and Coster 2001).

Although the web-based methodology has been discussed amongst psychologists for over 10 years, it is only very recently that it has been seriously considered by hearing researchers. This is no doubt largely due to technical difficulties in the reliable delivery of audio stimuli to web-users who may be using software that is several product cycles out-of-date and who could only recently be expected to have Internet connections with adequate bandwidth. Nevertheless, the increased ease and precision with which audio-based experiments can be conducted is evidenced by the rapidly growing interest among musicologists (Honing 2006; Honing and Ladinig 2008; Kendall 2008; Lacherez 2008), audiologists (Choi *et al.* 2007; Bexelius *et al.* 2008; Seren 2009; Swanepoel *et al.* 2010) and, of particular relevance for the current chapter, the speech technology community (Blin *et al.* 2008; Kunath and Weinberger 2010; Wolters *et al.* 2010; Mayo *et al.* 2012).

The first major web-based psychoacoustic experiment, published in 2008, studied the impact of visual stimuli on unpleasant sounds (Cox 2008). In this study, participants were asked to listen to a sound while observing an image and were then asked to rate the 'horribleness' of the sound on a 6-point scale. The experiment was run from a simple Flash-enabled website and was accessible to anyone with a web browser and a computer with audio output capabilities (e.g. loudspeakers or headphones). Given that attitudes to sound are highly individual, the study required a large and demographically broad subject base in order to produce meaningful conclusions. The study could not have been conducted using a traditional lab-based methodology. However, as Cox notes, simply placing an experiment online does not guarantee a large number of participants, regardless of how well designed the web interface is. In order to recruit participants some form of media campaign is required. By making use of the local, national and international press, Cox was able to collect more than 1.5 million votes. Clearly, success here rests on being able to capture the public imagination and being fortunate in conducting a study to which people could easily relate. Participants were not paid for their time: once engaged with the experiment the main motivating factor was that in return for each response, the response of the general population was revealed, allowing participants to compare their view with that of others.

As the existence of the current volume testifies, crowdsourcing has also recently been recognised as a useful methodology within the speech research community. In contrast to Cox's voluntary crowd, speech researchers have largely employed crowds that have been financially compensated, usually through the use of Amazon's Mechanical Turk (MTurk). Annotation of speech corpora was the first problem to be addressed in this way. The key concern, clearly articulated by Snow *et al.* (2008), is 'Cheap and fast – but is it good?' Their conclusion, echoed in the title of a similar study 'Cheap, fast and *good enough*' (Novotney and Callison-Burch 2010), is that if crowdsourcing output is suitably handled, many large labelling tasks can be completed for a fraction of the cost of using highly paid experts and, crucially, with no significant loss in quality. Crowdsourcing has also been used for read-speech corpus collection. McGraw *et al.* (2009) employ an online educational game to generate a 'self-annotating' corpus. Similarly, the VoxForge project (Voxforge 2012) asks visitors to their site to read prompted sentences with the aim of collecting a quantity of transcribed speech sufficient for training robust acoustic models for use with free and open-source speech recognition engines.

The success of crowdsourcing in the context of corpus collection, transcription and annotation is encouraging, but does not by itself demonstrate the suitability of the methodology for the study of *speech perception*. In labelling tasks, human judgement is not being recorded as a means to judge the human perceptual system, but rather as a means to generate data that will be used to either bootstrap or evaluate learning algorithms. Error in human judgements is a source of noise in the labels that may lead to suboptimal machine learning, but it will not lead directly to false experimental conclusions. Further, labelling tasks have a high degree of inter-listener agreement, allowing outlying data to be filtered. Perceptual tasks, in contrast, are often concerned with the distribution of judgements or small statistical differences between conditions that are more likely to be masked in the event of a lack of experimental control.

Examples of crowdsourced speech perception studies can be found in the *speech synthesis* community. The annual Blizzard speech synthesis challenges (e.g. King and Karaiskos 2010) use human judges to rank the quality of competing synthesis systems. The judges include a mix of expert listeners and a contingent of naive listeners recruited via e-mail and social networking sites. Listeners typically perform tests in their offices over the Internet using headphones, though judgements are supplemented and validated by extensive lab testing. In the context of the Blizzard challenges, Wolters *et al.* (2010) have recently tested the validity of using a purely crowdsourcing methodology for evaluating the intelligibility of speech synthesis systems. Using listeners recruited via MTurk they find that although absolute intelligibility is much worse than in laboratory testing (a finding echoed in other crowdsourced speech perception studies, e.g. Cooke *et al.* 2011; Mayo *et al.* 2012), crucially, the MTurk listener scores reflect the *relative* intelligibility of the systems fairly well. If the task is to compare a new system against the current state of the art then reliable relative judgements may be all that is required.

6.3 Challenges

Despite their many clear advantages, web-based studies are not without their problems. Reips' early review of web-based experiments identified a carefully considered list of disadvantages (Reips 2002). The extent to which these issues invalidate the web-based methodology has been much debated in the intervening years (Skitka and Sargis 2006; Kendall 2008; Honing and Reips 2008). Despite strongly polarised views, it is clear that the validity of a web-based methodology is highly dependent on the nature of the experiments being conducted. In this section, we will re-examine the key difficulties with a specific focus on the requirements of speech perception experiments.

The most commonly cited problem for the web-based methodology is the comparative lack of experimental control. In fact, most of the difficulties discussed by Reips (2002) can be seen as symptoms of this underlying problem, and the challenges of experimental control feature prominently in studies such as Wolters *et al.* (2010). Generally speaking, in all web-based experiments, there is a trade-off: the experimenter accepts a reduced amount of control, but hopes that this may be compensated by the opportunity to recruit a very large numbers of subjects; that is, the added measurement noise due to nuisance variables is, it is hoped, more than countered by the increase in the number of data points. Nevertheless, a large number of data points cannot protect against systematic biases and even if subjects are plentiful it is bad practice to waste resources through poor experimental design. It is, therefore, worth considering how to minimise the potentially damaging consequences of the reduced experimental control inherent in web-based experimentation.

In considering a speech perception experiment, the factors that we wish to control can be broadly categorised under three headings: (i) *environmental factors* that describe external conditions that might affect a subject's responses; (ii) *participant factors* that describe how listeners are selected; and (iii) *stimulus factors* that describe how the sounds that are heard will be controlled. We consider each factor in turn.

6.3.1 Control of the Environment

The loss of environmental control is perhaps the most obvious difficulty facing the web-based methodology. For speech perception studies the *acoustic* environment is clearly very important. Laboratory listening tests are typically conducted in sound-attenuating rooms with state-of-the-art equipment for audio reproduction. In contrast, web-based tests may be performed by listeners sitting at computers at home or at work, in rooms with different amounts of environmental noise and with uncalibrated headphones of unpredictable quality.

A first consideration is whether the experiment is likely to be sensitive to environmental noise. Clearly, it would be unwise, for example, to attempt to measure signal reception thresholds in a web experiment: even quiet offices typically contain significant ambient background noise as well as the possibility of intermittent audio distractions (e.g. incoming calls, visitors). However, if the experiment involves processing speech at an adverse signal-to-noise ratio (SNR) then the additional environmental noise may not be significant if its peak intensities lie below the level of the experimental masker. Alternatively, it may be possible to reduce the unpredictability of the noise background. One practical technique that may be appropriate in some tasks is to add a fixed noise floor to the stimulus to mask variation caused by differing *low* levels of external noise.

Variance caused by differing headset quality is a separate issue. Headsets may well possess significant differences in frequency response. Subjects could be asked to provide information about their headset, but this would serve little purpose as cheap consumer headsets are uncalibrated and variation may be present even between headphones of an identical make and model. However, for many experiments, these sources of variation are of little real significance. Consider in particular that there is natural variation in the spectral shaping of speech caused by room acoustics, and further variation in the audiograms of even supposedly 'normal hearing' listeners. Depending on the details of the study design, it might be argued that it would be unusual for the result of a speech perception experiment to be heavily dependent on factors that the perceptual system itself works hard to minimise or ignore.

A broader factor is the degree to which the environmental context affects the attention of a participant. Consider that in a traditional experiment a subject has been brought to a laboratory where they are placed in an environment designed to be free of distraction. The subject will have given up time in their day to perform the experiment and can generally be expected to be focused on the task. This is in stark contrast to the web-based situation where the participant, even if highly-motivated, situated in a quiet office and wearing good-quality headphones, is far less likely to be devoting their full attention to the experiment. They may have other applications running on their computer, they may be receiving e-mail alerts or instant-messages; they are likely to be at work and generally in possession of a multi-tasking mindset. There is little that can be done to control these factors, but two points are worth noting: first, these factors are not totally excluded from traditional experiments – subjects unfamiliar with listening experiments will find the unfamiliarity of a hearing laboratory a distraction in itself, and differences between the mental stamina of participants will lead to varying degrees of attention throughout an experimental session. Second, it can be argued that results that are obtained in a natural environment have greater external validity; that is, they are more likely to be representative of the type of hearing performance achieved in day-to-day life. As we noted earlier, it has been argued that web-based findings are likely to generalise better to a greater range of real-world situations (e.g. Laugwitz 2001).

6.3.2 Participants

When conducting a behavioural experiment it is normal to seek a homogeneous group of participants meeting some well-defined selection criteria. Relevant criteria in listening experiments might include factors such as gender, age, language history and normality of hearing. It is the experimenter's responsibility to ensure that the selection criteria are met when recruiting participants. The opportunity for face-to-face interaction between the experimenter and the participant allows for a robust selection process. In the web-based case, participant selection is more problematic. Criteria can be made explicit, allowing participants to self-select, or information can be gathered from participants using online forms, which allow non-conforming participants to be filtered out subsequently. Here, two problems arise: selection depends on participants' trustworthiness, and certain selection criteria – such as possession of normal hearing – may be difficult or impossible to apply remotely, at least with current technology.

Trust

The 'trustworthiness' of participants is an oft-cited problem with web-based experiments (McGraw *et al.* 2000). How can it be guaranteed that the participants are providing correct information? Moreover, how can it be guaranteed that they are providing meaningful responses during the experiment itself? However, proponents of web experimentation point out that this question applies generally to all behavioural testing whether web-based or laboratory-based. Further, it has been argued that participants in a web-based experiment may generally have less motivation to give deceptive responses: if they are taking time to complete the experiment it will be because they have a genuine interest (they are 'highly' voluntary) rather than because they have been drawn in by the promise of financial reward and are being coerced to complete the session while being overseen by an experimenter (Honing and Ladinig 2008). Of course, this reasoning only applies to those applications of crowdsourcing where participants are not being paid and provides an argument other than cost savings for not making crowdsourced experiments financially rewarding. Further, the ease with which a web-based subject can drop out ensures that participants completing the experiment will generally be highly motivated and providing good-quality data. (Ironically, high drop-out rate is often something that web experimenters discuss as a concern.)

If participant trustworthiness is considered to be a serious issue then precautionary checks and measures can be put in place to screen out untrustworthy data. This can often be achieved through careful experimental design. For example, if the test requires a certain level of hearing acuity which in turn necessitates good equipment and low levels of background noise, experimenters may set a threshold with tokens that must minimally be identified correctly to ascertain that the required conditions are being met. If listeners need to have a particular linguistic background (e.g. regional, native, non-native accent, L1 of origin), which will be determined by a questionnaire, experimenters can also add criterion tokens designed to filter out participants who are not being frank about their profile (see also the use of what we call 'anchor tokens' in Section 6.5.6). When membership of a specific group is sought, appropriate slang vocabulary presented orally can be a useful means to determine affiliation. More generally, to increase the quality of participant data, web forms should not be pre-filled with default values (something the current authors are somewhat guilty of in the case study presented later in this chapter!) and participants should be compelled to complete the form before commencing the experiment proper. It may be advisable to avoid disclosing the selection criteria to avoid participants supplying dishonest data in order to gain access to the experiment. This is particularly true if respondents are motivated by financial reward.

Even valid participants – that is, those meeting the experiment's selection criteria – may need to be screened out if they are not sufficiently engaged in the task and are simply providing arbitrary responses in order to complete the work with minimum effort. It may be possible to detect such participants by monitoring response timings and checking that they fall in a normal range. Another simple and commonly employed technique is to intersperse the genuine trials with number of dummy trials that have a highly predictable 'correct' response, where 'incorrect' responses to these trials is then a sure indicator that the participant is not cooperating

(Sawusch 1996). Finding reliably correct dummy trials is itself something that is aided by large listener samples in a crowdsourcing study, as we show in Section 6.5.6.

Hearing Impairment

The impossibility of measuring participants' audiograms is a major limitation of the web-based methodology. Many people with mild or even moderate hearing loss do not realise that they have a deficit, especially if the loss has been progressive and not associated with trauma (as is typically the case with age-related hearing loss). Despite remaining undetected, a hearing deficit can easily lead to a measurable and significant effect on speech perception, particularly in noise, and render a potential participant unsuitable for a wide variety of speech perception experiments. Robust solutions to this problem are not obvious. Wolters *et al.* (2010) screened participants using a standard hearing questionnaire based on the hearing handicap inventory for adults (HHIA) (Newman *et al.* 1990), but HHIA scores are only weakly correlated with audiometric measures and Wolters *et al.*'s conclusions appear to cast doubt on the efficacy of this approach. It is perhaps more reliable to identify abnormal subjects directly from the statistics of their stimulus responses and apply a *post hoc* filtering to the results. This may be made easier if 'diagnostic' stimuli can be inserted into the experiment. Also, as with other issues of poor subject control, the difficulties can be reduced by designing experiments that rely on within-subject rather than between-subject comparisons.

Linguistic Background

Speech perception tasks in general and accent judgements in particular are usually carried out by naive native listeners of the target language (Major 2007), although expert judgements (Bongaerts 1999) and non-native data (Riney and Takagi 2005; MacKay *et al.* 2006; Major 2007) have also been collected, depending on the experimental aims and listener availability. Despite their predominance, the reliability of native judges has been questioned (Van Els and De Bot 1987; Major 2007). Dialectology and L2 studies find native listeners to be far from homogeneous as a group: listeners vary in their ability to judge accents and to some extent in their perceptual performance depending on their history of exposure to different varieties and languages, as well as other individual variables such as metalinguistic awareness, age, hearing, and, for some tasks, personality and educational factors. In the case of crowdsourcing, familiarity with technology and computer interfaces can also introduce variability in the results.

One of the issues that needs to be handled carefully is the acquisition of indicators of a participant's linguistic background. In crowdsourcing, this monitoring has to be done indirectly, since the experimenter is not usually able to ascertain a participant's linguistic competence by means of observations of their speech. In principle, notwithstanding additional technical and ethical concerns, web-based collection of a speech sample is possible, though its analysis would be expensive in time and effort and perhaps difficult to justify in the context of a speech perception experiment. Careful questionnaires are needed that clarify which languages are spoken by listeners, to which level and from what age (i.e. which are native languages, second languages or foreign languages). In this respect, care should be taken with the terms used to describe multilingual situations (e.g. 'bilingualism' and 'second language learners'; see Garcia Lecumberri *et al.* 2010, for a review). However, the thoroughness of the questionnaire and, therefore, its length need to be weighed against the possibility of discouraging participation.

To ensure data reliability, as mentioned above, participants' self-descriptions can be correlated with performance on criterion tokens, which may provide useful indicators of the trustworthiness of questionnaire responses. Kunath and Weinberger (2010), exploring the use of MTurk listeners for perception tasks, establish a baseline pre-test to determine listeners' accuracy. They also propose for future studies a more demanding and comprehensive 'qualification test' that will screen listeners before selecting them as participants in the main perception task.

6.3.3   Stimuli

In all listening experiments it is clearly important that a reproducible stimulus can be delivered undistorted to the subject. In the early years of web-based experimenting, technological constraints were such that the web was a poor substitute for the laboratory if experiments required delivery of reliable, high-quality audio and/or video. Web browsers would not always provide media support without the installation of non-standard 'plug-ins' and file sizes could require excessive download times on narrowband connections. However, with the more widespread availability of higher bandwidth connections to the Internet and with the advent of technologies such as MPEG4, HTML standardisation and client-side web scripting, these problems have largely vanished for most users.

A number of software frameworks for constructing and hosting web experiments have emerged in recent years. Systems such as WEXTOR (Reips and Neuhaus 2002), NetCloak (Wolfe and Reyna 2002) and DEWEX (Naumann *et al.* 2007) perform processing on the server side to avoid client-side compatibility issues. A fully server-side approach, however, is unsuitable for speech perception experiments that require controlled delivery of stimuli on the client. In contrast, the WebExp package employs a Java applet running on the client (Keller *et al.* 1998, 2009) that allows sophisticated control but at the expense of requiring that Java has been installed in the client's browser, something that cannot be guaranteed, especially on mobile devices. A potential solution is demonstrated by the Percy framework (Draxler 2011) that makes use of the latest HTML specification, HTML5, which provides multimedia tags to control the presentation of audio. This technology permits the development of web experiments that will run in any compliant browser with no need for external media players, plug-ins or additional client-side software.

Despite ongoing technological advances a few issues remain worth noting. First, although increased bandwidth means audio stimuli can be continuously streamed over the Internet, careful software design is needed to ensure that the participant receives stimuli in a predictable fashion. In some experiments even tiny uncertainties in the start time of a stimulus can impact the result. Therefore, stimuli need to be downloaded or buffered so that their onset times can be controlled with millisecond precision. Pre-downloading an entire experiment's stimuli may take appreciable time and participants may have little patience for watching a download bar. A solution calls for good software design, for example, using buffering and asynchronous downloads that occur in dead time while the participant is reading instructions or processing the previous stimulus.

A further issue concerns the quality of client-side audio hardware. Although a digital signal can be delivered with fidelity and in a timely manner to the participant, its reproduction can be compromised by poor-quality sound cards, variability of headphone frequency responses and interference from noise in the surrounding environment as discussed earlier.

Finally, *audio-visual* speech perception experiments require precise audio-visual synchronisation. Participants can be sensitive to asynchronies of as little as 40 ms if the audio arrives in advance of the video. For television broadcasting, for example, the Advanced Television Systems Committee recommends that audio should lead video by no more than 15 ms and audio should lag video by no more than 45 ms. The commonly employed MPEG encoding can ensure close synchronisation but only if care is taken during preparation by, for example, inserting presentation time stamps into the MPEG metadata field. Even with due care, data can potentially become desynchronised if there is significant mismatch in the video monitor and audio processing circuitry after decoding. However, such timing errors would be unusual in modern hardware and this problem can be expected to disappear in the near future for most users.

6.4   Tasks

Choosing a task depends mainly on the aims of the data collection and on the stimuli that will be used. In principle, we can classify speech perception tasks broadly according to what they are aiming to measure: signal properties such as speech intelligibility, quality and naturalness, speaker aspects (e.g. accent evaluation), and listeners' perceptual abilities and phonological systems.

6.4.1   Speech Intelligibility, Quality and Naturalness

Speech intelligibility measurement is the object of a great many studies in speech perception, motivated by investigation of factors such as speech style and listener characteristics as well as the effects of maskers, vocoders and synthesis procedures. Intelligibility is normally quantified objectively[1] by having listeners report what they have heard either orally or in writing. Thus, intelligibility is measured with tasks that take the form of oral reports (e.g. repetition of speech, answers to questions, utterance completion) or written reports (orthographic or phonetic transcriptions) or selection from response alternatives presented programmatically. In the case of studies that use crowdsourcing, the latter option is the only feasible modality at the present time. Participants can be asked to type what they have heard (Wolters *et al.* 2010) or choose their response using a custom-designed interface (Garcia Lecumberri *et al.* 2008).

Another measure that is sometimes grouped with intelligibility is 'comprehensibility' – how easy it is to understand a particular utterance or speaker. As opposed to intelligibility, comprehensibility is a subjective measure, since it depends on a listener responding based on their impression rather than on quantifiable data (e.g. number of words/segments understood). Comprehensibility, like many other subjective listener-derived judgements, is typically measured by means of Likert scales. In crowdsourcing, comprehensibility of synthetic speech has been assessed with a version of the mean opinion score, a 5-point Likert scale (Blin *et al.* 2008).

Naturalness, alongside intelligibility and comprehensibility, is the main criterion by which synthetic and other forms of generated or coded speech are routinely judged. One of the

---

[1]Note that the term *subjective intelligibility* is also frequently used in this context to distinguish between measures derived from listeners on the one hand and predictions made by so-called *objective intelligibility models* on the other.

pioneering applications of crowdsourcing has been in the annual Blizzard Challenge (e.g. King and Karaiskos 2010), which also employs Likert scales similar to the ones mentioned above.

The measures outlined above are frequently used in conditions that simulate some elements of everyday speech perception, usually by presentation in the presence of competing sound sources or under cognitive load. The latter is especially relevant for accent judgements (see Section 6.4.2). Formal studies have corroborated the intuition that foreign accents (Munro and Derwing 1995) and unfamiliar regional accents (Floccia *et al.* 2006) can make special demands on the part of the listener so that the cognitive effort required from listeners is higher than when listening to a familiar accent. One direct way of measuring cognitive effort is to monitor response times or latencies. Latencies can be calculated as a global quantification of overall accent effects or at a finer level of detail, in terms of segmental or featural variables (e.g. Bissiri *et al.* 2011).

From a technical perspective, reaction time monitoring in a crowdsourced experiment requires careful design, particularly with respect to the balance of responsibilities for client and server components, but the use of suitable client-side software enables reliable reaction monitoring (e.g. as demonstrated in Keller *et al.* 2009). However, a less tractable set of issues comes from the need for a commitment on the part of participants to focus on the task to the best of their abilities during its time span and to avoid distractions. A related issue is that it may be difficult for the remote experimenter to measure web-respondent fatigue effects that sometimes accompany tasks involving cognitive load.

6.4.2 Accent Evaluation

Outside the strict communicative confines of intelligibility as measured by narrow criteria such as the number of keywords identified correctly, speech provides a wealth of other information. For instance, speech provides cues to the geographical and/or social origin of speakers and also conveys affect. In turn, a talker's speech provokes attitudinal responses in listeners. Accent research, for both native and foreign accents, has addressed all these areas.

Accents may be analysed according to speakers' geographical or linguistic origins. Within this broad field, some researchers are concerned with straightforward accent classification. Clopper and Pisoni (2005) suggest a perceptual regional accent classification task in which listeners are asked to indicate on a map where a particular speech sample belongs to in geographical terms. In some contexts, regional accent judgements are linked to opinions of social class (Wells 1982; Trudgill and Hannah 2008). Kunath and Weinberger (2010) used crowdsourcing to classify foreign-accented English samples according to three possible first language origins as well as in the degree of foreign accent present in each sample. The magnitude or degree of accent corresponds to the extent to which it differs from a particular norm or standard. The notion of distance from a reference is flexible. In the case of synthetic or manipulated speech, an evaluation might be designed to measure the extent to which speech differs or conforms to a 'standard' set by natural speech or even a specific voice. Conformity evaluations used for synthetic speech and foreign/regional accents typically employ similar response measures as for studies of naturalness (i.e. rating scales).

Accents can cause attitudinal reactions in listeners, who may feel charmed, soothed or interested when listening to certain voices or conversely may get irritated, anxious or bored when listening to accents that differ from their own. Negative reactions are often present in the case of foreign accents (Brennan and Brennan 1981; Fayer and Krasinski 1987), probably due to communication breakdowns or the extra effort listeners need to make in order to repair phonetic deviations (Fernandez Gonzalez 1988). Paradoxically, foreign speakers who achieve near-native accents may also provoke unusual reactions such as suspicion or envy. Again, attitudinal style judgements are usually carried out by means of Likert scales.

Speech conveys paralanguistic and extralinguistic information and as such, listeners develop constructs about speakers' personalities and other characteristics such as intelligence, education, profession, trustworthiness and socioeconomic status, which may become generalised for particular combinations of speaker/listener groups and according to certain stereotypes. There are connections between accent and physical qualities of their place of origin (Andersson and Trudgill 1990). Thus, regional accents will be judged as euphonic if they belong to a scenic location, whereas accents from industrial areas tend to be considered uglier. Research on second language acquisition has shown that for foreign accents too listeners judge speech differently depending on the perceived origin of the speakers (Hosoda *et al.* 2007) and these judgements extend to beliefs about economic level, status, job suitability and professional competence (Boyd 2003; Dávila *et al.* 1993).

Since most of these accent evaluation tasks are based on speech tokens evaluated along Likert scales, they are certainly appropriate targets for crowdsourcing studies, just as Kunath and Weinberger (2010) have done for degree of foreign accent, and as in the evaluations of speech naturalness cited earlier.

6.4.3 Perceptual Salience and Listener Acuity

Exploring the perceptual salience of speech features and listeners' perceptual abilities in, for example, detecting just-noticeable differences (JNDs) is often the object of basic research or an adjunct to other speech perception tests. Discrimination tasks are typically used here. Discrimination tasks normally give no information as to listeners' phonological or grammatical classification of the stimuli. Stimuli may be presented in one or two pairs (AX or AA–BX), or in triads (ABX) in which a listener has to decide if X is the same or differs from the other tokens in the sequence. In the latter case, to avoid short-term memory biases towards the stimulus closest to X, the triad AXB is the preferred alternative (Beddor and Gottfried 1995).

Discrimination tasks require a very high degree of stimulus control to avoid ascribing perceptual performance to a confounding variable. It seems unlikely that crowdsourcing will be suitable (or indeed necessary) for the estimation of psychoacoustic distinctions such as JNDs in pitch or duration. It is perhaps surprising then that crowdsourcing has been used in speech-based discrimination tasks. For instance, Blin *et al.* (2008) developed a system to support ABX discrimination tasks used to compare manipulated and natural speech. Notwithstanding the fact that these two examples involve relatively high-level categorisations and hence to some extent prior information, the basis for any decision originates in part from low-level stimulus differences. One would certainly expect to find differences in sensitivity between formal and web-based discrimination tasks.

6.4.4 Phonological Systems

One of the main aims of speech perception research has been to discover something of the structure of a listener's phonological system, addressing questions such as what are the phoneme categories in an inventory, what is the internal structure of those categories, how is the phonological system organised, and how does it relate to other phonological systems. For these purposes, sound identification tasks have been widely employed. In identification tasks, the stimuli may correspond to phonemic categories or their realisations, but dialectology also uses this type of task to classify accents. Here, the task for the listener is to provide a label (e.g. phonetic symbol, spelling, word, accent name) for the stimulus.

Identification tasks may be totally open so that the listener comes up with the label. However, this can make it difficult to compare results across listeners because of a potential profusion of labels. Additionally, the task may be too difficult, particularly for speakers with low metalinguistic awareness. In that case, individual abilities introduce a great deal of response variability. To avoid these risks, experimenters frequently provide ready-made label choices. The number of choices may be restricted or open (e.g. presenting labels for all the language's phonemes or for all the dialectal areas). While limited response sets typically lead to easier tasks, they carry the danger of leaving out the label that the listener would actually choose in response to a stimulus (Beddor and Gottfried 1995).

In phoneme labelling studies, particular care needs to be taken with the labels chosen since orthography may play an important and often confounding part, particularly in cross-linguistic research (Beddor and Gottfried 1995). A study that compared English consonant perception across listeners from eight different L1 backgrounds (Cooke *et al.* 2010) found that for naive listeners, orthography has a strong influence. The alternative of using phonemic symbols restricts the participants to populations which are familiar with them, or runs the risk of providing unwanted perceptual training during symbol familiarisation.

In order to explore the internal structure of listeners' phonological categories and how different realisations are classified as exemplars of the same phonemic category, categorical discrimination tasks may be used. These tasks resemble straightforward discrimination tasks except that in categorical discrimination all stimuli are physically different (ABC) and listeners have to classify as 'same' those belonging to the same phonemic category (e.g. A and B). This task can also be extended to accent studies in order to group different speakers into accent groupings defined by regional origin or L1, for example.

Category Goodness Rating is a metric designed to explore the internal structure of phonemic categories at a more detailed level than is possible with categorical discrimination. Listeners rate individual stimuli based on how good an exemplar it is of a particular phonological category. This task usually accompanies either an identification task or a categorical discrimination task. Ratings can use Likert-like scales (Kuhl 1991) or continuous scales (Gong *et al.* 2011).

The types of task outlined above may well be suitable for crowdsourcing – there have been few studies in this domain to date – but their hallmark of multiple response alternatives that convey what might be quite subtle categorical distinctions raises a broader issue for web-perception experiments: how to instruct the participant and how to determine if the instructions have been adequately understood. In formal laboratory situations, the human–human interaction between experimenter and participant is rich, flexible and rapid, providing an immediacy of feedback both for the participant who may be unsure of what is required, and for the experimenter, who can form a judgement about whether instructions have been understood. The experience for a web-based participant is monochrome by comparison. Instructions are usually presented in textual form, perhaps with audio examples. There is generally no personalised interaction, nor an opportunity to ask questions. One positive aspect is that instructions are the same for all participants, although there is no guarantee that instructions are followed or even read!

6.5 BIGLISTEN: A Case Study in the Use of Crowdsourcing to Identify Words in Noise

We now describe in some detail a recent crowdsourcing exercise in which listeners attempted to identify words presented in noise via an online application. This study, which we call the BIGLISTEN, is an example of the *crowd-as-filter* approach, where the numerical advantage inherent in the crowd is used to screen a potentially vast number of stimuli to find tokens that meet some criterion, which are subsequently presented to listeners under a traditional controlled laboratory regime. In part, the BIGLISTEN web application was developed to pilot ideas in crowdsourcing for speech perception and in particular to enable comparisons between formal and web test results in order to evaluate the merits of the approach.

In this section, we describe the problem that motivated the BIGLISTEN and argue that crowdsourcing is a natural solution, before explaining the design decisions taken during development of the web application. We go on to highlight some of the principal findings and discuss the lessons from the pilot approach. More details of the BIGLISTEN can be found in (Cooke 2009; Cooke *et al.* 2011).

6.5.1 The Problem

A better understanding of how listeners manage to communicate effectively using speech in realistic environments – characterised by the presence of time-varying degradations resulting from competing sound sources, reverberation and transmission channels – will enable the development of more robust algorithms in speech and hearing technology applications. One key ingredient is a detailed computational model that describes how listeners respond to speech presented in noise. At present, we have what have been termed *macroscopic* models that make objective predictions of subjective speech intelligibility (e.g. ANSI 1997; Christiansen *et al.* 2010) and quality (Rix *et al.* 2001). By contrast, the study of *microscopic* models that attempt to predict what listeners hear at the level of individual noisy tokens is only just starting (see Cooke 2006). At the heart of the microscopic modelling approach is the need to discover *consistent* responses to individual speech-in-noise tokens across a sufficient sample of listeners, and to uncover a large enough corpus of such examples to allow comparative evaluation and refinement of microscopic models.

While less-sophisticated microscopic models might be expected to respond like listeners when tokens are correctly recognised, they are less likely to make the same errors as listeners unless the model successfully captures in some detail the processes involved in human speech perception. Therefore, while consistently reported *correct* responses in noise are useful in model evaluation, unexpected responses common to many listeners are particularly valuable for the microscopic modelling enterprise.

The main requirement, then, is to collect a corpus of individual noisy speech tokens, each of which induces a high degree of consistency in listener responses for both correctly heard and misheard cases. More generally, we are interested in measuring the response distribution for each noisy token. Low entropy distributions, characterised by one, or perhaps two, clear concentrations of responses, are the goal of token screening. Robust estimation of response distributions demands the availability of a large number of *different* listeners, and hence makes this an ideal application for crowdsourcing.

6.5.2   Speech and Noise Tokens

Users of the BigListen application identified one or more blocks of stimuli. Each block contained 50 monosyllabic English words mixed with one of 12 types of noise. Words came from an existing list (Cara and Goswami 2002) using selection criteria designed to encourage confusability (e.g. high spoken and written frequency and the possession of a large set of phonological neighbours) and screened to remove obscenities. Five native British English speakers, four males and one female, each recorded the subset of more than 600 words that met these criteria.

A variety of noises were used to encourage different kinds of confusions, resulting, for example, from foreground-background misallocation of patches of spectro-temporal energy or masking of target speech components. Maskers included speech-shaped noise, multitalker babble for a range of talker densities (including a single competing speaker), envelope-modulated speech-shaped noise and factory noise. Each block of stimuli contained words from a single target talker and a single type of masker. The SNR was set based on pilot tests to a range low enough to create potential confusions but not so low as to lead to near-random responses. In practice, the SNR decreased within a narrow range (SNR_{max} to SNR_{min}) within each block of stimuli. The first five stimuli in the block acted as practice tokens. Their SNRs decreased linearly from +30 dB (i.e. almost noise free) to SNR_{max}, after which the SNR decreased linearly for the remaining 45 tokens to SNR_{min}. Different maskers used different SNR ranges to reflect the finding that listeners' ability to reach a criterion intelligibility level varies with noise type (Festen and Plomp 1990). The purpose of using a decreasing SNR during the block was to test a range of noise levels where consistent confusions might be expected to occur and also to provide the user with a more challenging and perhaps engaging task experience with time. Users could complete as many blocks as they wished. More details of the task and stimuli are provided in Cooke (2009).

6.5.3   The Client-Side Experience

Visitors to the BigListen home page saw a single web page containing a small amount of motivational text, instructions and the test itself. The page also included clickable examples of words in noise, which had the dual purpose of illustrating the types of stimuli in the test and allowing the volume control to be set to a comfortable level. The test interface ran via a Java applet. The applet initially displayed a form to collect a small amount of information from the respondent and to seek their consent to take part in the test (Figure 6.1). Once the form was filled in and consent given, the main experimental interface – essentially a text input box – replaced the form (Figure 6.2). After completing a block, users received immediate feedback on their performance, expressed as a ranking based on their score of words correctly identified within the subset of listeners who had heard the same test block.

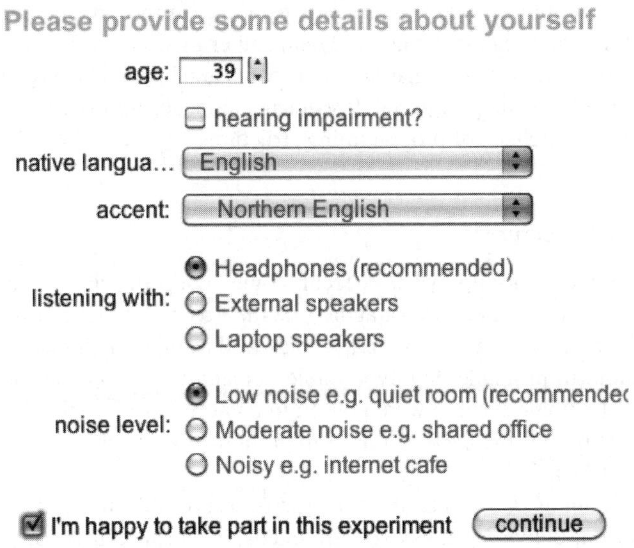

Please provide some details about yourself

age: [39] [↕]

☐ hearing impairment?

native langua... [English   ↕]

accent: [Northern English   ↕]

listening with:
◉ Headphones (recommended)
○ External speakers
○ Laptop speakers

noise level:
◉ Low noise e.g. quiet room (recommendet
○ Moderate noise e.g. shared office
○ Noisy e.g. internet cafe

☑ I'm happy to take part in this experiment   (continue)

Figure 6.1   The initial page of the test interface, showing the questionnaire filled in.

6.5.4   *Technical Architecture*

General Considerations

Previous sections highlighted those aspects of a web-based experiment that are largely outside an experimenter's control. However, the impact of many of these factors can be mitigated to a large extent through a careful consideration of software architecture and design. Key technical goals include minimising the impact of network delays (e.g. through buffering), maintaining precise control over data such as audio signals and user responses transferred between client

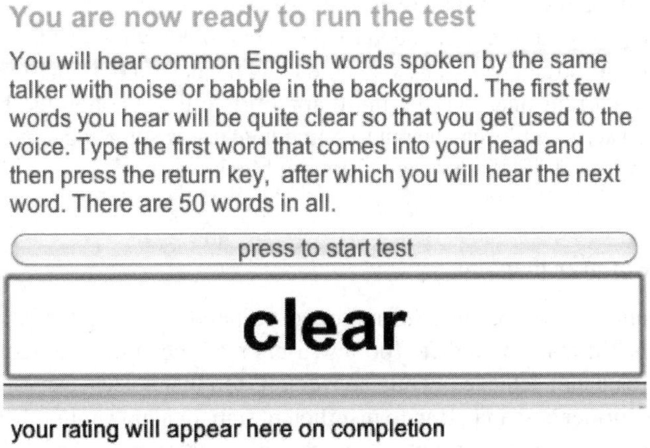

You are now ready to run the test

You will hear common English words spoken by the same talker with noise or babble in the background. The first few words you hear will be quite clear so that you get used to the voice. Type the first word that comes into your head and then press the return key, after which you will hear the next word. There are 50 words in all.

(press to start test)

clear

your rating will appear here on completion

Figure 6.2   The main experiment screen.

and server, robust handling of spikes in user interest (e.g. via resource pooling), encouraging task completion through a seamless and rapid data gathering process, and by accommodating as far as possible differences in client hardware, software and location. Consequently, technical solutions are favoured that support portability, localisation, scalability and client–server load sharing in addition to a rich set of programming structures.

The BIGLISTEN Architecture

The BIGLISTEN application employs Java technologies coupled with a back-end relational database. Java provides good support for audio (via the `javax.sound.sampled.*` package) and user interfaces (via `javax.swing.*`) as well as multiple threads of execution and database integration. In principle, highly variable demand can also be accommodated using Enterprise Java technologies. These were not felt to be necessary for the initial version of the BIGLISTEN application but the ease of future migration to a scalable web application is an attractive feature of Java.

A Java applet running on the client's browser is responsible for collecting respondent-provided information, delivering noisy speech tokens, gathering participants' responses and providing feedback at the end of each test block. A further applet supports the inclusion of buttons on the web application's introductory page to provide examples of stimuli and also to allow the user a convenient means to check the volume setting.

A Java servlet mediates all information flows from and to the applet. The servlet is responsible for all communication with the database and filestore, in addition to one-time initialisation of common resources such as connection pools. The applet–servlet design pattern permits full abstraction of implementation details (e.g. no database language code is present in the applet, nor any direct links to other back-end resources), facilitating rapid reconfiguration of the back-end without affecting the user view of the application and without requiring recoding at the applet level.

Information about test blocks as well as homophone, language and accent lists is held in a relational database in the BIGLISTEN application. The database also stores participant-supplied information, word responses and timing data. For efficiency, complete blocks of 50 test stimuli are bundled into single files stored on the server. To enable delay-free presentation during the test itself, a block of stimuli is downloaded to the client applet while the user fills in the form. Intermediate buffering strategies, such as downloading the next or next-but-one stimulus while the user hears the current one, may be more appropriate than monolithic block transfer in situations where a user's results can be put to immediate use in selecting stimuli for successive users. Here, the overhead of transferring a 50-word block was not high.

Making best use of user demand

In many crowdsourcing applications, the number of respondents using the system in any given time period can be difficult to predict. Too few users may mean that the required number of responses per token is not achieved, while conversely, too many users can rapidly exhaust the supply. While the former case may lead to insufficient statistical power in subsequent analyses, the latter represents a missed opportunity. A number of techniques can be used to address these issues. A low response rate can still result in valuable data if stimuli are rationed with the aim of maintaining a given number of responses per token. A lower limit on the number of tokens available at any instant might be based on the maximum expected number of tokens screened by a single individual, given that users should not in general hear the same token twice. A higher-than-expected usage can be accommodated either by dynamic generation of new stimuli to meet demand, or by overgeneration of tokens.

The BIGLISTEN application adopts a rationing approach. Blocks of stimuli progress through three states – 'unused', 'active' and 'exhausted'. At any time, a small number of blocks are active. When a block has been screened sufficiently, it is moved to the exhausted state and replaced by an unused block. Sufficiency of screening is defined in the BIGLISTEN based on reaching a criterion number (here set to 20) of 'high-quality' listeners (the definition of high quality here is approximately the same as the 'subj' category described in Section 6.5.6).

6.5.5 Respondents

Here, we examine quantitative aspects of the BIGLISTEN experiment as well as the information provided by respondents themselves.

Raw Response Statistics

Two adverts placed 11 days apart via the University of Sheffield's internal announcement service (which has the potential to reach more than 20,000 staff and students) led to 2120 respondents filling in the initial applet form within the first 20 days of the first advert. Of these, 1766 (83.4%) went on to complete the task (i.e. respond to at least one block of stimuli). Note that since respondents were not required to register to use the system, no user-tracking between page visits was possible, so what we call respondents here are actually separate page visits. Predictably, most of the activity occurred on the days of the adverts themselves, with a rapid decrease over time (see Figure 6.3). Clearly, peaky demand is a consequence of the method used to garner interest in the web experiment. Ideally, publicity measures that produce

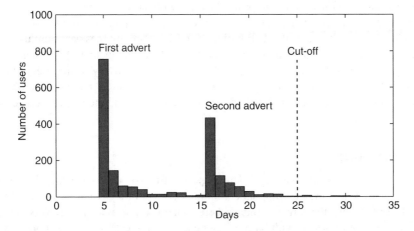

Figure 6.3   Number of responses per day.

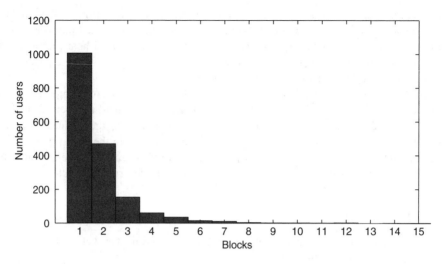

Figure 6.4   Number of stimulus blocks identified per respondent.

a more uniform demand over time are preferred, although in this case the level of demand was not problematic for the tool.

Between them, respondents heard 157,150 individual noisy tokens, corresponding to 3143 blocks, a mean of 1.78 blocks per respondent. Figure 6.4 demonstrates that while most listeners identified stimuli from a single block, a significant proportion went on to complete several blocks. The number of additional blocks screened gives an indication of how engaging the task was for listeners. An additional 0.78 blocks per listener perhaps suggests that while many respondents were curious enough to carry out the task once, most did not feel it sufficiently engaging to continue. Here, it seems likely that the relatively sparse feedback provided (essentially just a user ranking) and the lack of any reward – monetary or otherwise – was responsible for the relatively low task engagement. In practice, task designers can use this kind of quantitative information to improve the web application.

The mean response time per block was 155 seconds, corresponding to just over 3 seconds per stimulus. Figure 6.5 shows the distribution of mean response times per stimulus.

Tables 6.1, 6.2, 6.3 and 6.4 summarise data supplied by respondents about their first language, accent, listening conditions and audio hardware, respectively, while Figure 6.6 plots their age distribution. In addition, 58 respondents (3.3%) reported some degree of hearing impairment. Figures are based on the 1766 respondents who completed at least one block.

First Language (L1)

More than four out of every five respondents reported English as an L1, while the remaining native languages reflect the multilingual community typical of a UK university. While native English listeners were the target audience here, our experience with later versions of the BigListen application tested with large L2 populations suggests that robust confusions can also be harvested from non-native listeners, particularly from homogeneous samples such as

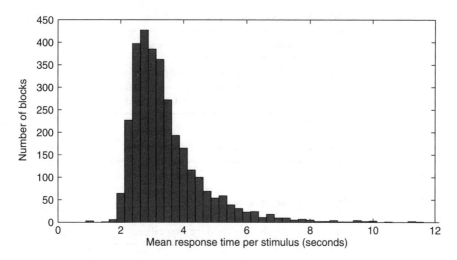

Figure 6.5   Response time.

Table 6.1   Respondents' self-reported first language.

N	Percent	L1
1442	81.65	English
70	3.96	Chinese
39	2.21	German
31	1.76	Spanish
27	1.53	Bulgarian
19	1.08	Arabic
16	0.91	Hindi
11	0.62	Greek

Note: Here and elsewhere categories with fewer than 10 respondents are omitted.

Table 6.2   Accents reported by respondents with English as L1.

N	Percent	Accent
746	42.24	UK and Republic of Ireland
317	17.95	Not supplied
265	15.01	Northern English
162	9.17	Southern
104	5.89	Midlands
55	3.11	Received Pronunciation
12	0.68	Scottish
10	0.57	West Country
10	0.57	Welsh
10	0.57	Northern Irish

Table 6.3 Respondents' listening conditions.

N	Percent	Noise level
1541	87.26	Low (e.g. quiet room)
207	11.72	Moderate (e.g. shared office)
18	1.02	Noisy (e.g. Internet cafe)

Table 6.4 Respondents' audio hardware.

N	Percent	Audio delivery
815	46.15	Headphones
577	32.67	External loudspeakers
374	21.18	Laptop speakers

advanced learner groups with the same L1. For L2 listeners confusions appear to be dominated by L1 influences rather than masking.

Accent

Table 6.2 lists the dominant accents of English amongst respondents. Knowing a listener's linguistic origins within the native population can – in principle – help to make sense of their responses. One issue is the granularity at which to define accents. A detailed classification can lead to problems in finding an appropriate category for the many listeners who have moved around, producing the potential for confusion on the part of users as to the desired response.

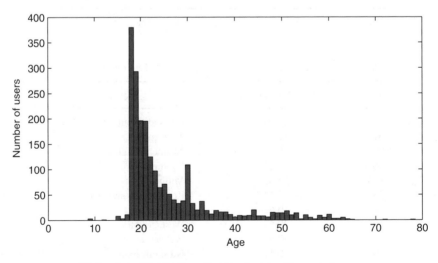

Figure 6.6 Distribution of respondents' self-reported ages.

The problem is more acute for bilinguals or individuals with mixed accents. In the BIGLISTEN listeners could choose from 10 options within the United Kingdom, 7 each for Oceania and North America, and around 5 each for other English-speaking countries. A design decision was taken to also permit null responses, or one of several less-specific categories such as 'UK and Republic of Ireland', or 'General American'. The aim was to enable respondents to get through the questionnaire rapidly in order to encourage completion of the whole task. A better approach might be to forego self-classification of accent and instead to embed accent-diagnostic words within the main test, along the lines of SoundComparisons (2012). As we will see later, certain word confusions reveal something of the likely broad accent region of the listener and provide an indirect way to classify a respondent's accent.

Listening Conditions

Crowdsourced listening tests will inevitably contain many responses from users listening under nonideal acoustic conditions. This aspect of crowdsourcing is one of the most difficult to control (but see Section 6.6 for some suggestions). Part of the problem stems from the robust nature of human speech perception: listeners are so capable of tracking a target source in the presence of reverberation or other sound sources that their tolerance for extraneous sound is high, and what is subjectively a quiet environment may well contain a significant level of noise. A very high proportion of respondents in the BIGLISTEN claimed to take the test in a quiet environment, a figure perhaps influenced by the availability of such spaces for a university population and not necessarily representative of a wider audience. On the other hand, the test itself demands a certain degree of quietness. We later introduce a method for selection of responses based on performance on near-universally correct stimuli, which can be expected to identify those respondents listening in reasonably quiet conditions.

Audio Hardware

Extraneous noise is attenuated by headphone listening. As for listening conditions, the fidelity of audio delivery is one area where a large amount of variability can be expected. Here, perhaps surprisingly, the majority of respondents did not use headphones but instead listened through external or laptop speakers, the latter in particular being clearly sub-optimal for speech in noise tasks.

Age

Due to factors such as the possibility of age-related hearing loss, knowing a respondent's age can be valuable for later subsetting or rejection of responses. Here, the age profile (Figure 6.6) probably says more about that of the group who received the invitation to participate than it reveals of any age-related predilection for online tests. Note that the peak at age 30 stems from this being the default choice on the questionnaire, again resulting from a design decision to facilitate rapid test completion. In a large-scale test, it would make more sense to force respondents to choose an age. Even so, it is interesting to observe that all but an estimated 4% of respondents did indeed go to the trouble of selecting an age rather than using the default.

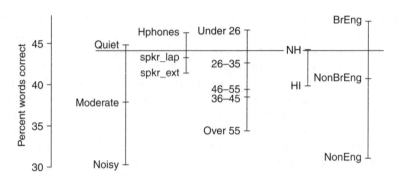

Figure 6.7 Mean word identification scores for each level of respondent-supplied factors. (Figure reproduced from Cooke *et al.* 2011.)

6.5.6   Analysis of Responses

In this section, we examine the responses supplied by users of the BigListen and go on to compare them to those of a group tested using the same task and materials under traditional laboratory conditions (for details see Cooke 2009). Since not all blocks heard by the formal group were exhausted by the web group (in the sense defined in Section 6.5.4), the following analysis is based on a subset of the exhausted web data, corresponding to material spoken by one of the male talkers in each of the 12 noise conditions.

Effect of Self-Reported Factors on Recognition Rates

While the principal purpose of the BigListen web experiment is to discover interesting word confusions, most of the time in formal tests listeners reported the correct answer, so it is of interest to explore how the information supplied by respondents (e.g. first language, age) correlated with overall recognition scores. Figure 6.7 shows mean scores for each level of the factors gathered from participants.

This figure needs to be interpreted with some care. These are univariate scores, that is, computed over all other factors, and thus it is important to note that control variables are not independent. For example, a correlation can be expected between those respondents who reported hearing impairment and those in the older age brackets. For a sufficiently large sample, a full conditional dependency analysis between factor levels could be carried out, but the relatively small scale of the current sample precludes this kind of analysis here. Also note that the distribution of respondents across levels for some of these factors is non-uniform. This caveat aside, we include the data to give some idea about the likely average effect of participant factors on performance.

Ambient noise in the test environment had a large effect, as did having a first language other than English. More surprisingly, the performance of listeners having as their L1 a variety of English other than British English (NonBrEng) was substantially lower than the level obtained by native British English speakers (BrEng). Predictably, older listeners fared less well than younger, and similarly users with headphones outperformed those relying on internal or external speakers. Listeners who reported hearing impairment showed relatively

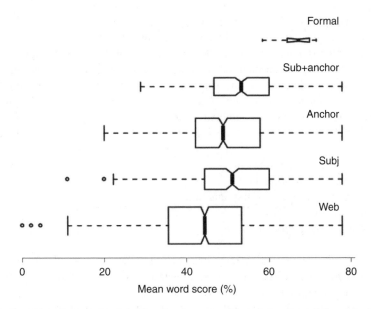

Figure 6.8 Boxplots of scores for formal and web groups. Lines extend to 1.5 times the inter-quartile range, circles indicate outliers, box thickness is proportional to the number of listeners in the group and notches depict 95% confidence intervals. (Figure from Cooke *et al.* 2011.)

little degradation, although it is likely that listeners with moderate or severe HI either did not attempt the task or used a hearing aid.

While the ranking of levels within each factor is almost as expected (the exception being the poorer performance of external loudspeakers compared with internal loudspeakers), the cross-factor comparisons afforded by this type of plot are revealing. The difference between means for quiet and noisy conditions is of a similar size as the difference in performance between British English and non-native listeners. The benefit of headphone listening is, by comparison, not so large.

Web versus Formal Listening Tests

The upper and lower boxplots of Figure 6.8 depict word score statistics for the crowdsourced (WEB) and traditionally tested (FORMAL) groups prior to any type of respondent-filtering. The intermediate boxplots (SUBJ, ANCHOR, SUBJ+ANCHOR) describe scores for subsets of web respondents selected on the basis of subjective and objective criteria defined below.

This figure demonstrates that mean scores obtained via unfiltered crowdsourcing are very significantly reduced – here, by well over 20 percentage points – compared to those obtained under traditional testing procedures. This outcome has been found in other web-based speech perception studies (e.g. Wolters *et al.* 2010; Mayo *et al.* 2012). For instance, in Mayo *et al.* (2012), MTurk listeners had an absolute performance level of around 75% of that measured in a traditionally tested group.

Nevertheless, Figure 6.8 suggests that individual web listeners are capable of high scores. Indeed, some WEB participant scores are higher than those obtained in the FORMAL group, although it should be noted that the latter employed far fewer participants (which also accounts for the wider confidence intervals for the FORMAL group).

Clearly, the WEB group includes data from respondents whose first language is not English, or who reported hearing-impairment, or might be expected to suffer from age-related hearing deficits, less-than-ideal listening conditions or audio delivery hardware. As a first post-filtering step, respondent-supplied criteria were used to select a subjectively defined subset of web respondents (SUBJ). This subset contained only those respondents who satisfied *all* of the following criteria:

  (i) Listening in a quiet environment
 (ii) Audio delivery via headphones
(iii) British variety of English as first language
 (iv) Aged 50 or under
  (v) No reported hearing problems.

Around 31% of web listeners satisfied the intersection of these constraints. As anticipated, the mean score for this group (Figure 6.8) is significantly higher [$p < 0.01$] than the unfiltered WEB group, although still far below the level of the FORMAL group.

Taking respondents' information at face value, subjectively defined criteria go some way to matching conditions in traditional testing environments, where more control over the listener population can be exercised. However, they retain responses from those listeners who, for whatever reason, performed very poorly on the test compared with others in the cohort (see the outliers in Figure 6.8). These listeners may have given up at some point during a block of stimuli and then entered arbitrary responses in order to receive feedback at the end of the test, for example. For this reason, it is useful to seek objective criteria to select well-motivated respondents. In the crowdsourcing scenario, one approach is to examine response consistency across listeners. In general, many techniques are possible based on measuring the likelihood of a response sequence by comparing the response to each token with the distribution of responses from all other listeners who screened that token. In the BigListen we adopted an approach based on first identifying a type of criterion token (see Section 6.3.2) that we call an 'anchor token' – an individual stimulus that satisfies the joint criteria of (i) having been screened by many listeners and (ii) having a very high rate of correct identification. Once anchor tokens are identified, they can be used to filter out those respondents who failed to reach a criterion score on these stimuli. Here, anchor tokens were defined as those stimuli heard by at least 30 listeners and which, as individual tokens, resulted in scores of at least 80% correct. Since not all listeners heard the same blocks of stimuli, different sets of anchor tokens are used in each case. Fortunately, many anchor tokens meeting the above criteria were present in the response set.

Respondents who achieved mean scores of at least 90% on anchor tokens made up this objectively defined ANCHOR subset. Around 63% of all web listeners met this rather strict criterion. Subjective and objective respondent filtering approaches can also be combined to produce a SUBJ+ANCHOR group. In this case, the dual criteria retained only 23% of web respondents.

The use of anchor tokens has the desired effect of removing outliers, and produces an increase in mean score, although by less than the application of subjective criteria [$p<0.05$]. Combination of the two criteria leads to higher scores, at the cost of removing more than three out of every four respondents from the analysis. However, a 13 percentage points gap still remains between the traditionally tested and best web subset.

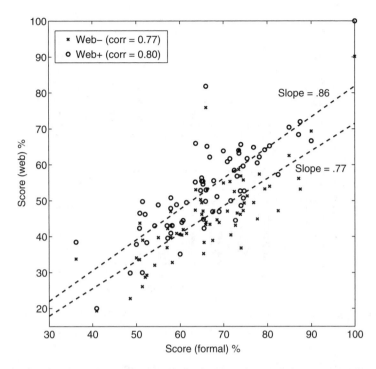

Figure 6.9   Mean scores in each masker and SNR condition for the formal and web groups. (Figure from Cooke *et al.* 2011.)

In subsequent analyses, the responses of the formal group are compared with the best-performing web subset SUBJ+ANCHOR and its complement (i.e. the set WEB-(SUBJ+ANCHOR)). For brevity, these web groups are denoted web+ and web−.

Score Correlations across Masker and SNR

The degree to which the different listener groups pattern in a similar way as a function of noise type and SNR is shown in Figure 6.9. Each point represents responses from a single noise type in a narrow SNR range (quantised to 1 dB). The strong correlation that exists between formal and web scores suggests that both the varying difficulty in identifying word subsets at a given SNR as well as the challenge produced by each of the masker types leads to the different listener groups being affected to a very similar degree.

An even larger correlation of 0.96 in intelligibility scores across five different speech styles was reported in a comparison of MTurk and laboratory-tested listeners in Mayo *et al.* (2012), strengthening the view that even when absolute scores differ, the pattern of scores across conditions can be remarkably similar in web-based and formal speech perception tests.

Response Consistency

Another way to measure similarity in responses is to look at the proportion of words where listeners reached a certain level of consensus in their decisions. Figure 6.10 shows how many

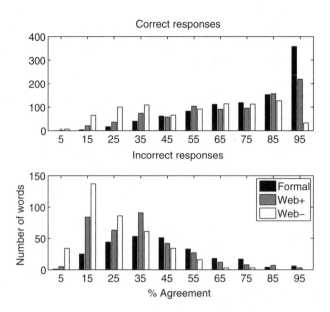

Figure 6.10   Agreement levels for correct and incorrect responses. (Figure from Cooke *et al.* 2011.)

words were identified correctly (upper panel) or misidentified, but in a consistent way (lower panel) as a function of the degree of agreement.

The right-most bars in both plots depict a very strict level of agreement, with more than 90% of listeners providing the same response to a given stimulus. For the formal group, more than 350 words were identified correctly on the basis of this criterion, with rather fewer for the web+ subset of crowdsourced listeners. Here, there is a clear difference between the web+ and web- groups, the latter showing far lower degrees of response consistency.

In the middle of the range, for 50% agreement upwards, we have the weaker criterion of *majority agreement*. For correct responses, the majority agreement levels are similar for each group. However, for incorrect responses, where majority responses identify the robustly perceived confusions that we are mainly interested in, the formal group shows a greater degree of consistency than the web+ group, while the web- group discovered relatively few consistent confusions. In fact, the formal group discovered 129 majority confusions compared with 85 and 44, respectively, for the web+ and web- groups. This suggests that although the web-based procedure leads to lower overall scores, it is still effective in finding potentially interesting word confusions in noise if both subjective and objective listener selection procedures are followed. Some of these confusions are shown in Figure 6.11.

An unexpected outcome was the finding that the web+ group's majority confusions were not simply a subset of those discovered by formally tested listeners. In fact, only 33 were common to both groups, while the remaining 96 from the formal group were not majority confusions for the web group. Intriguingly, the reverse was also the case: the web+ group crowdsourced 52 exemplars that were 'missed' by the formal group. The reasons for this finding are unclear. It is possible that the lower quality audio equipment likely to have been used by the web group led to consistent response biases. For instance, if significant high-frequency attenuation was

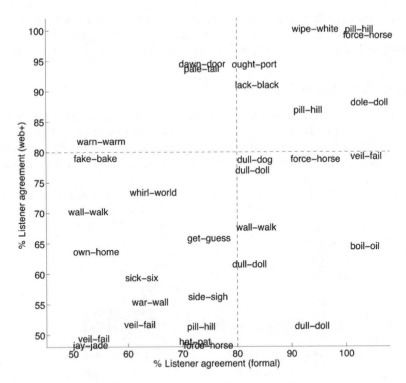

Figure 6.11 Majority confusions in common for formal and web+ listeners. Dotted lines show 80% agreement levels. (Figure from Cooke *et al.* 2011.)

more likely to be present in the web group, confusion between certain fricatives might be more frequent, and may have led to a tendency to pattern in similar ways across the web group. This is an area that demands further investigation.

Typical Confusions

The relatively small scale of the BIGLISTEN experiment means there is insufficient data to support a comprehensive discussion of confusions. However, to date we highlight some tendencies we have observed in the data:

 (i) Most confusions involve consonants rather than vowels (although the reverse was true in a subsequent unpublished study with non-native Spanish listeners). Most vowel confusions (mainly /ʌ/-/ɒ/) are likely to be caused by an accent mismatch between the speaker and listener.
 (ii) Labial plosives and fricatives are often involved in onset confusions. Sometimes, the confusions are inter-labial (/f/ to /p/ or /b/) involving fricative/plosive errors (Hazan and Simpson 1998), but we often observe a labial to /h/ confusion, which highlights the lack of salience of the labial gesture in acoustic/perceptual terms.
 (iii) Nasals are frequently substituted or deleted, especially in coda position (Benki 2003).

(iv) Some confusions involve consonant insertion in both coda and onset position, perhaps due to incorporation of background energy fragments (e.g. 'pea-peace').

(v) Other confusions suggest an effect of word familiarity (e.g. 'veil-fail' and 'whirl-world').

Example Response Distributions

We end this case study with a look at some of the response distributions to individual noisy speech tokens. Each panel of Figure 6.12 plots the number of times a given word was reported in response to the presented word and noise type indicated. To keep the response distributions manageable and relevant, only those responses that were reported by at least three listeners are

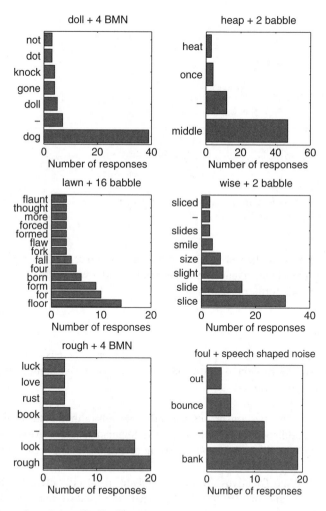

Figure 6.12 Example response distributions. A hyphen indicates a null response. The stimulus in each case is shown as the combination of word and noise type. 2/16 babble indicates natural babble made up of 2 or 16 voices. 4 BMN is speech-shaped noise with envelope modulations from 4-talker babble.

retained. These examples have been chosen both to illustrate facets of the task and to highlight some of the issues that need to be considered when using crowdsourcing to gather responses in speech perception tasks. While we present some conjectures, the underlying mechanisms that create the response patterns are still far from understood.

(i) **'Doll' in 4-talker babble-modulated noise (BMN):** This is a classic case of a very robust confusion with a high degree of listener agreement. Respondents identifying this stimulus as 'dog' outnumbered those reporting the correct answer by 6-to-1 here. It is possible that energetic masking of the final consonant followed by misallocation of a suitable brief noise burst from the background masker was responsible for this confusion. The vowel in this and many other examples was correctly reported. As noted above, vowels tend to be robust and survive masking at the SNRs used in BigListen.

(ii) **'Heap' in 2-talker babble:** Complete background words typically remain audible in 2-talker maskers and in this case nearly all listeners identified the word 'middle' instead of the target 'heap'. What is surprising about this example is that a two-syllable word was reported in spite of listeners receiving instruction that all target words were monosyllabic. This highlights a methodological difference between crowdsourcing and traditional testing: there is no guarantee that participants bother to read the instructions, and there is less opportunity to emphasise experimental factors such as this compared with a formal testing situation where the experimenter is physically present. If necessary, immediate and automatic feedback could be provided to correct the misunderstanding.

(iii) **'Lawn' in 16-talker babble:** We speculate that high-entropy confusions like this are symptomatic of energetic masking, where parts of the target word are swamped by noise. How listeners fill in the inaudible segments will depend on respondent-specific 'language model' factors, leading to some variety in responses. However, this example illustrates that useful information can be obtained even in the case of a relatively high-entropy response distribution. While no respondents reported the word actually present in the noisy stimulus, nearly all agreed on the attachment of a word-initial fricative /f/, presumably recruited from the masker. Again, the vowel was reported correctly in all cases. Note also the need to handle homonyms (e.g. 'floor' and 'flaw') in open-response tasks of this sort.

(iv) **'Wise' in 2-talker babble:** This is a similar example to 'lawn': a surviving target diphthong surrounded by a largely consistent initial consonant cluster and some variation in coda consonant. The different coda consonants presumably reflect both non-uniform stimulus ambiguities which favour some interpretations over others, as well as lexical constraints. It is worth noting that while misspellings were present (though infrequent), respondents in the main did not invent words, that is, use nonwords to identify their response. The lesson here is that, unlike some tasks in speech perception (such as those mentioned in Section 6.4.4), a task demanding words as responses is highly-appropriate for naive listeners.

(v) **'Rough' in 4-talker BMN:** What is interesting about this example is not the fact that a small majority reported the correct response, but that some of the incorrect responses reveal something about the accent or linguistic environment of the respondents. Within the United Kingdom, regional variation in pronunciation of words is rife, with words such as 'look' and 'book' being produced with either /uː/ or /u/. In principle, these diagnostic responses might be used to corroborate respondent-supplied information on accent. This example also illustrates that homonym handling needs to be sensitive to accent; for example, 'look' and 'luck' are not homonyms for all listeners.

(vi) **'Foul' in speech-shaped noise**: Here, no listener reported the correct answer and many produced a null response, but there was enough evidence in the noisy stimulus for 19 listeners to report the word 'bank'. This is an interesting case, because the phonological transformation from 'foul' to 'bank' is not at all obvious (to say the least!) and yet the background noise type is supposedly uninformative (it does not contain speech), nor was it temporally modulated. This concluding example demonstrates one of the primary benefits of carrying out speech perception tasks with open-response sets and large numbers of listeners, namely the emergence of intriguing and unexpected outcomes.

6.5.7   Lessons from the BIGLISTEN Crowdsourcing Test

While small scale in nature, the BIGLISTEN experiment suggests that crowdsourcing is capable of eliciting response distributions that are of potential interest in speech perception studies. Quantitative estimates (e.g. from Figure 6.10) of the rate at which even formally tested groups make consistent mis-identifications of noisy stimuli indicate that robust confusions are rare, and motivates the use of crowdsourcing as an initial sieve prior to formal confirmation tests.

The BIGLISTEN also demonstrates that both respondent-provided information and internally generated anchor tokens can contribute to the selection of listeners who better match the levels of homogeneity and motivation that we aim for in laboratory-based tests. Nevertheless, coherent subsets of web respondents never matched scores seen in the laboratory. This finding echoes other studies of crowdsourcing with speech and/or noise stimuli (Wolters *et al.* 2010; Mayo *et al.* 2012). Clearly, crowdsourcing in speech perception is not suitable for those tasks that seek to estimate absolute performance levels of listener samples. The reasons for this discrepancy have yet to be pinpointed – in itself not an easy task – but seem likely to include differences in the overall audio delivery path from a client's computer to their auditory system: digital-to-analogue conversion, amplification, connectors, leads and headphones are all candidates for signal degeneration relative to a typical speech perception laboratory setup.

The BIGLISTEN benefitted from a surprisingly high rate of voluntary participation, estimated at around 7–10% of all those receiving one of two e-mail invitations. While careful timing of the invitations, just following the annual influx of new students, no doubt contributed to this level of involvement, it is also possible that the promise of a rapid, hassle-free and anonymous experiment requiring no user registration appealed to many respondents. Designing a test that could generate useful data with an end-to-end time of under 3 minutes per user was a primary design goal, even at the expense of permitting null responses in the elicitation of user data (e.g. default values for age and accent). In hindsight, allowing default responses is not to be recommended as best practice due to its potential to invalidate sample-wide estimates of the desired factor.

One of the advantages of a large-scale listening test with a relatively unconstrained response set is the possibility of finding unforeseen yet robust responses with non-trivial explanations. For example, the BIGLISTEN has, for us, motivated a change in the way we think about the effect of noise on speech, with the notion of masking giving way to a more complex sequence of speech-noise 'reactions' that result in a given word confusion. The lesson here is that while it is possible in principle to find similar outcomes with traditional test procedures, the use of large and somewhat uncontrolled samples seems to encourage unexpected outcomes. Control of everything that can be controlled, from participants to instructions, is the official ethos in most formal tests (although it need not be), but may well be counter-productive in tasks which seek to discover 'interesting' specimens.

The finding that formal and web tests differ not only in absolute scores but also in the patterns of majority confusions suggests that additional care needs to be exercised in preparing for a web experiment. One implication is that pilots carried out in a formal setting may give a biased picture of what can be expected in a crowdsourced test.

6.6   Issues for Further Exploration

Further and more extensive use of crowdsourcing in speech perception seems inevitable. Some of the driving forces for greater use of non-formal testing procedures include the increasing use of spoken language output technology that calls for large-scale comparative evaluations, for which crowdsourcing enables ranking of systems; the online delivery of simple hearing tests; and the need for more speech perception tests to better understand hearing and to develop more robust speech technology. Here, we raise some of the issues in crowdsourced speech perception that deserve further study and highlight some technological developments that might enable better control of web experiments.

Matching Traditional Levels of Performance

Currently, as we have seen, the absolute level of performance in web-based speech perception tests falls short of that obtained in formal settings, perhaps restricting the use of this methodology to crowd-as-filter approaches and rank ordering of conditions, assuming in addition that appropriate formal validations are carried out. What can be done to raise the performance baseline? Here are six areas to focus on:

(i) **Better listener selection procedures**: Pre-tests, using criterion tokens, might help to select listeners suited to the target language, for example.

(ii) **Automatic determination of audio delivery hardware.**

(iii) **Automatic sampling of a listener's acoustic environment**: While the technology already exists to make client-side estimates of, for example, the background noise spectrum and level, its use raises important privacy concerns and could only be employed based on informed consent.

(iv) **Improved procedures for task explanation, including mechanisms to check for correct interpretation**: A short instructional video could better simulate the oral interaction typical of a laboratory-based experiment.

(v) **Improving respondent motivation**: Many opportunities exist to incorporate the collection of speech-based judgements into more entertaining applications. The provision of timely and relevant feedback is an additional facet of motivation.

(vi) **Options to cope with client-side disruptions during the task**: A participant with the best will in the world will find it more difficult to prevent disruptions – caused by such things as visitors or telephone calls – than an experiment under laboratory control. While response time monitoring is a passive means to identify disruption, an approach that allows participants to signal 'unreliable' trials would permit better identification of reliable data.

Decreasing Variability

A key issue is how to reduce response variability, which has many of the same origins as those speculated to cause lower absolute performance – listeners, equipment, environment. Targeted advertising in special-interest communities or forums might lead to increased listener homogeneity, if this were a desirable outcome in any given web experiment, at the expense of a reduced rate of participation. More stringent respondent questionnaires are likely to produce the same trade-off.

Equipment variability is one area that should be more easily controlled in the future. Experiments can be aimed at users of specific devices whose audio characteristics are well-understood. For more limited sample sizes (perhaps involving longer or more intensive testing), headphones could be mailed out as a gift to participants, providing an incentive to participate. Introduction of a gaming/competitive element may motivate certain types of user to undertake the test using the best equipment at their disposal.

While the BIGLISTEN made no use of IP addresses (e.g. to estimate participant location/ language, or cross-session tracking), this information could be employed to increase the likelihood that different sets of responses originate from different individuals.

One source of variability in applications like the BIGLISTEN which have open-set response alternatives stems from user input errors, for example typos and misspellings. If the user responds with another valid word, little can be done. Otherwise, the participant could be passively alerted to the possibility of an input error, perhaps adopting the commonly used method of input underlining. Handling input errors is best done on the client side and is likely to become easier to integrate into crowdsourcing in the future, for example, through the use of spellchecked forms in HTML5.

Ethics and Safety Concerns

Two related issues we have barely touched upon are the ethical and safety dimensions of crowdsourcing in speech perception. The BIGLISTEN required explicit consent to be given before commencement of the main test, but it is not clear that this will be sufficient in all tasks or jurisdictions. Ethical and especially safety concerns involve many distinct questions, some of which have been covered in other chapters and are common to the domain of speech perception. We focus on those most relevant to the speech domain here.

First, there is the issue of possible temporary or permanent hearing damage caused by the delivery of intense stimuli. Here, we would suggest that while there are numerous examples of web-based audio delivery (e.g. online videos or music samples) and that there is very little that can be done to control the final sound intensity level at which stimuli are reproduced, deployment of crowdsourcing in speech research requires high standards to minimise user risks. Techniques include issuing warnings about setting the output level via examples prior to reaching the main test; requiring a user to identify correctly practice examples chosen to distort at high volume levels; monitoring performance in the main test and curtailing the experiment if a performance threshold on easily recognised tokens is not reached; preventing overlong exposure to the experiment by fixing a maximum number of repeated listens from a given IP address in a fixed time period; ensuring that output levels are fixed across stimuli and tested at high volume settings on commonly used computer hardware.

Second, detailed questionnaires, particularly those permitting complete linguistic histories that might be solicited in speech and hearing studies, should not compromise user anonymity where this has been promised. This concern applies most acutely for smaller samples that might result from targeted recruitment.

Third, feedback should be relevant, accurate and useful. In tests involving the perception of speech signals, it is essential to make clear to respondents that they are not undertaking an online hearing test, and to stress in any feedback given that the results cannot be interpreted in ways which relate to their individual hearing sensitivity. The provision of useful feedback needs careful consideration in applications such as the BigListen that actively seek confusions and typically lead to low scores from listeners who are performing quite normally. Here, other feedback metrics might be required, such as the degree of listener consistency rather than raw accuracy.

6.7 Conclusions

- Crowdsourcing in speech perception can be a valuable adjunct to traditional testing methods.
- For tasks such as those which require calibration of presentation levels, or involve the reporting of fine distinctions or estimates of absolute levels of intelligibility, traditional tests remain the method of choice.
- For evaluative tasks such as accent judgements or speech synthesis quality assessment, where ranking of alternatives is the desired outcome, web-based testing is an option that merits consideration.
- In domains where the availability of a large listener sample is an essential element of experimental design, crowdsourcing may be the only practical approach.
- Further studies are required to validate the application in any new task or domain and in particular to test for the existence of consistent biases in responses from the crowd.
- Methodological innovations will be needed to enable objective confirmation of subjective wisdom.

References

Andersson L and Trudgill P (1990) *Bad Language*. Basil Blackwell, Oxford.

ANSI (1997) *S3.5-1997: American National Standard Methods for Calculation of the Speech Intelligibility Index*. American National Standards Institute, New York.

Beddor PS and Gottfried TL (1995) Methodological issues in cross-language speech perception research with adults in *Cross-language Studies of Speech Perception: Issues in Cross-language Research* (ed. W Strange). York Press.

Benki J (2003) Analysis of English nonsense syllable recognition in noise. *Phonetica* **60**, 129–157.

Bexelius C, Honeth L, Ekman A, Eriksson M, Sandin S, Bagger-Sjoback D and Litton J (2008) Evaluation of an Internet-based hearing test: comparison with established methods for detection of hearing loss. *Journal of Medical Internet Research* **10(4), e32**.

Bissiri MP, Garcia Lecumberri ML, Cooke M and Volín J (2011) The role of word-initial glottal stops in recognizing English words. *Proceeding of Interspeech*, pp. 165–168.

Blin L, Boeffard O and Barreaud V (2008) Web-based listening test system for speech synthesis and speech conversion evaluation. *Proceedings of the 6th International Language Resources and Evaluation (LREC'08)*, pp. 2270–2274, Marrakesh.

Bongaerts T (1999) Ultimate attainment in L2 pronunciation: the case of very advanced late L2 learners, in *Second Language Acquisition and the Critical Period Hypothesis* (ed. D Birdsong). Lawrence Erlbaum.

Boyd S (2003) Foreign-born teachers in the multilingual classroom in Sweden: the role of attitudes to foreign accent. *International Journal of Bilingual Education and Bilingualism* **3-4**, 283–295.

Brennan EM and Brennan JS (1981) Accent scaling and language attitudes: reactions to Mexican American English speech. *Language and Speech* **3**, 207–221.

Cara BD and Goswami U (2002) Similarity relations among spoken words: the special status of rimes in English. *Behavior Research Methods, Instruments, and Computers* **34**, 416–423.

Choi J, Lee H, Park C, Oh S and Park K (2007) PC-based tele-audiometry. *Telemedicine and e-Health* **13**, 501–508.

Christiansen C, Pedersen MS and Dau T (2010) Prediction of speech intelligibility based on an auditory preprocessing model. *Speech Communication* **52**, 678–692.

Clopper CG and Pisoni DB (2005) Perception of dialect variation, in *The Handbook of Speech Perception* (ed. DB Pisoni and RE Remez). Blackwell Publishing Ltd.

Cooke M (2006) A glimpsing model of speech perception in noise. *Journal of the Acoustical Society of America* **119**, 1562–1573.

Cooke M (2009) Discovering consistent word confusions in noise. *Proceedings of Interspeech*, pp. 1887–1890, Brighton, UK.

Cooke M, Barker J, Garcia Lecumberri M and Wasilewski K (2011) Crowdsourcing for word recognition in noise. *Proceedings of Interspeech*, pp. 3049–3052.

Cooke M, Garcia Lecumberri M, Scharenborg O and van Dommelen W (2010) Language-independent processing in speech perception: identification of English intervocalic consonants by speakers of eight European languages. *Speech Communication* **52**, 954–967.

Cox T (2008) The effect of visual stimuli on the horribleness of awful sounds. *Applied Acoustics* **69**, 691–703.

Dávila A, Bohara A and Saenz R (1993) Accent penalties and the earnings of Mexican Americans. *Social Science Quarterly* **74**, 902–915.

Draxler C (2011) Percy: An HTML5 framework for media rich web experiments on mobile devices. *Proceedings of Interspeech*, pp. 3339–3340.

Fayer JM and Krasinski E (1987) Native and non-native judgments of intelligibility and irritation. *Language Learning* **37**, 313–327.

Fernandez Gonzalez J (1988) Reflections on foreign accent. *Issues in Second Language Acquisition and Learning*, Universitat de Valencia, Valencia.

Festen J and Plomp R (1990) Effects of fluctuating noise and interfering speech on the speech-reception threshold for impaired and normal hearing. *Journal of the Acoustical Society of America* **88**, 1725–1736.

Floccia C, Goslin J, Girard F and Konopczynski G (2006) Does a regional accent perturb speech processing? *Journal of Experimental Psychology: Human Perception and Performance* **5**, 1276–1293.

Garcia Lecumberri ML, Cooke M and Cutler A (2010) Non-native speech perception in adverse conditions: a review. *Speech Communication* **52**, 864–886.

Garcia Lecumberri ML, Cooke M, Cutugno F, Giurgiu M, Meyer B, Scharenborg O, van Dommelen W and Volin J (2008) The non-native consonant challenge for European languages. *Proceedings of Interspeech*, pp. 1781–1784.

Gong J, Cooke M and Garcia Lecumberri ML (2011) Towards a quantitative model of Mandarin Chinese perception of English consonants, in *Achievements and Perspectives in SLA of Speech* (eds M Wrembel, M Kul and K Dziubalska-Kowaczyk). Peter Lang.

Hazan V and Simpson A (1998) The effect of cue-enhancement on the intelligibility of nonsense word and sentence materials presented in noise. *Speech Communication* **24**, 211–226.

Honing H (2006) Evidence for tempo-specific timing in music using a web-based experimental setup. *Journal of Experimental Psychology* **32**, 780–786.

Honing H and Ladinig O (2008) The potential of the Internet for music perception research: A comment on lab-based versus web-based studies. *Empirical Musicology Review* **3**, 4–7.

Honing H and Reips U (2008) Web-based versus lab-based studies: A response to Kendall (2008). *Empirical Musicology Review* **3**, 73–77.

Horswill M and Coster M (2001) User-controlled photographic animations, photograph-based questions, and questionnaires: Three Internet-based instruments for measuring drivers risk-taking behavior. *Behavior Research Methods* **33**, 46–58.

Hosoda M, Stone-Romero E and Walter J (2007) Listeners' cognitive and affective reactions to English speakers with standard American English and Asian accents. *Perceptual and Motor Skills* **1**, 307–326.

Keller F, Corley M, Corley S, Konieczny L and Todirascu A (1998) WebExp. *Technical Report*, HCRC/TR-99, Human Communication Research Centre, University of Edinburgh, Edinburgh.

Keller F, Gunasekharan S, Mayo N and Corley M (2009) Timing accuracy of web experiments: a case study using the WebExp software package. *Behavior Research Methods* **41**, 1–12.

Kendall R (2008) Commentary on the potential of the Internet for music perception research: A comment on lab-based versus web-based studies by Honing & Ladinig. *Empirical Musicology Review* **3**, 8–10.

King S and Karaiskos V (2010) The Blizzard Challenge 2010. *Blizzard Challenge Workshop.*

Kuhl PK (1991) Human adults and human infants show 'perceptual magnet effect' for the prototypes of speech categories, monkeys do not. *Perception and Psychophysics* **50**, 93–107.

Kunath S and Weinberger S (2010) The wisdom of the crowd's ear: speech accent rating and annotation with Amazon Mechanical Turk. *Proceedings of the NAACL HLT 2010 Workshop on Creating Speech and Language Data with Amazon's Mechanical Turk*, pp. 168–171. Association for Computational Linguistics.

Lacherez P (2008) The internal validity of web-based studies. *Empirical Musicology Review* **3**, 161–162.

Laugwitz B (2001) A web-experiment on colour harmony principles applied to computer user interface design, in *Dimensions of Internet Science* (eds UD Reips and M Bosnjak). Pabst Science Publishers Lengerich, Germany, pp. 131–145.

MacKay IRA, Flege JE and Imai S (2006) Evaluating the effects of chronological age and sentence duration on degree of perceived foreign accent. *Applied Psycholinguistics* **27**, 157–183.

Major RC (2007) Identifying a foreign accent in an unfamiliar language. *Studies in Second Language Acquisition* **29**, 539–556.

Mayo C, Aubanel V and Cooke M (2012) Effect of prosody changes on speech intelligibility. *Proceedings of Interspeech.*

McGraw I, Gruenstein A and Sutherland A (2009) A self-labeling speech corpus: collecting spoken words with an online educational game in *Tenth Annual Conference of the International Speech Communication Association.*

McGraw K, Tew M and Williams J (2000) The integrity of web-delivered experiments: Can you trust the data? *Psychological Science* **11**, 502.

Munro M and Derwing TM (1995) Processing time, accent, and comprehensibility in the perception of native and foreign-accented speech. *Language and Speech* **38**, 289–306.

Naumann A, Brunstein A and Krems J (2007) DEWEX: A system for designing and conducting web-based experiments. *Behavior Research Methods* **39**, 248–258.

Newman C, Weinstein B, Jacobson G and Hug G (1990) The hearing handicap inventory for adults: psychometric adequacy and audiometric correlates. *Ear and Hearing* **11**, 430.

Novotney S and Callison-Burch C (2010) Cheap, fast and good enough: automatic speech recognition with non-expert transcription. *Human Language Technologies: The 2010 Annual Conference of the North American Chapter of the Association for Computational Linguistics*, pp. 207–215. Association for Computational Linguistics.

Reips U and Neuhaus C (2002) WEXTOR: a web-based tool for generating and visualizing experimental designs and procedures. *Behavior Research Methods* **34**, 234–240.

Reips UD (2000) The web experiment method: advantages, disadvantages, and solutions. *Psychological Experiments on the Internet*, pp. 89–114.

Reips UD (2002) Standards for Internet-based experimenting. *Experimental Psychology (formerly Zeitschrift fur Experimentelle Psychologie)* **49**, 243–256.

Riney T, Takagi, N and Inutsuka K (2005) Phonetic parameters and perceptual judgments of accent in English by American and Japanese listeners. *TESOL Quarterly* **39**, 441–466.

Rix A, Beerends J, Hollier M and Hekstra A (2001) Perceptual evaluation of speech quality (PESQ) – a new method for speech quality assessment of telephone networks and codecs. *Proceedings of ICASSP* **2**, 749–752.

Sawusch J (1996) Instrumentation and methodology for the study of speech perception, in *Principles of Experimental Phonetics* (ed. N Lass). Mosby St. Louis, MO, pp. 525–550.

Seren E (2009) Web-based hearing screening test. *Telemedicine and e-Health* **15**, 678–681.

Skitka L and Sargis E (2006) The Internet as psychological laboratory. *Annual Review of Psychology* **57**, 529–555.

Snow R, O'Connor B, Jurafsky D and Ng A (2008) Cheap and fast – but is it good? Evaluating non-expert annotations for natural language tasks. *Proceedings of the Conference on Empirical Methods in Natural Language Processing*, pp. 254–263. Association for Computational Linguistics.

SoundComparisons (2012) www.soundcomparisons.com (accessed 4 October 2012).

Swanepoel W, Clark J, Koekemoer D, Hall JI, Krumm M, Ferrari D, McPherson B, Olusanya B, Mars M, Russo I and Barajas J (2010) Telehealth in audiology: the need and potential to reach underserved communities. *International Journal of Audiology* **49**, 195–202.

Trudgill P and Hannah J (2008) *International English: A Guide to the Varieties of Standard English*, 5th edn. Hodder Education, London.

Van Els T and De Bot K (1987) The role of intonation in foreign accent. *The Modern Language Journal* **71**, 147–155.

Voxforge (2012) http://www.voxforge.org/ (accessed 4 October 2012).

Wells JC (1982) *Accents of English I: An Introduction*. Cambridge University Press, Cambridge.

Wolfe C and Reyna V (2002) Using NetCloak to develop server-side web-based experiments without writing CGI programs. *Behavior Research Methods* **34**, 204–207.

Wolters M, Isaac K and Renals S (2010) Evaluating speech synthesis intelligibility using Amazon's Mechanical Turk. *Proceeding of 7th Speech Synthesis Workshop (SSW7)*.

7

Crowdsourced Assessment of Speech Synthesis

Sabine Buchholz[1], Javier Latorre[2], and Kayoko Yanagisawa[2]

[1] SynapseWork Ltd, UK
[2] Toshiba Research Europe Ltd, UK

7.1   Introduction

Speech synthesis is the artificial generation of human speech, usually by a computer. Since the most common input for a speech synthesizer is text, these systems are often called text-to-speech (TTS) systems. The two main requirements of any TTS system are intelligibility and naturalness. For the first synthesizers created in the 1940s and 1950s, just achieving sufficient intelligibility over normal sentences was a huge task. With the development of computer-based TTS systems and especially after the appearance of data driven systems (waveform-concatenation and parametric-based speech synthesis) in the late 1990s, the quality improved dramatically to the point that intelligibility over normal sentences is mostly a given. At the same time, researchers' focus shifted from intelligibility toward naturalness and those speech features related to it such as similarity to the original speaker and expressiveness (the ability to convey emphasis, emotions, attitudes, irony, character, etc.). However, these features are highly subjective and thus evaluation of such features is more complex and requires more listeners than for intelligibility tests.

This chapter is mainly about crowdsourcing *assessment* of TTS systems, but it also presents some results from attempts to use it for other purposes related to TTS development. Assessment is an important part of the research and development cycle for any type of speech technology. For speech recognition, assessment is usually performed by automatically computing the word error rate (WER) against manually transcribed reference texts. For speech synthesis, some automatic methods have been used that compute the differences between the synthesized

Crowdsourcing for Speech Processing: Applications to Data Collection, Transcription and Assessment, First Edition.
Edited by Maxine Eskénazi, Gina-Anne Levow, Helen Meng, Gabriel Parent and David Suendermann.
© 2013 John Wiley & Sons, Ltd. Published 2013 by John Wiley & Sons, Ltd.

sentence and a reference recording of the same sentence by a human speaker. In most cases, this comparison is based on the root mean square error between some acoustic parameters, such as the fundamental frequency, the phone duration, or the cepstral coefficients. However, none of these distances correlates well enough with human judgments of the *overall* naturalness or quality of the TTS output. Hinterleitner *et al.* (2010) analyzed several more complex approaches for automatically predicting naturalness. Although their results are promising, they are still not very reliable. Therefore, overall assessment of TTS systems is still carried out by asking human subjects to take part in *listening tests*, which involves listening to synthesized speech samples and answering questions about them. The nature of samples as well as the types of questions/answers vary according to the *type of listening test*; these are discussed in more detail in Section 7.2.

Traditionally, formal listening tests have been conducted under controlled conditions in the laboratory, with external subjects. However, this is time-consuming and costly. First, TTS systems do not perform equally well for all sentences. Therefore, a TTS assessment needs to include as many different sentences from each system as possible. Second, TTS assessment is based on subjective judgments, and different subjects often have different opinions. This means that in order to obtain results that can be generalized over the whole population, the test should be taken by as many and as varied subjects as possible. Less formal tests typically involved researchers asking their laboratory colleagues to perform the test just at their desks. However, this is suboptimal due to the limited and nonnaive subject pool, especially when comparing the laboratory's own system against another, or for languages that are not the researchers' native language. Nowadays, tests are increasingly conducted over the Internet, but finding subjects can still be a problem (Black and Tokuda 2005). This is where the varied and almost unlimited pool of subjects available through crowdsourcing provides a solution.

While some of the aspects of crowdsourcing for TTS are similar to those for other speech applications, there are other aspects that are different. First, assessment of TTS does not only pose the common problems of dealing with speech files on the Internet, but entails some additional difficulties due to the researchers' desire to work with speech files in WAV format, a format less well supported than the more common MP3 format. This will be discussed in Section 7.3.2. Second, assessment of TTS is mostly a subjective task. This means that many quality control mechanisms used for other applications of crowdsourcing cannot be used.

This chapter is a much extended version of Buchholz and Latorre (2011), which analyzed 127 crowdsourced listening tests, of one test type, for one language, using one crowdsourcing platform and interface design. In this chapter, that analysis is built on and extended to other languages, test types, platforms/labor channels, interface designs, and over a longer time span. Section 7.3 reports which approaches worked and which did not.

In contrast to Chapter 6, this chapter deals with paid, not voluntary, crowdworkers, and as such, needs to take the detection and prevention of spamming seriously. Section 7.4 reviews related work on this topic and Section 7.5 develops dedicated metrics for a specific case. Finally, Section 7.6 provides a conclusion and discussion on future directions.

7.2 Human Assessment of TTS

There is a vast body of literature about the assessment of TTS; see Bech and Zacharov (2006) for an overview. Here only the most relevant types of listening tests are discussed.

One way to categorize listening tests is according to the unit of assessment. Segmental tests typically aim to assess performance at the level of individual phones. For example, in the modified rhyme test (MRT; House *et al.* 1963), listeners have to indicate which word they heard in the sample. The word can be embedded in a carrier phrase, but the phrase is always the same, only the one word differs. The words are chosen among minimal pair sets such that mishearing one phone results in a different word. The most common level of TTS assessment nowadays is the sentence level (as most TTS systems' unit of synthesis is still the sentence). Sentence-level listening tests will be discussed in more detail below. Finally, while listening tests involving units beyond the sentence are not new (e.g., ITU 1994), there has been renewed interest for them lately (e.g., Hinterleitner *et al.* 2011; King 2012), due to a desire to move beyond the synthesis of isolated sentences toward the synthesis, and therefore assessment, of longer texts.

Going back to sentence-level assessment, there is a broad division into intelligibility and nonintelligibility listening tests. In intelligibility tests, subjects are asked to type in the words they hear in the synthesized sentence. To prevent listeners from guessing some words by their context, semantically unpredictable sentences (SUS; Benoit and Grice 1996) are often used. In nonintelligibility tests, subjects answer multiple choice questions about the samples. These tests can be further subdivided by the type of multiple choice answers. In mean opinion score tests (MOS; or Absolute category rating, ACR; ITU 1988a), listeners are asked to *rate* one speech sample, typically on a Likert scale (Likert 1932) from 1 to 5. An example is a MOS *naturalness* test, in which listeners have to rate the naturalness of the speech, with 1 being "Completely Unnatural" and 5 being "Completely Natural," (see Fraser and King 2007). It is important to note that MOS only refers to this rating scale of the answer and is largely independent of the exact question asked. For example, ITU (1994) recommends MOS with seven different questions, ranging from "Listening Effort" to "Overall Quality."

In *AB* tests (also called paired or pairwise comparison tests), listeners are typically asked to indicate which of the *two* speech samples has *more* of some property, with the property being defined by the question. For example, in *preference* tests, the question is simply "Which one do you prefer?" For the assessment of expressive synthesis, the question could be "Which one sounds happier (sadder/angrier)?" *AB* tests can be *forced choice*, or allow for a "No preference" option.

In *AB* tests specifically designed to collect data for multidimensional scaling (MDS) analysis, listeners indicate whether two speech samples are the *same* or *different* with respect to some property such as naturalness (see Clark *et al.* 2007).

Orthogonally to the above subdivision of test types, there is another dimension of whether a reference sample is given. For example, while in a normal MOS test only one sample is given, a "degradation" or "differential" MOS (DMOS; ITU 1988b) test in addition provides a reference sample and listeners are asked to rate the sample with respect to this reference. DMOS tests are often used to assess speaker similarity, for example, for speaker conversion or adaptation approaches, and the scale is then labeled "1—Sounds like a totally different person" to "5—Sounds like exactly the same person" (see Fraser and King 2007).

Likewise, while in a normal *AB* test two samples are given, an *ABX* test additionally provides a reference and asks which of the two samples under test sounds more like the reference with respect to the property stated in the question, such as speaker similarity or emotion. Table 7.1 gives an overview of the sentence-level test types discussed.

Table 7.1 An overview of sentence-level listening test types.

Task	Number of samples	Question	Without reference	With reference
Type in	1		Intelligibility e.g., SUS	n/a
Rate	1		MOS e.g., naturalness	DMOS e.g., similarity
Compare	2	Which one more ...?	AB e.g., preference	ABX e.g., expressiveness
Compare	2	Same or different?	For MDS e.g., naturalness	n/a

It is important to note that possibly with the exception of intelligibility tests, all the other types of tests discussed above can only be used to measure *relative* differences between (versions of) TTS systems. In other words, they cannot yield an absolute measure of subjective quality that can be compared across different tests with different systems being compared. For comparison tests, this means that the results are not necessarily transitive. In other words, when comparing three systems A, B and C, it is possible to find that $A > B$, $B > C$ but $C > A$, as have been observed in some of our experiments. This may happen when the differences between systems are not large, for example, $\leq 10\%$, and/or when the tests are not conducted simultaneously by the same group of subjects. In the case of MOS tests, a certain inter-test comparability can be achieved by including natural speech in all the tests to anchor the upper limit of the scale (Taylor 2009), or by always including a benchmark system (Karaiskos *et al.* 2008 onward) but even then, the interpretation of further levels on the scale depends on the systems being compared. As listening tests involve humans and humans cannot be standardized, it is not rare that the same test repeated twice yields different *absolute* results, especially when the group of subjects involved are different. This is relevant for crowdsourcing, as one is almost guaranteed not to get exactly the same set of listeners (workers) on different occasions.

Finally, different test types and instantiations (i.e., test questions) measure different aspects and there is not a single "best" test for TTS assessment. For example, although the final goal of both MOS and *AB* tests is the same, that is, to rank systems according to some subjective criterion, the way they work is different. In MOS tests, the exact score might depend on the previous stimulus the subject has judged (hence randomization is important). However, scores refer to the overall impression over the whole stimulus. Therefore, local differences are usually ignored. In an *AB* test, subjects listen to two versions of the same sentence, which in many cases are very similar to each other. As a result, the overall impression is often the same and what really counts is local differences/errors. This means that paired tests can provide a much finer ranking than MOS tests can. It is not unusual to find significant differences in preference for systems with very similar MOS values. Of course, the disadvantage of *AB* tests is the cost. To evaluate S samples over N systems using *AB* tests $S \cdot N \cdot (N - 1)$ stimuli have to be judged, whereas with MOS tests $S \cdot N$ stimuli are sufficient. This disadvantage can partially be counteracted by the abundance of listeners and lower costs of crowdsourced tests.

Sityaev *et al.* (2006) conducted several MOS tests with different test questions and SUS intelligibility tests with three TTS systems. They observed a high correlation between the "listening effort," "comprehension," and "articulation" MOS tests recommended in ITU (1994), but opposite results for the other tests. While one system scored highest on acceptability, another's voice was liked the most. Also, one system scored the highest in the intelligibility test, but lowest in the MOS naturalness and "overall quality" tests. When comparing different TTS systems, the de facto standard of the Blizzard Challenges—an annual open evaluation of TTS systems—is to at least test intelligibility as well as naturalness.

7.3   Crowdsourcing for TTS: What Worked and What Did Not

7.3.1   Related Work: Crowdsourced Listening Tests

A sizable number of papers have been published in the last 2 years whose main focus is TTS research but which use crowdsourcing as a means to perform the evaluation part thereof (Watts *et al.* 2010; Zen and Gales 2011; Latorre *et al.* 2011; Black *et al.* 2012; Parlikar and Black 2012).

As one example, Hashimoto *et al.* (2011) used MTurk to perform several types of evaluations in the context of speech-to-speech translation (S2ST): intelligibility of TTS, naturalness of TTS, as well as adequacy and fluency of the output of the whole S2ST system and just the machine translation part. Intelligibility was measured by WER while naturalness, adequacy and fluency used MOS. One hundred and fifty people participated in the evaluations, a number well above traditional laboratory experiments. However, crowdsourcing is not the main focus of the paper and no further details of this aspect are given.

So while there is TTS research *using* crowdsourcing, as shown in Chapter 2 there are only a few published studies *about* using crowdsourcing for the assessment of speech technology and even fewer specifically about the assessment of TTS.

The Blizzard Challenges (Black and Tokuda 2005 and newer) evaluate corpus-based TTS systems on common datasets. From the start, they have run evaluations online, and crowd-sourced parts of their listeners through an open call for unpaid volunteers and members of participating teams (alongside paid undergraduates in the laboratory). Over the years, slightly different sets of test types have been run, including intelligibility (MRT, SUS), naturalness (MOS, MDS), speaker similarity (DMOS), and appropriateness in dialog context (MOS) (King and Karaiskos 2009). While not mentioned in King and Karaiskos (2010), the organizers of the 2010 Blizzard Challenge repeated the evaluation of one of the main tasks (EH1) on MTurk, and concluded that this is a usable approach (S. King, personal communication).

Wolters *et al.* (2010) conducted tests to measure the intelligibility of four TTS systems using SUS. They repeated the same experiments in two settings: a controlled environment in the laboratory involving 20 university students, and crowdsourced via MTurk involving 159 workers from a variety of age groups and backgrounds. While absolute WERs were worse in the crowdsourced experiments, relative differences between systems were preserved and the much larger number of listeners in the crowdsourced experiments resulted in more statistically significant differences. They conclude that crowdsourcing is viable for TTS intelligibility tests.

Although not for TTS, Chen *et al.* (2009) describe a crowdsourced listening test framework for evaluating algorithms that process multimedia files, such as audio or video codecs. For evaluating n algorithms, the framework runs a task consisting of $\binom{n}{2}$ paired comparisons (*AB* tests). In each comparison, users see/hear the output of one algorithm or the other depending on whether they press or release the space bar. They then submit a forced choice preference between the two states (pressed/released). The assignment between algorithms and states is randomized. Tests were run in three settings: the laboratory, MTurk, and "an Internet community." In most but not all cases, the laboratory participants yielded higher quality; however, all three groups were reasonably consistent. In terms of cost and participant diversity, the crowdsourced approaches had a clear advantage. Note that the "spacebar switching" presentation method described in Chen *et al.* (2009) is not suitable for TTS listening tests (except potentially for vocoding research) as the outputs of two different (versions of) TTS systems will typically have different timing structure and cannot simply be switched mid-sample.

Ribeiro *et al.* (2011) present an approach to crowdsource MOS naturalness tests for TTS using MTurk. To validate their approach they run MOS tests on part of the 2009 Blizzard Challenge (King and Karaiskos 2009) samples, and show that (1) two subsequent runs produce highly correlated results, that is, good repeatability, and (2) scores from these runs also correlate highly with those of the Blizzard-paid laboratory students, that is, are reliable.

Buchholz and Latorre (2011) ran five TTS *AB* preference tests laboratory-internally as well as on CrowdFlower (http://crowdflower.com/; accessed 17 October 2012) using MTurk workers and compared results. They found that relative preferences were very similar between the two settings and that results in terms of whether or not observed differences are statistically significant were identical. They also observed that the crowdworkers more often expressed a preference (rather than choosing the answer "No preference").

Eyben *et al.* (2012) used crowdsourcing to evaluate the expressiveness of speech, both human and synthetic. They applied unsupervised clustering to the sentences of an audiobook read in often expressive styles and then evaluated that clustering by running *ABX* expressiveness tests on CrowdFlower/MTurk. Listener were asked which of two sound files (*A* or *B*) was more similar in emotion or speaking style to the reference (*X*). The reference was either from the same cluster as *A* or *B*. The authors then used the data from the best performing clustering to build expressive TTS voices and again evaluated the expressiveness using crowdsourced *ABX* tests. While the crowdsourced tests were not formally verified by comparison to noncrowdsourced ones, both the cluster as well as the TTS evaluation showed a clear "winner," indicating that the tests worked.

Nearly all the research mentioned so far was conducted on English TTS voices (American or British). The only exception are the 2008–2010 Blizzard Challenges, in which one of the voices was Mandarin Chinese (Karaiskos *et al.* 2008; King and Karaiskos 2009, 2010).

Other Crowdsourcing for TTS

Apart from in the *assessment* of TTS, crowdsourcing has also been used to prepare resources for the development of the text normalization part of TTS systems (in which "nonstandard tokens" such as abbreviations, numbers, and symbols are expanded into the corresponding words). For example, Pennell and Liu (2010) and Liu *et al.* (2011) asked MTurk workers to list the standard English form of abbreviations found in a Twitter corpus.

7.3.2   Problem and Solutions: Audio on the Web

For all speech applications, one wants to make sure that the audio works at least for the majority of users (operating systems/browsers/audio players). However, as also described in Chapters 4 and 6, delivering audio over the web is not without problems. For TTS assessment in particular, there are additional complications. First, it is imperative that the audio is played uninterrupted; that is, the whole file is downloaded before playing can start. Otherwise, listeners might mistake a temporary lag for a synthesis fault and score the sample much worse than it ought to be. Second, the output of a TTS system is uncompressed audio, such as the PCM WAV format. Therefore, for assessing the quality of a TTS system, one wants to assess the uncompressed audio, rather than a lossily compressed version of it, such as MP3. (Note that as synthesized speech differs in various ways from human speech, and those differences change with changes in the synthesis method, it would be difficult to prove that the choice of WAV vs. MP3 would never influence listening test outcomes for future research versions of a synthesis system.) Third, for listening tests that involve more than one audio sample, one wants to ensure that these are played sequentially, in the intended order. Fourth, one ideally wants to ensure that workers can only submit answers after having played all audio files.

The last two requirements mean that a programmatic way, such as JavaScript, is needed to control the playback of the audio, and a callback from the audio player confirming when a file has played successfully. This, in turn, means that it is not enough to leave audio playback to whatever player happens to be installed in a worker's browser. A typical solution for such a problem involves Flash. However, while Flash supports the MP3 audio format commonly used on the web, it does not natively support WAV. The fourth requirement additionally means that one needs to somehow influence the mechanism by which workers submit their answers to the crowdsourcing platform. However, this part is typically beyond the control of the requester.

The problems described above meant that the first interface (optional playback) described in Buchholz and Latorre (2011) did not fulfill all requirements. Based on QuickTime, which natively supports WAV, it was not completely crossbrowser, and while it did use JavaScript to enable the button for a second audio file only once the first one had played, it could not actually enforce playing of all audio files before a worker could submit their answer. Figure 7.1 shows this interface for a preference test.

The second interface (mandatory playback), developed about a year later, was based on the only Flash-based WAV player (Fedorov 2012) that an extensive web search found. It had better crossbrowser support and through custom-made code also ensured not only that both audio files were played sequentially but also that the answer radio buttons only showed up once the last file had played. As this was marked as a required question, workers could not submit until they had answered it. This is shown in Figure 7.2.

Both interfaces have an additional checkbox that workers can tick if something is wrong with the audio. This was useful as it uncovered the fact that an initial version of the Quick-Time interface using a complicated combination of `<object classid="...">` and `<embed>` tags caused problems in about a third of the cases.

Even with the latest interface, there are still a number of technical issues. There is anecdotal evidence that having another Flash player open in another tab of the same Firefox browser can interfere with playing the audio of the listening test. There seem to be issues related to some sampling rates, and the worker forums also recently had reports of audio files taking much longer to load. This might be related to load on the server and needs to be investigated. Finally in rare cases workers still seem to be able to submit answers faster than should be possible when playing all audio. Due to the lack of direct contact between requesters and workers in crowdsourcing, especially when going through an intermediary such as CrowdFlower, knowing how much technical problems affect workers is very difficult. One solution would be

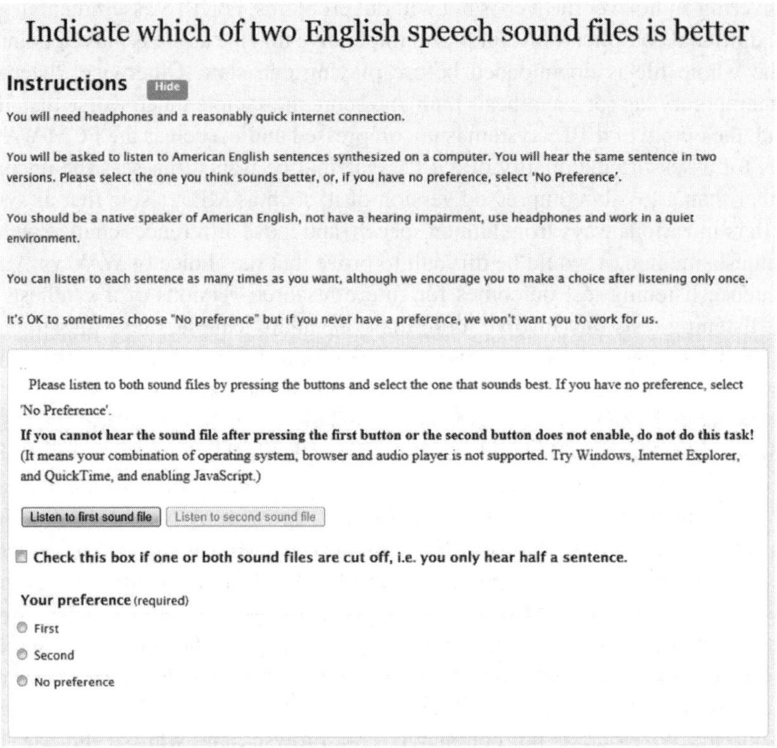

Indicate which of two English speech sound files is better

Instructions `Hide`

You will need headphones and a reasonably quick internet connection.

You will be asked to listen to American English sentences synthesized on a computer. You will hear the same sentence in two versions. Please select the one you think sounds better, or, if you have no preference, select 'No Preference'.

You should be a native speaker of American English, not have a hearing impairment, use headphones and work in a quiet environment.

You can listen to each sentence as many times as you want, although we encourage you to make a choice after listening only once.

It's OK to sometimes choose "No preference" but if you never have a preference, we won't want you to work for us.

Please listen to both sound files by pressing the buttons and select the one that sounds best. If you have no preference, select 'No Preference'.

If you cannot hear the sound file after pressing the first button or the second button does not enable, do not do this task! (It means your combination of operating system, browser and audio player is not supported. Try Windows, Internet Explorer, and QuickTime, and enabling JavaScript.)

`Listen to first sound file` `Listen to second sound file`

☐ **Check this box if one or both sound files are cut off, i.e. you only hear half a sentence.**

Your preference (required)

○ First
○ Second
○ No preference

Figure 7.1   The first "optional playback" interface for preference tests.

to include a feedback/comment box into every single HIT (as CrowdFlower does not support a general "exit" questionnaire).

Note that even with the latest development of audio on the web, the HTML5 `<audio>` tag, the problem is not solved. Up to present, different browsers support different subsets of audio formats, and in particular, Internet Explorer does not support WAV (see the overview at http://www.w3schools.com/html5/html5_audio.asp; accessed 17 October 2012).

7.3.3   Problem and Solution: Test of Significance

After running a subjective experiment, a test of significance is required to assess how likely it is for the results to be due to chance. This measurement involves comparing the probabilities obtained in the test with the probability of a purely random experiment that is called the null hypothesis. Experiments run in the laboratory are designed in such a way that a standard null hypothesis can be used. In most cases, these standard tests assume that the test is complete; that is, all the desired samples were evaluated. However, in crowdsourcing this cannot be guaranteed. Some files might never be evaluated, some subjects might experience problems, and yet other subjects might later be excluded by quality control measures. To cope with this in the significance test, some adjustments are required.

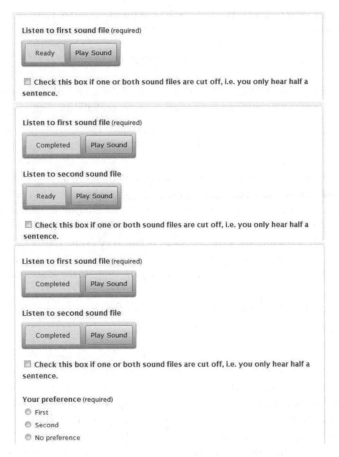

Figure 7.2 The second "mandatory playback" interface for preference tests. The three screenshots show successive stages of the interface as a worker plays one audio after the other.

Comparison Tests

In preference tests, the standard null hypothesis is that the samples are obtained from a Bernoulli distribution with $q = 0.5$. In other words, the null hypothesis is that both systems are equally preferred. In this way, the z-score is computed as the number of standard errors S between the observed system preference X_A and the expected one q:

$$Z = \frac{X_A - q}{S}, \tag{7.1}$$

where

$$S = \frac{\sigma_{\text{bernoulli}}}{\sqrt{N}} = \frac{\sqrt{q * (1 - q)}}{\sqrt{N}} \tag{7.2}$$

with N the number of judgments. The significance of the test is given by the two-tailed p-score that is computed as

$$p = 2 * (1 - \Phi(Z)), \tag{7.3}$$

where $\Phi(Z)$ is the cumulative distribution function of the standard normal distribution. In comparison tests with a "No preference" option, half the probability of that option is added:

$$X_A = P(A) + \frac{P(\text{None})}{2}. \tag{7.4}$$

This is justified, as the "No preference" option is an accumulator of those judgments that would otherwise be randomly assigned to A or B. Usually, the assumption is that $q = 0.5$ and to avoid any possible bias toward the sample played first or second altering it, stimuli are played the same number of times as AB as BA. However, in a crowdsourced experiment this cannot be guaranteed. To correct for this, the proper q value needs to be estimated by taking into account the preference for the first, second, or no sample, as well as the number of times stimuli were played as AB and as BA. In this way, the estimated $\hat{q}$ to compute the Z-score should be

$$\hat{q} = P(\text{first})P(AB) + P(\text{second})P(BA) + \frac{P(\text{None})}{2}. \tag{7.5}$$

MOS Test

For MOS tests, Ribeiro *et al.* (2011) proposed a method to compute the confidence score that tries to take into account the variability due to the system, the subject and the subjective test. Although this approach is appropriate for excluding workers, it cannot be used to compare whether a system is significantly better than another. The reason is that MOS is a Likert-type scale (Likert 1932), which means that although the MOS scale is ordered, the distance between two consecutive levels is not guaranteed to be equal. As a result, it is not appropriate to compare systems based on their mean MOS scores. This problem was discussed in Clark *et al.* (2007), in which the use of a Bonferroni-corrected Wilcoxon signed rank test was proposed. The problem with this type of test is that it assumes a set of matched results for the compared systems, but this condition is not always true for crowdsourced experiments. A possible way to solve this is to consider the MOS test as a kind of asynchronous comparison test and test the statistical significance of $P(A > B)$ versus the null hypothesis that the scores for A and B come from the same distribution. Assuming statistical independency $P(A = i, B = j) = P(A = i)P(B = j)$, therefore, for a 5-points MOS scale,

$$P(A > B) = \sum_{i=2}^{5} \sum_{j=1}^{i-1} P(A = i)P(B = j) \tag{7.6}$$

and

$$P(A = B) = \sum_{i=1}^{5} P(A = i)P(B = i). \tag{7.7}$$

Equating these probabilities to $P(A)$ and $P(\text{None})$ of a standard comparison test, respectively, the preference for system A can be estimated as

$$\bar{X}_A = P(A > B) + \frac{P(A = B)}{2}. \qquad (7.8)$$

Since in a MOS tests subjects listen to each stimulus independently and the presentation order is randomized, it is almost impossible to estimate the corrected $\hat{q}$. Therefore, it is reasonable to assume a null hypothesis with a default $\hat{q} = 0.5$. The standard error is then computed using Equation (7.2) and approximating N by

$$\bar{N} = \sqrt{N_A B_B}, \qquad (7.9)$$

where N_A and N_B are the number of collected judgments for systems A and B, respectively.

7.3.4 What Assessment Types Worked

As was seen in Section 7.3.1, most papers about crowdsourced assessment of TTS report about one or two experiments, performed at one point in time, for English only. At Toshiba Research, crowdsourced listening tests have been used routinely since the middle of 2010 to evaluate new TTS research and development. A variety of test types was run, for a variety of languages, although the focus has been on English. Table 7.2 gives an overview.

The listening tests in Table 7.2 were all run on CrowdFlower (another platform will be discussed in Section 7.3.6). CrowdFlower offers several "labor channels"; that is, they can post tasks on different platforms. Unless otherwise noted, the listening tests discussed in the remainder of this chapter used the MTurk labor channel; that is, CrowdFlower posted the tasks on MTurk where they were completed by MTurk workers. Officially, only people or institutions located in the United States can directly post work on MTurk (MTurk 2012a). This is the main motivation for using CrowdFlower, as they do not impose such a restriction. They also offer easy support for using a gold standard (see Snow *et al.* 2008) as quality control *during* the task. Workers are prevented from continuing work on the task as soon as they fail too many gold questions. However, unless otherwise stated, this functionality was not used, due to the subjective nature of listening tests. Finally, while MTurk offers the possibility to

Table 7.2 Numbers of listening tests (run via CrowdFlower), by test type and language.

Test\language	enUS	enUK	deDE	esUS	esES	frFR	frCA	All
AB	229	9	32	4	4	5	0	283
ABX	54	0	0	0	0	0	0	54
MOS	32	2	4	5	0	2	1	46
DMOS	12	0	0	0	0	0	0	12
total	327	11	40	9	4	7	1	400

enUS, American English; enUK, British English; deDE, German (Germany); esUS, American Spanish; esES, European Spanish; frFR, European French; frCA, Canadian French.

restrict access to a task to workers from specified countries, it does so on the basis of the country that the *worker* filled in when signing up. CrowdFlower, on the other hand, uses IP address geolocation to actively prevent access to tasks for workers who are not located in the countries specified by the requester. This is a welcome extra advantage.

However, there are also disadvantages: CrowdFlower charges a percentage on top of MTurk's price; did not initially offer the possibility to block specific workers from all future tasks; and does not offer functionality for rejecting workers, paying bonuses or external HITs (where the task is hosted on the requester's server rather than Amazon's). The latter is of particular relevance to listening tests. Wolters *et al.* (2010) ran each complete intelligibility test (involving several questionnaires and 50 SUS samples) as a single human intelligence task (HIT) on MTurk by using the external HIT interface. This gives complete control over the experimental setup and allows advanced test designs. For example, the Blizzard Challenges use a Latin square design for their listening tests (Bennett and Black 2006). This is not possible on CrowdFlower. All listening tests discussed here were, therefore, run in such a way that each rating (MOS, DMOS) or paired comparison (*AB, ABX*) was its own HIT. This setup probably speeds up completion rates as it is less risky for workers who do not yet know this type of task to sample a few (as they can just stop at any point and be paid proportionally). Wolters *et al.* (2010) reported having to re-release the task each day for at least 8 days, while the English listening tests discussed in this section mostly finish within a day.

The remainder of this section discusses in more detail how the individual listening test types were implemented.

MOS Tests: Naturalness, Quality, Diagnostic

A number of types of MOS tests have been run: MOS *naturalness*, MOS *quality,* and *diagnostic* MOS. The first two types are rated on a 5-point scale, as described in Section 7.2. MOS naturalness is rated on a continuum of "completely natural" – "completely unnatural," whereas the continuum for MOS quality is from "very good" to "very bad." These have been used to evaluate systems for research purposes as well as part of quality assurance in product development, whereby our own system is compared against those of the competitors. The third type, diagnostic MOS, is a modified form of MOS, in which the objective is to examine the extent of any problems in a large number of sentences, typically 2000 or more. The ratings are on a 3-point scale, and the points are labeled "No problem," "Minor problem," and "Major problem."

In ordinary MOS tests, the overall performance of the systems is our main interest, and as such, scores for individual sentences are irrelevant. In diagnostic MOS, however, scores for individual sentences are respected. This is used as a filter for a more detailed diagnostic evaluation, which is described in Section 7.3.5. Note that in order to draw the listener's attention to any linguistic or phonetic errors as well as synthetic artifacts, the text of the sentence is displayed in the interface for diagnostic MOS.

In all diagnostic MOS tests run on CrowdFlower, gold-standard items were included. They consisted of "bad" gold, in which problems were introduced artificially, as well as (in later tests) "good" gold, which were (1) unanimously judged to be problem free in a gold-selection diagnostic MOS test that preceded the main test and then (2) confirmed as being incontestably "good" by an in-house native speaker of the language. A small number of MOS quality and MOS naturalness tests were run with human speech samples as gold standard items to anchor the upper limit of the scale, but the majority were run without gold.

CrowdFlower/MTurk ensures that workers are served HITs in random order and that they are never served the same HIT more than once. However, as already pointed out in Section 7.3.4, it has not been possible to implement tests with a Latin square design using CrowdFlower, so a listener may hear the same sentence (by different systems) more than once.

Listeners need to listen to a few samples before they are able to make a call on the range of naturalness, quality, or extent of errors present in the evaluation samples. Therefore, the first three responses from each worker are normally excluded from the analysis of MOS tests.

Since June 2010, 64 MOS tests have been run on CrowdFlower/MTurk and Clickworker (http://www.clickworker.com/; accessed 17 October 2012), out of which nine are diagnostic MOS. The tests cover seven languages, completed or aborted with varying degrees of success. A worker is paid US$0.03 per HIT. This amount is based on the US federal minimum wage and the average time it is expected to take for a worker to complete a HIT.

AB Tests: Preference
As explained above, each HIT of an *AB* preference test consists of just one paired comparison. Workers see instructions and two buttons that they need to click to play the two audio samples, and then indicate their preference for the first sample, the second sample, or none (cf. Figure 7.1). To avoid presentation order bias (Wherry 1938), separate HITs are posted for each order (*AB* as well as *BA*) of each sample pair.

For each preference test, the test outcome is computed in terms of percentages of preference expressed for each system, *A* and *B*, and "No preference." To check whether any observed difference in preference for the systems is statistically significant, the two-tailed *t*-test described in Section 7.3.3 is applied to the results after splitting the "No preference" votes equally over the two systems, simulating a forced choice situation in which people who really have no preference can be expected to vote for *A* as often as for *B*.

The above method to crowdsource preference tests has been used within our TTS research group extensively. Since June 2010, 284 preference tests have been run on CrowdFlower. Research evaluated encompassed a wide range of topics, including prosody, acoustic modeling, and vocoding. Typically, all but the largest test groups finish overnight, which greatly increases researcher efficiency.

By default, a test consists of a minimum of 50 sample pairs played in both orders, and each pair is evaluated by at least 5 workers, yielding a total of 500 judgments. Usually each worker is allowed to provide a maximum of 40 judgments (functionality provided by CrowdFlower). Therefore, the minimum number of workers per test is 12.5. A worker is paid US$0.05 per comparison and the average sound sample is 8.9 seconds long. US workers so far came from all the 50 states and, therefore, should be more representative than any group one could hope to assemble in a traditional laboratory experiment.

DMOS and ABX Tests: Similarity
DMOS and *ABX* tests were used for testing the similarity to a given sample in terms of speaker identity by Zen *et al.* (2012) and Wan *et al.* (2012), and in terms of expression/speaking style by Eyben *et al.* (2012), Latorre *et al.* (2012), and Chen *et al.* (2012). As these tests involve listening to more audio samples than for an *AB* preference test, a worker is paid US$0.05 and US$0.07 per comparison, respectively. We have not carried out any specific analyses for these test types other than the standard analysis used for standard MOS or *AB* tests. In general, subjects in these types of tests find it difficult to abstract the factor for which they

are asked from the general speech quality of the samples. People who are used to listening to TTS systems may have learned to discriminate between the different factors, but this is not necessarily the case for naive listeners. Even in controlled experiments, when comparing speaker similarity in a DMOS test with a human speech sample as the reference, it is not unusual for listeners to give higher scores to the samples that "sound" better, even though the speaker identity might actually be very different. To avoid this, the question for these types of tests has to be formulated very carefully so as to be as unambiguous as possible. For crowdsourced experiments this is even more important, as the average naivety of the workers tends to be higher.

7.3.5   What Did Not Work

While the listening test types described above could be crowdsourced successfully, there were some types that did not work well. Here, we will describe two experiments that we failed to crowdsource successfully.

Diagnostic Evaluation

One such type is a diagnostic evaluation, which is designed to pin-point problems of quality in a set of sentences. The desired result is not a MOS or a preference, but a set of indications concerning problems of certain kinds such as synthesis artifacts, text normalization errors, mispronunciations, pause errors, and intonation errors. While also meant to identify segmental errors, the scope of a diagnostic evaluation is much broader than the segmental tests discussed in Section 7.2. It can serve as a tool for quality assurance purposes, highlight the areas that need improvement, and provide per-word diagnostics of sentences agreed upon by multiple listeners. By combining the diagnostic results with MOS scores, it may also be possible to obtain an indication of which error types have a greater impact on the overall quality than others.

Previously, attempts were made to collect such information from free-form comments. For example, in an experiment conducted with naive listeners in a controlled environment, Krstulović *et al.* (2008) encouraged listeners to give free-form comments for each sentence, in order to augment the information collected by means of MOS scores. The comments were then manually reduced to a set of semantic tags to summarize the problem in a standardized manner. However, this involved a potentially costly phase of manual lexical and semantic analysis, and would be impractical with a large number of utterances and listeners. While one option would be to crowdsource this tagging phase, we opted for an approach in which the listeners' judgments are obtained in a more direct manner, and in finer detail.

A pilot experiment for a diagnostic evaluation was run on CrowdFlower/MTurk with 20 evaluation sentences and 10 gold-standard sentences. The gold-standard sentences consisted of "problematic" sentences, in which problems were artificially introduced. Problems included mispronunciations (segmental and suprasegmental), pause insertion, cutoff, flat prosody, and overlaying artifacts such as echo, noise, and beeps. These were intended to serve as training examples as well as to exclude spammers. CrowdFlower allows, upon request, for a certain number of Gold samples to be presented at the beginning of the evaluation for each worker, so that any workers who fail these are barred from proceeding further.

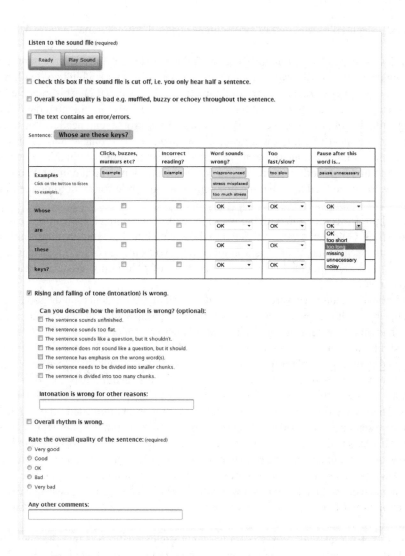

Figure 7.3   A screenshot of the diagnostic evaluation interface.

Figure 7.3 shows an interface similar to the one used. Listeners were presented with the unnormalized text of the sentence as they listened to the corresponding audio file. They were asked to indicate any problems at the word level, by ticking the checkbox or selecting an option from the dropdown menu in the intersection cell for the word and the error type. Word-level errors included artifacts, text normalization, mispronunciation, lexical stress, sentence stress, rhythm, and pause. They were also asked some questions about sentence-level errors such as overall intonation contour, chunking, and overall sound quality. Each error type was described in a nontechnical manner and an explanation was given in the instructions. Finally, listeners were also asked to indicate the overall quality of the sentence on a scale of 1 (very bad) to 5 (very good). Each HIT consisted of two sentences and there was a payment of US$0.10 per HIT.

The task was suspended automatically soon after it was started, due to a large number of gold units being missed. CrowdFlower automatically suspends jobs with a high gold failure rate, in order to let the requester review their gold units (CrowdFlower 2012). This indicated that the questions asked were not appropriate for naive listeners, and also highlighted the difficulty of selecting suitable gold items. Gold items with intonation errors (sentence level or word level) were missed the most often. CrowdFlower allows workers to "contest" the integrity of any gold items they miss, but none of the missed gold items was contested in this pilot. It may be inferred from this that workers were able to detect the problem with the missed gold samples once they were given feedback, but the problem was too subtle for them to notice in the first instance.

A second pilot was then run with the wording revised and with each error type exemplified with audio files. Gold items were restricted to those which had obvious synthesis artifacts and one item with a segmental error (mispronunciation). No examples of incorrect sentence stress or lexical stress were included as gold. The purpose of gold items, this time, was solely to reject spammers.

The 15 nongold sentences were a mixture of "problematic" sentences and "good" sentences as selected by multiple expert listeners within our group using the same interface. Each sentence had on average nine unnormalized words per sentence.

At least nine workers evaluated each sentence. They varied widely on their judgment of the position and nature of problem for each sentence, and even on whether there was a problem or not. There was not a single sentence that all the workers unanimously found to be without a problem, not even those that had been judged as being problem free by internal expert listeners. There were four sentences for which 90% of workers indicated no problems. Some of them were sentences that had been judged to be problem free in the internal test; others had problems that expert listeners had detected but which only one crowdworker picked up on.

There was one sentence that had been deemed problem-free internally, but for which 80% of the workers reported some problem or other. Upon close examination, the sample was found to contain one word with a debatable vowel quality. This is an interesting case that exemplifies the power of crowdsourcing, but there was little agreement on the nature or the position of the problem.

One of the problems with crowdsourcing diagnostic evaluations is that the questions need to be asked in such a way that the responses can be reduced to a technically meaningful set of qualities, but they also need to be intuitive for naive listeners. This sometimes means the categorization of errors that makes sense from a technical point of view may not make sense from a layman's point of view. Questions asked in our evaluation may have been too detailed for naive listeners, and a more simplified interface may provide useful results. For example, testing for different error types separately would reduce the cognitive load for the workers. However, this would lead to an increase in costs as a greater number of tests would need to be run in order to collect information about different aspects of speech.

Another difficulty is in the subjective nature of the questions being asked. MOS scores are also subjective, but they can be averaged out over a large number of sentences. With diagnostic evaluations, problems pertaining to individual sentences are often of interest, as well as the overall score. Even in the laboratory-internal test, there was some level of disagreement as to the nature or the position of the problem. This is undoubtedly partly due to the fact that many error types are interrelated. For example, errors in stress will affect the rhythm as well as vowel quality and a listener may mark one or the other, or any combination of the above, if at all.

Creating or selecting gold samples presents another issue, as they need to be incontestable. If "bad" gold samples are artificially created, there is a risk of biasing listeners as to the types of errors they pay attention to. The gold-selection task itself may be crowdsourced, but depending on the quality of the system being evaluated, one may end up submitting thousands of sentences to select a relatively small number of gold samples, which could be costly. Selecting "good" gold is even more difficult, as there is a good chance a phonetically trained listener will detect a problem with a sentence judged to be problem free by 10 naive listeners. Still, one could accumulate gold samples through repeated crowdsourced evaluations.

The evaluation contained a MOS element, and this was the only part from which relatively reliable information could be drawn. As described in Section 7.3.4, we currently crowdsource modified (diagnostic) MOS tests in which the workers' attention is drawn to specific error types and they are asked to rate a large number of sentences on a scale of 1 "Major problem" to 3 "No problem." This is used as a rough quality filter to identify the most problematic sentences, which are then submitted to further analysis by phonetically trained listeners using the interface described above (Figure 7.3).

Emotion Labeling

This section will describe another pilot study into crowdsourcing for TTS that has not resulted in a workable procedure. Contrary to the other sections, it is not about *assessment of* TTS but about *labeling for* TTS.

Proper use of expression makes speech more interesting and aids understanding of content by adding nuances and information beyond the pure text content. Expressiveness in speech includes emotions (e.g., angry and sad), speaking style (e.g., whisper and boasting), and the "character voices" often used in story reading (e.g., "old man"). While state-of-the-art TTS systems can achieve high intelligibility and naturalness for reading isolated sentences in a relatively neutral style, the synthesis of expressive speech is still a challenge. The first key components in expressive TTS synthesis are the speech corpus used and the annotation of "expression" in that corpus. It is this latter aspect that we tried to crowdsource.

The speech corpus used consisted of one chapter each of 18 audiobooks publicly available from LibriVox (http://librivox.org/; accessed 17 October 2012). These had previously been manually labeled with an open set of emotions by one in-house labeler. As this way of creating annotations is quite slow, and Snow et al. (2008) had previously shown that it is possible to crowdsource emotion labels for *text*, crowdsourcing seemed worth a try. For the pilot study, we chose the 10 emotion labels that the in-house labeler had used most often (annoyed, anxious, confused, defiant, desperate, excited, happy, sad, surprised, upset). For each, 10 sentences that she had labeled as such were selected at random. Via CrowdFlower/MTurk, crowdworkers were asked to listen to a sentence and indicate the emotion(s) that they thought were expressed in the sentence. The set of emotion labels they could choose from were the top 10 labels used by the annotator plus 3 more missing from this set but used in Snow et al. (2008) (and much

Table 7.3 Interannotator agreement (IAA) for the 13 emotion labels, expressed in α_κ, sorted descending.

Emotion	Excited	Angry	Sad	Happy	Confused	Annoyed	Surprised
IAA (α_κ)	0.39	0.29	0.25	0.24	0.24	0.22	0.21

Emotion	Other	Desperate	Upset	Defiant	Disgusted	Anxious
IAA (α_κ)	0.16	0.16	0.11	0.09	0.08	0.06

of the preceding literature): afraid, angry, disgusted. Also, they could indicate "no emotion," or "other" and supply their own label. Finally, for each emotion label they could choose from 3 grades of intensity, for example, "slightly angry," "angry," or "extremely angry."

Five judgments were collected per audio sample. Ten workers participated in the task. Two of those listened to only two samples each and were, therefore, ignored for the analysis. One further worker was ignored as their answers looked like *random spamming*, often indicating more than six emotions for a sample. The final analysis, therefore, includes seven workers.

Interannotator agreement (IAA) was computed as α_κ (Artstein and Poesio 2008). This metric is suited for annotations with multiple labelers, different amounts of labeling per labeler and weighted classes. Table 7.3 shows the IAA for all 13 emotion labels. All α_κ were lower than 0.4 ($\alpha_\kappa > 0.8$ is considered to be good agreement).

The initial interface was quite complex. Also, emotions might be hard to judge without context (but note that for the purposes of training an expressive TTS system, the emotion *should* really be audible in the audio, not just inferable from the context). Various simplifications and extensions were tried, such as showing a few sentences of context, asking for dimensional (negative/positive; active/passive) instead of categorical emotion annotations, and in the extreme case just asking a binary question whether the sample sounded happy (angry/sad/...) or not. However, even in informal laboratory-internal trials, no setup gave convincing results. It should be noted that we were trying to tackle a very hard problem. Emotions annotation, in general, is quite subjective, and the emotions expressed in audiobooks are typically not as strong as those in acted emotion recordings.

While the pilot study did not result in a workable annotation method, it did show how quick and easy it is with crowdsourcing to try out a variety of approaches with a sizable number of native speakers.

7.3.6 Problem and Solutions: Recruiting Native Speakers of Various Languages

Are Our Workers Native Speakers?

Assessment of TTS usually requires participants whose native language is the language of the system being tested. In a controlled experiment in the laboratory, participants are often recruited with native language skills as a required criterion. They may even be required to take a pretest to ensure they possess the skills necessary. In crowdsourced tests, this is more difficult to control.

Ideally, a qualification test to assess basic language skills should be administered. MTurk allows requesters to filter workers by means of qualification tests, but Ribeiro *et al.* (2011) found that HITs requiring qualifications attract fewer workers. Furthermore, different types of tests may require different levels of language skills. For example, preference tests for vocoding experiments may require linguistic skills to a lesser degree than a diagnostic MOS test. The qualification test would thus need to be designed appropriately for each test.

Short of administering tests, self-reported information about the worker's native language could be utilized. Wolters *et al.* (2010) asked US-based MTurk workers about their country of birth and dialect of English and *post hoc* excluded those workers who were not native speakers of US English or not born in the United States. This excluded 3.6% of the participants. However, one can of course not rule out the possibility that workers provided false information when filling out the demographic questionnaire. Chapter 6 describes an approach in which linguistic background information provided through a questionnaire is tested against performance on "anchor tokens" in order to ensure data reliability.

On CrowdFlower, it has not been possible to run qualification pretests to filter workers, so we rely on geolocation information based on IP address. CrowdFlower introduced *restricting* workers based on IP address geolocation in late 2010. Out of the 2158 MTurk workers who contributed to our tests prior to this, 164 actually participated from outside the United States, as evidenced by their IP address and geolocation information provided by CrowdFlower. These could have been US citizens living outside the Unites States[1] or using a proxy service or a virtual private network (VPN), but there is no way to verify this. As shown in Table 7.4, the greatest number of non-US-based workers were located in the United Kingdom, so they may have been native speakers of English, but not of the variant being evaluated. Worse still, many of the workers were based in non-English speaking countries, so it is likely that some of our results were skewed by nonnative speakers.

Since the introduction of geolocation restriction by CrowdFlower, our evaluations are made available only to workers in the country where the language variant is spoken as the primary language. However, geolocation restriction is not sufficient to ensure that the workers are native speakers of the language being evaluated, as they may be foreigners living in the country of interest. Although the instructions are localized to each language and workers are asked to only participate in the test if they are a native speaker of the language, there is nothing to prevent nonnative speakers from participating in the task, except the thought of missing too many gold items and thus being rejected, which would affect their reputation and the number of HITs available to them.

Availability of Native Listeners in Non-English Languages

According to Ipeirotis (2010), 46.8% of the MTurkers come from the United States, followed by 34.0% from India. A problem encountered when crowdsourcing tests with languages other than English is the availability of native listeners. For example, while Irvine and Klementiev (2010) successfully used MTurk to collect annotations for translation lexicons for 37 of 42 less commonly used languages, Novotney and Callison-Burch (2010) reported having trouble finding workers for a Korean transcription task. They eventually solved this by posting a special HIT where the task was to recruit such workers.

---

[1]Officially, US citizens living outside the United States are currently not allowed to work on MTurk, due to tax reasons (MTurk 2012c).

Table 7.4 Breakdown of the country of workers who worked on our American English evaluations prior to introduction of geolocation restriction by CrowdFlower.

Country	Number of workers
United States	1965
United Kingdom	29
Unidentified	27
Serbia	22
Macedonia	11
Canada	10
India	9
Dominica, El Salvador	7
Belgium	6
Philippines, Romania	5
Indonesia, Russia	4
Japan	3
Italy, Lithuania, Netherlands	2
Australia, Brazil, China, Colombia, Luxembourg, Pakistan, Singapore, South Africa, Turkey	1

As of June 2012, US-based MTurk workers have the option of receiving their earnings through a bank account or in the form of an Amazon.com gift certificate. Workers based in India may receive bank checks or Amazon.com gift certificates. All other workers can only disburse their earnings to an Amazon.com gift certificate (MTurk 2012b). This undermines the incentive of workers outside the United States, as they can only spend the voucher on Amazon.com and not on its local versions such as Amazon.co.uk, often incurring shipping costs on items purchased.

CrowdFlower provides other labor channels, some of which have a large user base in countries outside the United States, for example CrowdGuru in Germany. Using the right labor channel can have a dramatic effect. Before the CrowdGuru channel was available, a German preference test requiring 1500 judgments (i.e., workers × HITs) was canceled after 3 weeks, when only 950 judgments had been collected via MTurk (and the internal interface). More recently, a German preference test needing 900 judgments completed within 5 hours, and nearly 90% of the judgments came via CrowdGuru.

The most difficult native speakers to find in our experience so far have been those of Canadian French. The reason might be that there are "only" about 7 million of them to begin with, and workers can only be paid in Amazon.com vouchers. A Canadian French MOS test on CrowdFlower/MTurk requiring 2400 judgments was canceled after 3.5 months, when only 139 judgments had been collected. This is despite CrowdFlower automatically refreshing the task on MTurk so they appear toward the top of the HIT list. Later, another Canadian French MOS test was run via Clickworker, who were initially confident about quick delivery, but in the end it took nearly 3 months to collect just over 1200 judgments.

Even European Spanish tests have taken 20 days and European French over a week to collect 1000 judgments on CrowdFlower. American English tests with the same number of judgments usually complete within a couple of hours. With Clickworker, twice as many judgments can be collected in the same time span for French; and for Spanish, usually twice as many judgments can be collected in a day.

With less common languages, the situation can be more dire. Irvine and Klementiev (2010) reports that out of the 42 languages for which they crowdsourced lexical translations on MTurk, only three completed within 20 hours, and they only required 30 judgments (3 workers × 10 HITs per language). The three languages were Hindi, Spanish and Russian. The rest were less common languages.

Knowing the user base of each platform helps in recruiting workers as quickly as possible. Under-resourced languages are often spoken in economically under-resourced areas and the number of speakers in the online community may be lower than in developed countries. However, Samasource has workers concentrated in areas of high linguistic diversity, so it may be a useful platform for some minority languages.

It is definitely worth exploring various channels, but attention needs to be drawn to the possible bias that may arise from using particular channels. First of all, one needs to bear in mind that crowdworkers are not representative of the general online community. For example, many of the workers on MTurk are younger and with a higher education level than the overall population (in their respective countries). They also have a lower income level (Ipeirotis 2010). Channels that are likely to attract specific types of workers will introduce additional skew. For example, CrowdFlower used to offer tasks through Gambit, whose workers were paid in the form of social gaming credits which could be used in online games. No tests or analyses have been conducted to compare results obtained through different channels.

7.3.7   Conclusion

While simple listening tests such as (D)MOS and AB(X) can be successfully and efficiently crowdsourced, more demanding assessments and labeling tasks have not been so successful, and even for the simple listening tests, one has to be careful when dealing with audio, applying standard statistical significance tests and choosing crowdsourcing platforms/labor channels for non-English languages.

7.4   Related Work: Detecting and Preventing Spamming

As in crowdsourcing in general, it is important to think about quality control when using crowdsourcing for TTS assessment.

There are clear incentives for people to spam. Not listening to samples before submitting an answer means workers can earn money faster. Listening without proper attention, for example, while simultaneously surfing the Internet, requires less concentration and is less boring.

Quality control is easiest with intelligibility tests because, as in the case of a transcription task, there is a "ground truth," that is, the correct word sequence. Wolters *et al.* (2010) simply excluded the two workers with a mean WER above 0.9. The next worst worker had a WER of only 0.61, making thresholding straightforward. One of the excluded workers had indicated problems with playing the audio stimuli, again indicating that not all poor workers are so intentionally. Bennett (2005) also reports excluding some participants of the intelligibility test who typed in comments, generic answers like "don't know," or the words belonging to the previous sentence.

Quality control becomes harder when dealing with more subjective tests.

In the case of MOS tests, it is possible to include the original human samples, which should always receive relatively high scores, therefore providing something akin to a gold standard. Indeed, Bennett (2005) excludes some listeners who score the human voice very low in the MOS test part. However, note that a proper gold standard should contain very good as well as very bad cases. If human samples acting as gold standard were the only quality control, malicious spammers could simply score all samples high.

Chen *et al.* (2009) checks the transitivity of the results. If *A* is preferred over *B* and *B* over *C*, then *A* should also be preferred over *C*. Workers are rejected, that is, not paid, if less than 80% of the triplets satisfy this transitivity rule. Although interesting, this method would present several problems for TTS listening tests. First, it assumes that there are at least three systems in the comparison, which is typically not the case for a TTS preference test. Second, workers' subjective criterion sometimes evolves over the test so that the same feature that was annoying at a certain point is ignored later or vice versa. Actually, when the differences between systems are not obvious, it is often found, even in controlled experiments, that subjects who preferred one system when the stimuli were presented as AB chose the other one when the same stimuli were presented as BA. In a controlled experiment, researchers can make sure that each subject compares all the systems for the same sentence with both presentation orders. However, to guarantee this in a crowdsourced experiment, all the comparisons between all the systems with both presentation orders would need to be collected into one large HIT.

Ribeiro *et al.* (2011) experimented with requiring workers to complete a simple pre-qualification (scoring two audio samples, "one obviously good, the other obviously bad") but found that that requirement dramatically limited uptake of the task. They also present a novel metric for post-experiment quality control of crowdsourced MOS tests. They compute the correlation coefficient between the MOS values of one worker and those of all workers. For a MOS tests comparing *K* systems, the correlation coefficient for worker *n* is computed as

$$r_n = \frac{\mathrm{cov}(v_n^1, \ldots, v_n^K; \mu^1, \ldots, \mu^K)}{\sqrt{\mathrm{var}(v_n^1, \ldots, v_n^K)}\sqrt{\mathrm{var}(\mu^1, \ldots, \mu^K)}}, \tag{7.10}$$

where v_n^k is the average MOS score assigned to system *k* by worker *n*

$$v_n^k = \frac{\sum_{m=1}^{M_n^k} s_{m,n}^k}{M_n^k} \tag{7.11}$$

and μ^k is the average MOS score for system *k* over all workers

$$\mu^k = \frac{\sum_n^N M_n^k v_n^k}{\sum_n^N M_n^k} \tag{7.12}$$

with $s_{m,n}^k$ the MOS score assigned to sample *m* of system *k* by worker *n*, M_n^k the total number of samples from system *k* evaluated by worker *n* and *N* the total number of workers. If $r_n < 0.25$ all the HITs from worker *n* are excluded. In addition, they reject individual answers that were submitted quicker than the time needed to listen to the samples. Finally, they encouraged "good" work by awarding bonuses to those workers with the highest r_n.

Buchholz and Latorre (2011) developed metrics for post-experiment quality control of preference tests. As this chapter is an extension of that work, those metrics will be introduced in more detail in Section 7.5.

There is much research on quality control for crowdsourcing outside of the speech technology literature. However, most focus on techniques that are not directly applicable to listening tests, such as using gold-standard samples (Snow *et al.* 2008), interworker comparisons that assume that there is a "true" label for each sample (Sheng *et al.* 2008; Ipeirotis *et al.* 2010; Hirth *et al.* 2010) or using a second crowd to assess the performance of the first (Hirth *et al.* 2010). One relevant paper from the field of crowdsourcing relevance judgments for query-document pairs is Vuurens *et al.* (2011). They classify crowdworkers into *proper workers*—who follow instructions and produce good judgments, *sloppy workers*—who have good intentions but deliver poor judgments (through poor understanding or incapability), *random spammers*—who intentionally cheat by giving random answers, and *uniform spammers*—who intentionally cheat but mostly repeat the same answer. They show through simulations that the standard approaches of using majority voting and gold-standard examples are susceptible to spam, and especially at higher spam rates are outperformed by proposed new metrics that specifically target random spammers, uniform spammers, and sloppy workers. However, the application of these metrics requires an iterative approach in which the worst spammers are removed first, and new judgments are collected to replenish the discarded ones. Such an iterative approach is not possible on CrowdFlower, and requires a separate crowdsourcing management tool on top of the MTurk API.

Finally, Rzeszotarski and Kittur (2011) introduce a technique based on *task fingerprinting*, which involves logging user events such as clicking, scrolling, pressing a key or moving the mouse, while workers perform tasks on MTurk, and using those to accurately predict the quality of the work. While this is an extremely interesting avenue for the future, it is beyond the scope of this chapter.

7.5   Our Experiences: Detecting and Preventing Spamming

As described in Section 7.4, Ribeiro *et al.* (2011) rejected workers whose system correlation coefficient between their MOS values and those of all workers was less than 0.25. We applied a similar approach to some of our diagnostic MOS tests (3-point scale) using a sample-based correlation $\hat{r}_n$ instead of a system-based one,

$$\hat{r}_n = \frac{\text{cov}(\mathbf{s}_n; \bar{\mathbf{s}}_n)}{\sqrt{\text{var}(\mathbf{s}_n)\text{var}(\bar{\mathbf{s}}_n)}}, \tag{7.13}$$

where $\mathbf{s}_n$ is the vector formed by all the samples evaluated by worker n, and $\bar{\mathbf{s}}_n$ the vector formed by the average scores assigned by all the workers to the samples evaluated by worker n. The workers rejected based on $\hat{r}_n$ were compared with those rejected by excluding those who responded with a recurring pattern such as 1,2,3,1,2,3 or 1,1,1,1,1,1. The latter type of spammers are not necessarily easy to detect by automated spam detection, but are easily detected when manually inspected. For all four tests (comprising, on average, 84.25 workers and 10,000 judgments per test), all workers identified as uniform spammers were excluded by using the sample correlation approach. This approach also rejects workers other than uniform spammers, but no further analysis has yet been carried out to detect whether any of the other rejected workers are actually "trusted" workers. Our experience with detecting spammers in MOS tests has so far been limited, and thus in this section, we report on our experience with preference tests only. As seen in Table 7.2, this is also the test type for which we have most data.

As explained in Section 7.3.2, the optional playback interface could not prevent workers from submitting answers without playing the audio. While this might seem a major drawback, it was also an opportunity to easily identify spamming workers and to develop dedicated metrics for their detection. This is similar to the "honeypots" used in computer security (Spitzner 2002). The work was originally reported in Buchholz and Latorre (2011) and is reproduced in a slightly extended and edited version in Section 7.5.1.

After publication of that work, a mandatory playback interface was implemented, which forces workers to play the audio before submitting answers. However, playing the audio does not necessarily entail listening to it with proper attention, so quality control is still important. Section 7.5.2 reports on an updated analysis on this topic.

7.5.1   Optional Playback Interface

The work in this section reports on 125 listening tests of the *AB* preference type, for evaluating American English TTS, performed between June 2010 and beginning of March 2011, on CrowdFlower using the MTurk labor channel, restricted to US-based workers. The optional playback interface did not force workers to play the audio. However, every time a worker clicked a button to play an audio sample, a piece of JavaScript incremented a hidden counter for that button. For each HIT, the final values of the two counters were stored together with the worker's preference. Given that information, it was clear that any worker who did not play both samples before answering was spamming *on that HIT*. For each test group,[2] the *not-played percentage* could be computed, that is, the percentage of submitted answers for which one or both samples were not played by the worker. Figure 7.4 shows the not-played percentages for all 73 test groups, ordered chronologically. Apart from the wide variation of values, for which no clear explanation was found, it is interesting to note the low percentages of not-played HITs in the earlier tests, sometimes below 1%. More recent tests never show such low percentages, which further analysis revealed is due to *returning* spammers, that is, workers who spam on more than one test group. At that time CrowdFlower did not allow directly blocking or rejecting workers, and these jobs were easy to identify because most of them shared the same title. Therefore, some workers might have concluded that they could spam without any risk to their reputation and were consequently looking for these jobs.

Gold-Standard Data

In a subjective comparison test, no ground truth or trusted annotations are available for the systems under test, as typically one or both are the result of recent research. However, one can construct crude gold questions by pairing a sample from the human voice used to train the

---

[2]Sometimes researchers wish to compare not just two but three or more versions of a TTS system. To facilitate this, preference tests can be grouped together. In other words, all the HITs for all the tests in the group will show up in the same HIT group on MTurk. The 125 listening tests formed 73 test groups.

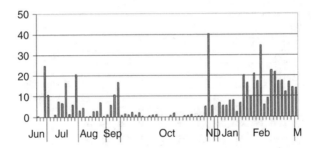

Figure 7.4 Percentage of submitted answers for which one or both samples were not played by the worker, for each of 73 test groups ordered chronologically, with months indicated (June 2010—March 2011).

TTS with a sample from a TTS system, typically an older version of the TTS versions in the main comparison. It can be expected for people to prefer the human over the TTS and we will henceforth say that they "fail" the gold question if they do not. To confirm whether that type of gold samples work as expected, a preference test was conducted in which all 103 questions were gold questions. Out of the 20 participating workers 12 always preferred the human; 2 always preferred the human except once; 2 preferred the human 92% and 78% of the time, respectively; and 4 preferred the human and the TTS about equally often. From these numbers, it seems obvious that the last 4 workers were spammers and their answers should be excluded. However, this is not so clear for those who failed only one question, as excluding them would also exclude many correct answers. As a result, it was left to researchers to decide whether they wanted to use gold samples in their tests, and only 46 out of the 73 test groups used them.

An obvious reason for failing a gold question is if the samples were not even played; however, those are the uninteresting cases. A further analysis was conducted on those gold questions where both samples *were* played. The percentage of (played) gold questions that were failed per test group did not increase over time in the same way as the number of submissions by spammers. One possible explanation is that the spamming "method" of playing both samples and then choosing the response randomly requires more time and effort than simply not listening at all. Consequently, it is not unexpected for the latter to spread faster over the web. We later found some posts on worker forums referring to our HITs as being lucrative. The average percentage of failed gold was 7.8.

New Dedicated Metrics

The previous paragraphs introduced two ways to detect spamming in preference tests. Using that information, the following two subgroups of workers can now be defined:

Spammers: Never play both samples of a HIT.

Reliable workers: Always play both samples, were asked at least four gold questions and do not fail a single gold question in the whole test group.

Note that for these definitions, a person who took part in multiple test groups is counted multiple times. For example, a person who participated in two test groups is considered as

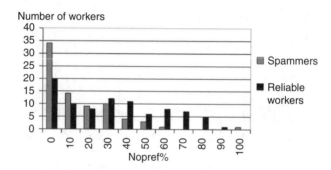

Figure 7.5   Counts (*y*-axis) of workers that answered **"No preference"** *x*% (rounded to nearest multiple of 10) of the time for workers who did at least 25 HITs and (a) never played samples (known spammers; 76 workers) or (b) played everything, did at least five gold and passed all gold (reliable; 88 workers).

two different workers. So the same person could be a reliable worker for one test group but a spammer for another, as they might indeed have changed their behavior. By comparing these two groups, new metrics that differentiate well between the two groups can be identified. Note that although spammers did not even play the samples, their answer pattern should be similar to those of workers who play but do not listen. Therefore, the effect of the optional playback interface was to allow "recruiting" a group of "genuine" spammers whose behavior could later be studied in order to better identify and exclude similar forms of spamming in the future.

Most of the metrics described in this study depend on the number of HITs evaluated by the worker. Therefore, only those spammers and reliable workers that undertook at least 25 relevant HITs per test group were analyzed.

As shown in the instructions in Figure 7.1, workers were discouraged from answering "No preference" too often. Figure 7.5 shows the percentage of times that spammers and reliable workers gave that answer. The histogram of the spammers is much more skewed toward the smaller values. In other words, they tend to avoid "No preference" more than reliable workers. In particular, no spammer had a "No preference" percentage over 56%, except one outlier who *always* answered this way. Unfortunately, while there are clear differences between the two groups, this information cannot be used to identify spammers, only very indecisive reliable workers.

Presentation order bias is the tendency of people to prefer the second of two samples if there is no clear difference between them (Wherry 1938). However, spammers can be expected to have either no preference for a specific position, as they choose their answer basically at random (Vuurens *et al.*'s (2011) *random spammers*), or a very clear preference, due to always choosing the same answer (Vuurens *et al.*'s (2011) *uniform spammers*). This is confirmed by the data shown in Figure 7.6. The peak of the spammers' distribution is at 50% while that of the reliable workers is at 60%. Also, all the data points at the extremes are from spammers. By looking at the values prior to rounding, an "extreme" value was defined as $x < 30$ or $x > 90$.

A spammer exclusion method proposed by Ribeiro *et al.* (2011) consists of looking for workers whose answers deviate widely from the average of all workers. This approach is based on the tacit assumption that spammers display outlier behavior, and conversely that outliers are probably spammers. In Ribeiro *et al.* (2011), MOS values are considered as

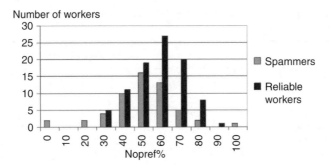

Figure 7.6   Counts (*y*-axis) of workers who had a **preference for the second sample** *x*% (rounded to nearest multiple of 10) of the time that they had a preference at all, for workers who had a preference in at least 25 HITs and (a) never played samples (spammers; 55 workers) or (b) played everything, did at least four gold and passed all gold (reliable; 91 workers; raising the threshold to five gold resulted in too few data points).

standard numeric values and used to compute the sample correlation coefficient. Since the answers of a preference test are not numeric, the **disagreement** metric was used instead. This is the deviation in system preference for a worker computed as the mean distance between the worker's answers and the answers of all the other workers for the same stimuli pair. The distance between two answers was defined as 0 if the answers were identical, 1 if one of the answers was "No preference" and 2 if different systems were preferred. Figure 7.7 shows that the two worker groups have different distributions for this value. It was found that no reliable worker in this subset had a disagreement higher than 1.03.

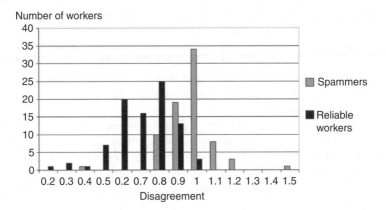

Figure 7.7   Counts (*y*-axis) of workers who had a **disagreement** of *x* (rounded to nearest multiple of 0.1) between the worker's answer and everybody else's answer for those sample pairs, for workers who did at least 25 HITs and (a) never played samples (spammers; 76 workers) or (b) played everything, did at least five gold and passed all gold (reliable; 88 workers).

Table 7.5 Effect of spammer exclusion metrics. "+" means named metric plus exclusion of "Not played" answers.

Exclude	% Answers excluded	Average p-score
None	0.0	0.137
"Not played" answers	6.4	0.131
+ Fail gold	14.6	0.133
+ Preference for the second sample < 30 or > 90	14.7	0.126
+ Disagreement > 1.03	10.6	0.124
+ Both above	18.5	0.122
+ Random	14.5	0.134

Spammer Exclusion Metrics and Their Effects

The previous section presented several spammer detection mechanisms and metrics. Next, the effect on the test outcome when they are used to exclude answers/workers is investigated. Spammers can be considered as noise, whose main effect is to "dilute" the preference of the test outcome. In other words, they increase the probability of obtaining the test results by chance, that is, higher p-score. Therefore, it can be expected that excluding spammers results in lower average p-scores in the outcome. The first row of Table 7.5 shows the simplest case, in which no answers or workers were excluded. The second row shows the effect of excluding answers for which the worker did not play both samples. This alone already excludes 6.4% of the answers and, as expected, decreases the average p-score. The remainder of the table reports the effect of combining "Not played" answer exclusion and other additional metrics. When all workers who fail at least one gold question are excluded, the mean p-score increases. This seems to indicate that this metric is not helpful in excluding spammers, either because the percentage of gold question per test-group, usually less than 10%, was too small to be effective; or because it excludes all the answers of any reliable workers who might have failed one gold question by mistake; or because spammers who played the samples were still able to identify them with a minimum or attention. The next two rows show results for the two new metrics developed in the previous section: preference for the second sample and disagreement. Both yield lower mean p-scores than just excluding the "Not played" answers. When applied together it results in a further boost, showing that they complement each other. As a sanity check, the last row of the table shows that simply excluding a percentage of workers at random does result in a larger mean p-score.

From the comparison of known spammers and reliable workers, it should be clear that these metrics can only hope to exclude *some* spammers and that there is a remaining risk of excluding genuine workers. Therefore, these metrics should not be used for withholding payment for workers, but only for *post hoc* exclusion of workers from results (see also Chapter 11 about the ethics of crowdsourcing). In addition, the metrics can be used for monitoring changes in spammers prevalence.

Of the two metrics, the one based on disagreement gives a slightly lower mean p-score while excluding far fewer answers. However, it also risks biasing the results by excluding genuine workers who happen to have a preference different from the majority. The metric based on the preference for the second sample introduces no such bias but it might exclude some workers simply because they completed too few judgments and therefore the percentages are skewed, in the extreme case, 0% or 100% if they have only completed one judgment. In later experiments, this metric is only applied to workers who have completed a minimum of 10 judgments.

The metrics can be used after the experiment to exclude spammers' answers from the results. However, they do not directly help in preventing spamming from happening in the first place. For this purpose, the mandatory playback interface described in Section 7.3.2 was developed. This new interface prevented the spamming method of submitting the answer without listening, by only showing the answer radio buttons once both samples had been played. The new interface was used in listening tests from mid-2011 onward. At the same time, 25 clear spammers identified by the analyses presented earlier were blocked from further tests through a new functionality added by CrowdFlower around that time.

7.5.2   Investigating the Metrics Further: Mandatory Playback Interface

The values of the thresholds for disagreement and preference for the second sample were set based on analyses performed on preference tests with American English-speaking MTurk workers spanning from June 2010 to March 2011 using the optional playback interface. Therefore, it is interesting to verify whether the same settings still work at a later time with the mandatory playback interface, and for different worker populations. The questions are:

- Does application of the antispamming filters defined for the optional playback interface still result in improved test outcomes, that is, lower p-scores, with the mandatory playback interface? And for other worker populations?
- Would different thresholds in the filters result in even lower p-scores, that is, were the chosen thresholds optimal?
- Is there any explanation for these thresholds?
- Does spamming increase over time? Do spammers return?

New Mandatory Playback Interface, and Other Languages

Table 7.6 shows the results obtained for different languages and MTurk worker populations after the interface change. For US MTurk workers, the proposed metrics still have the desired effect of lowering the mean p-score, and in general, the proposed filters produce an improvement in the average p-score as compared with not applying any filter. The first point to be noted is that the filter based on the preference for the second sample is less effective. One reason for this is that in order to avoid this filter from excluding workers who simply submitted too few HITs, it is now applied only to those workers who completed a minimum of 10 judgments. It might be argued that with the exception of German and maybe UK workers, the number of tests per language is not enough to draw any conclusion. Nevertheless, it seems obvious that the thresholds originally proposed to exclude a worker, that is, disagreement > 1.03 and pref. 2nd< 30 or > 90, are not optimal in the sense of minimizing the average p-score.

Table 7.6 Test outcomes for different workers population and spamming exclusion after changing to the mandatory playback interface.

Language, Worker location	Exclude	Answers excluded	Average p-score
American English, US	None	0.0	0.152
	Disagreement >1.03	10.4	0.145
	Pref. 2nd$<$ 30 or $>$ 90	6.4	0.146
	Both above	16.1	0.144
British English, UK	One	0.0	0.176
	Disagreement >1.03	3.8	0.170
	Pref. 2nd$<$ 30 or $>$ 90	3.2	0.178
	Both above	6.7	0.174
German, Germany	None	0.0	0.141
	Disagreement >1.03	3.6	0.135
	Pref. 2nd$<$ 30 or $>$ 90	3.6	0.140
	Both above	6.7	0.138
Spanish, Spain	None	0.0	0.022
	Disagreement >1.03	3.9	0.013
	Pref. 2nd$<$ 30 or $>$ 90	6.6	0.031
	Both above	10.4	0.020
French, France	None	0.0	0.070
	Disagreement >1.03	4.1	0.064
	Pref. 2nd$<$ 30 or $>$ 90	4.8	0.075
	Both above	8.1	0.071

Finding the Antispamming Settings That Minimize the p-Score

In order to find out the settings that would minimize the p-score, an extensive search was conducted with pref. 2nd thresholds between 0 and 30, and between 70 and 100 for the lower and upper limits, respectively, and disagreement thresholds from 0.85 to 1.2 in steps of 0.05. Table 7.7 shows the ten combinations with the lowest average p-score for US English-speaking workers with the mandatory playback interface. Figures 7.8, 7.9, and 7.10 show the same results graphically for different values around the global optimum. It can be seen that there is a clear valley when excluding workers with disagreement > 0.95. The second most relevant factor is the upper pref. 2nd threshold, but only when used to exclude workers with extreme values (pref. 2nd$>$ 95). The variation of the average p-score with respect to the lower pref. 2nd threshold seems to be less important.

The p-score optimization analysis was also conducted for other languages. In general, the results showed that the optimum threshold and the percentage of excluded workers based on it are similar to those found for English and US-based MTurkers, though in each case some extra-optimization was required.

Understanding the Results

It is not surprising that the values that optimize the average p-score differ from those suggested previously, as those were not obtained by optimizing the p-score—actually the minimum

Table 7.7 Top ten threshold combinations in terms of the average p-score for English-speaking MTurk workers in the United States with the mandatory playback interface.

Disagreement	Pref. 2nd Lower	Pref. 2nd Upper	Average p-score	Excluded answers (%)	Excluded	Workers always excluded (%)	Always accepted
0.95	15	95	0.131	36.4	37.4	19.6	50.3
0.95	25	100	0.131	36.5	37.4	19.6	50.4
0.95	25	95	0.131	37.3	38.2	20.2	49.4
0.95	20	100	0.132	36.0	37.0	19.2	50.8
0.95	15	100	0.132	35.5	36.7	19.0	51.4
0.95	20	95	0.132	36.8	37.8	19.8	49.8
0.95	5	95	0.132	36.1	37.2	19.5	50.4
0.95	10	95	0.132	36.2	37.3	19.5	50.4
0.95	0	95	0.133	35.5	36.8	19.2	50.8
0.95	15	90	0.133	36.7	37.7	19.8	50.0

p-score would have been obtained with disagreement ≤ 1.0 and pref. 2nd < 10 or > 90—but with the criterion of excluding as many spammers and as few reliable workers as possible. For that reason, those thresholds were based on the distributions of the proposed metrics for "reliable" workers and clear spammers. For the tests using the optional playback interface, these distributions were relatively simple to obtain because workers who never listened to the given samples were easily identifiable as spammers. The mandatory playback interface eliminated this easy way of cheating but it also made it impossible to obtain the distributions for "obvious" spammers. Still, it is possible to obtain an approximation to it based on the distributions of

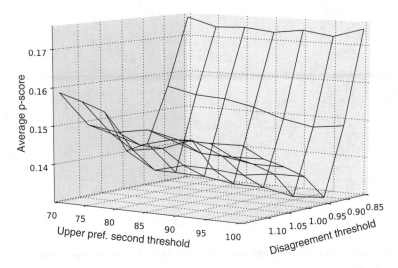

Figure 7.8 Average p-score for different disagreement and upper pref. 2nd thresholds with a fixed lower pref. 2nd threshold of 15%.

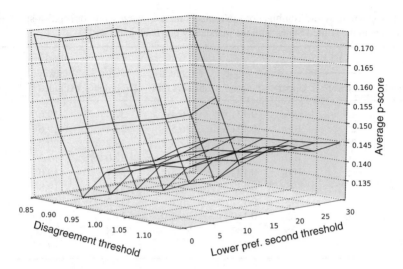

Figure 7.9 Average *p*-score for different disagreement and lower pref. 2nd thresholds with a fixed upper pref. 2nd threshold of 95%.

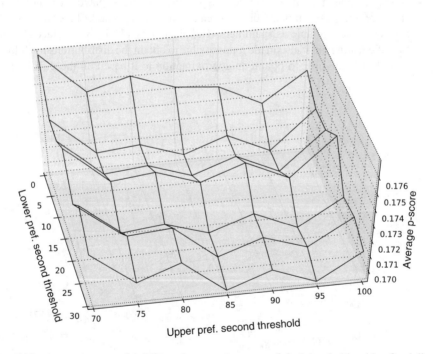

Figure 7.10 Average *p*-score for different lower and upper pref. 2nd thresholds with a fixed disagreement threshold of 0.95.

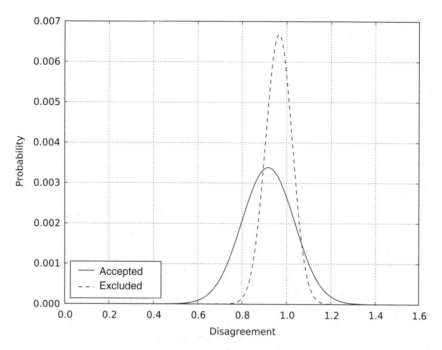

Figure 7.11   Gaussian approximation of the probability distribution of the system disagreement for accepted and excluded workers based on a pref.2nd threshold of < 15 or > 95 for those workers who evaluated a minimum of 25 samples.

accepted and excluded workers. To run this analysis, first one of the metrics is used to classify workers into two groups: accepted and excluded. Then the distributions for the other metric can be computed based on those groups. To analyze the disagreement, the pref. 2nd filter was first set to the optimized pref. 2nd thresholds obtained in the previous test; that is, workers were accepted if $15 \leq$ pref. 2nd ≤ 95. With these thresholds, workers who had evaluated a minimum of 25 samples were classified into "accepted" and "excluded." Figure 7.11 shows the Gaussian approximations of the probability distributions of both groups. Two interesting points are the crossover between both distributions for which the probability of the "excluded" workers equals that of the "accepted" ones. These points can be obtained analytically as the solutions of

$$x^2 \left(\frac{1}{\sigma_A^2} - \frac{1}{\sigma_R^2} \right) - 2x \left(\frac{\mu_A}{\sigma_A^2} - \frac{\mu_R}{\sigma_R^2} \right) + \left(\frac{\mu_A^2}{\sigma_A^2} - \frac{\mu_R^2}{\sigma_R^2} \right) + \log \left(\sigma_A^2 \right) - \log \left(\sigma_R^2 \right) = 0, \quad (7.14)$$

where μ_A and σ_A are the mean and standard deviation of the accepted workers distribution and μ_R and σ_R those of the excluded ones. For the system disagreement distributions, the interesting one is the right-side crossover, which corresponds to the greater of these solutions. Figure 7.12 shows the error functions for false exclusion and false acceptance associated with the above Gaussian approximations. In this case, the crossover point corresponds to the value for which the probability of false exclusion equals the probability of false acceptance.

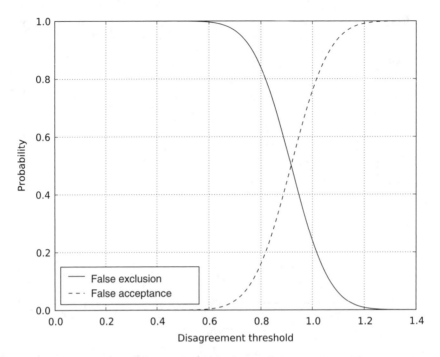

Figure 7.12 Gaussian approximation of the false exclusion and false acceptance probabilities for accepted and excluded workers based on the a pref. 2nd threshold of < 15 or > 95 for those workers who evaluated a minimum of 25 samples.

Still, nothing guarantees that the 15–95 thresholds for the pref. 2nd are the best to identify spammers. Therefore, the same type of analysis was performed for lower pref. 2nd thresholds between 0 and 30, and upper thresholds between 70 and 100. Figure 7.13 shows the variation of the average disagreement for both groups and the crossover points of the corresponding distributions and error functions with respect to various upper and lower thresholds for pref. 2nd. The mean and median of the distribution crossover are 1.036 and 1.032, respectively, which are roughly the values recommended in Buchholz and Latorre (2011). This seems to indicate that the real distributions of the disagreement for reliable workers and spammers with the mandatory playback interface are not so different from those obtained with the optional playback one. The mean and median of the error crossover are 0.94 and 0.942, respectively, which roughly agree with the disagreement threshold that was found to optimize the p-score. In other words, the disagreement threshold that optimizes the p-score is the one that minimized the workers' classification error. It should be noted that although a lower disagreement threshold might improve the p-score, this would be at the expense of excluding not just spammers but also "reliable" workers who disagree with the majority opinion. Figure 7.14 shows the estimated probability of excluding reliable workers for different disagreement thresholds. As expected, for the p-score "optimum" disagreement threshold of 0.95, the probability of excluding reliable workers is almost 40% regardless of the pref. 2nd thresholds. A similar analysis was carried out for the preference for the second sample, this time with workers classified into "excluded"

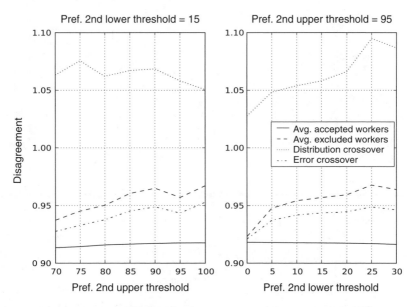

Figure 7.13 Average disagreement for accepted and excluded workers for a given pref. 2nd threshold for workers who completed at least 25 HITs.

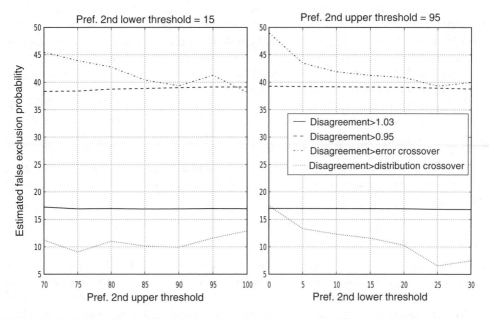

Figure 7.14 Estimated probability of excluding a reliable worker for different thresholds of disagreement and the distributions obtained by filtering workers based on the pref. 2nd metric.

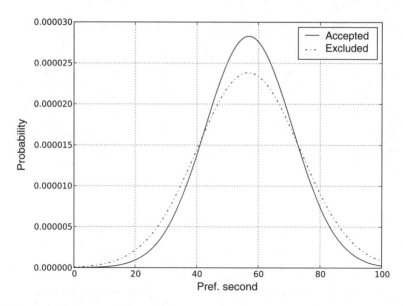

Figure 7.15 Gaussian approximation of the preference for the second sample for workers accepted and excluded based on a disagreement threshold of 0.95 for workers who evaluated a minimum of 25 samples.

or "accepted" according to their disagreement. Figure 7.15 shows the histogram for the 0.95 threshold. Except for very low pref. 2nd values, the distributions of accepted and excluded workers overlap. An analysis over different thresholds is presented in Figure 7.16 also with the upper and lower crossover points of the distributions. The average pref. 2nd is almost the same for both groups of workers regardless of the disagreement threshold and only the variances are different. In this case, the relationship between the data and the optimum pref. 2nd threshold is not so clear. Nonetheless, some insight can be obtained from the analysis of the crossover points of the distributions. The maximum upper crossover of the analyzed thresholds is at 92.14 and the minimum lower crossover is at 27.07, which more or less agree with the thresholds obtained with the optional playback interface. No possible explanation was found for the optimum lower threshold at 15 although this threshold is the less relevant parameter with respect to the p-score, as was shown in Table 7.7 and Figures 7.9 and 7.10.

Spamming Evolution Over Time

An important question is whether with the mandatory playback interface the number of spammers increases over time, as it did with the optional playback one (see Figure 7.4). An approximate answer to this question can be obtained by analyzing the evolution over time of the average disagreement and the preference bias. As Figure 7.17 shows, although these values fluctuate wildly, there does not seem to be any clear tendency.

As mentioned in Section 7.5.1, workers who are excluded are not automatically blocked from participation in other tests. Therefore, it is interesting to know whether those workers

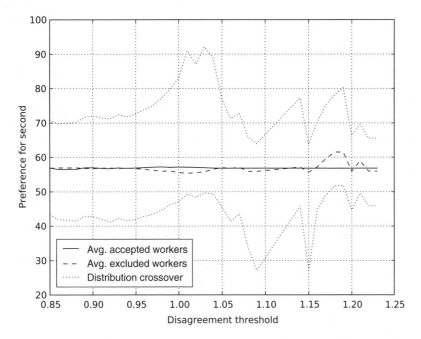

Figure 7.16 Average pref. 2nd and crossovers of the distributions of accepted and excluded workers classified based on the disagreement threshold.

who participate in more than one test were more likely to be or become spammers. Again, this question has to be answered by analyzing the tendencies of the average order bias and the disagreement with respect to the number of test groups on which workers participate. Figure 7.18 shows that the disagreement increases slightly for those who took part in more than 60 tests, but otherwise it remains largely constant. The order bias, on the other hand, has a clear tendency to decrease with the number of tests. The conclusion is that the more tests workers participate in, the more likely they are to be accepted. This can happen because they become more reliable workers, or because they become more efficient spammers. However, the probability of workers participating in more than one test follows the Zipf's law. Therefore, the more "experienced" a worker, the lower the probability to "recruit" them and the more negligible their expected impact on the test results.

Summary

The goal of the above analyses was to answer the questions posed at the beginning of the section. In that sense, it showed that the metrics proposed in Buchholz and Latorre (2011) do improve the general results of crowdsourced preference tests, not just for English-speaking MTurkers in the United States but also for other worker population. It also showed that the distributions of the metrics for spammers and reliable US workers do not seem to have changed much with the introduction of the mandatory playback interface. This means that the general behavior of the most reliable workers and the most obvious spammers is roughly the same. The analysis also showed that if the goal is to minimize the p-score, the thresholds have to

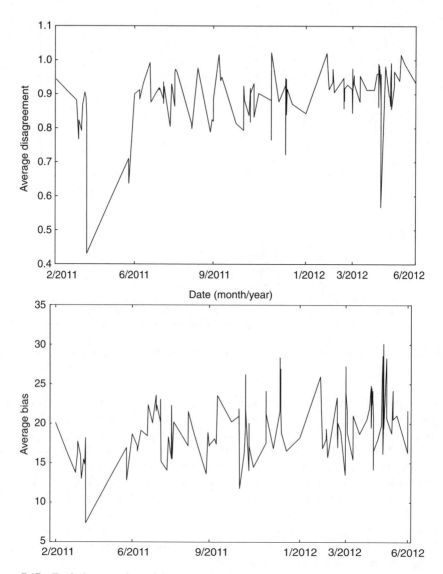

Figure 7.17   Evolution over time of the average disagreement and mean order bias($|$pref. 2nd $- 50|$).

be set much more aggressively. However, such "optimal" values would eliminate around half of the spammers at the cost of eliminating around 40% of "reliable" workers with an opinion different from the majority.

7.5.3   The Prosecutor's Fallacy

The goal of antispamming filters is to detect and exclude spammers. With the techniques described above this means excluding those workers for which the probability of observing

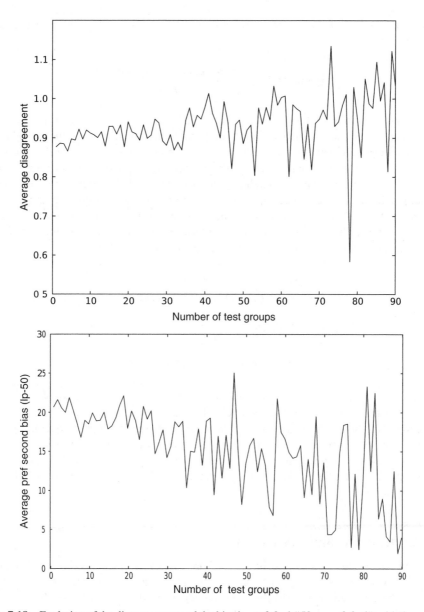

Figure 7.18 Evolution of the disagreement and the bias in pref. 2nd ($|50 - \text{pref. 2nd}|$) with the number of test groups in which workers participated for those workers who participated in more than 1 test group.

certain metrics $\bar{m}$ among spammers is high. In other words, exclude those workers for which the observation probability $P(\bar{m}|\text{Spammer}) > \text{Threshold}$. Although the experiments show that this approach works, it is nonetheless based on a flawed statistical reasoning commonly known as the "prosecutor's fallacy." The classification of a worker as a spammer should not be based on the generative model of the observation probability $P(\bar{m}|\text{Spammer})$ but on the discriminative model of the posterior probability of that worker being a spammer given the metrics, which is

$$P(\text{Spammer}|\bar{m}) = \frac{P(\bar{m}|\text{Spammer})P(\text{Spammer})}{P(\bar{m})}. \tag{7.15}$$

The main problem in using this discriminative model is how to estimate the prior probability $P(\text{Spammer})$. There are two main reasons that make this difficult. First, the probability of having spammers fluctuates continuously as can be seen in Figures 7.4 and 7.17 so that only a prediction of that prior would be available. Second, this approach requires a training set in which all workers are labeled as reliable or spammers, which is not easy to obtain. The optional playback interface made it possible to identify the most obvious spammers and to some extent the most reliable workers, but the majority of workers fell somewhere between these two categories. The mandatory playback interface does not even enable the identification of those clear cases. If the prior is assumed to be $P(\text{Spammer}) = 0.5$,

$$\frac{P(\text{Spammer}|\bar{m})}{P(\text{ReliableWorker}|\bar{m})} = \frac{P(\bar{m}|\text{Spammer})}{P(\bar{m}|\text{ReliableWorker})}. \tag{7.16}$$

Therefore, a worker would be excluded if

$$P(\bar{m}|\text{Spammer}) \geq P(\bar{m}|\text{ReliableWorker}). \tag{7.17}$$

Assuming that half of the workers are spammers is probably excessive. A possible alternative would be to use the spammer priors obtained with the optional playback interface. However, the cheating "methods" that can be applied with each interface are very different. Therefore, the prevalence of spammers is likely to be different as well. So far no proper method to estimate the spammers prior has been proposed; the best practices would consist of using different priors to filter the workers and reporting those values together with the results of the experiments.

7.6   Conclusions and Discussion

This chapter presented an overview of the standard assessment methods for speech synthesis (intelligibility, (D)MOS, $AB(X)$) and showed how all of them can be crowdsourced. It also described the problems one is likely to encounter when doing so and described possible solutions. One such problem concerns delivering audio in WAV format over the web, and ensuring that the audio is played by the workers. Two different interfaces for this were introduced and their implications for quality control described. Another problem concerns statistical significance tests. It was shown how these need to be adjusted to cope with the fact that some data points might be missing. The native language of the listener is important in the assessment of TTS systems, but this is not a factor that is easily controllable. It was shown how CrowdFlower's IP address-based geolocation helps to restrict workers to specific countries and how choosing the right crowdsourcing platform/labor channel can be crucial for languages other than English.

Finally, quality control for crowdsourced TTS assessments was discussed in detail. Prior work on quality control for intelligibility and MOS tests and relevant work from areas other than TTS was summarized. Novel parameterized metrics specifically for crowdsourced AB preference tests were then developed using a honeypot approach of identifying spammers and analyzing their behavior. These metrics were shown to be applicable for a variety of language and interface scenarios, and the parameter space was explored in detail. While the metrics will likely be useful to other researchers wishing to crowdsource their own AB preference tests, the honeypot approach in general should be of broader interest as a method to perform (some) quality control for a subjective task.

The main goal of the proposed techniques is to identify and filter out spammers. However, it would be even better to avoid getting them in the first place. The change of the interface from optional to mandatory playback was one step into that direction. However, workers might still have a greater incentive to cheat or work sloppily than to do the work properly. In that sense antispam filters such as the ones proposed should be better combined with incentives such as the bonus payments described by Ribeiro *et al.* (2011) (however, note that bonuses are currently not supported by CrowdFlower). The analysis about the most appropriate type and quantity of the incentives is left to psychologists and/or experts in game theory. For example, other applications such as image tagging have been successfully crowdsourced by turning them into an enjoyable game. To our knowledge, no such approach has yet been tried for TTS assessment.

As described above, crowdsourcing provides a reliable, efficient and economic alternative to controlled laboratory experiment for standard assessment methods of speech synthesis. As in other areas, such as the social sciences (Paolacci *et al.* 2010), this alternative is quickly becoming more and more common. However, this chapter also showed that more complex assessment types such as diagnostic evaluations still present a challenge. Ultimately, researchers want to know not only *which* TTS system/version is better or worse but also *why*. It will be interesting to see whether future research in crowdsourcing for TTS assessment can solve this problem.

For other aspects of TTS beyond *assessment*, crowdsourcing has not caught on in the same way. As shown at the end of Section 7.3.1, there is some work related to preparing resources for the development of the text normalization (TN) part of TTS. However, there may be potential for a much broader approach, in which a completely data-driven TN component is derived from a huge corpus annotated by the crowd, maybe along the lines of Moore *et al.* (2009, 2010). Likewise, it would be interesting to be able to crowdsource other types of annotations such as those needed for training TTS systems that display appropriate expressiveness. However, as shown in Section 7.3.5 for emotion labeling, this needs further research. Interestingly, there is a fully crowdsourced word-emotion association lexicon (Mohammad and Turney 2011), which might be of use for expressive TTS. There is already work to allow "users with no special training to provide phonemic transcriptions efficiently and accurately" (Ainsley *et al.* 2011). This could in the future pave the way for a fully crowdsourced pronunciation lexicon.

Furthermore, state-of-the-art approaches of data-driven TTS thrive on ever more training data. So far, crowdsourced speech acquisition (see also Chapter 3) has mainly been performed for automatic speech recognition (ASR) purposes, and likewise a huge average-voice model-based TTS has been trained on corpora originally collected for ASR (Yamagishi *et al.* 2010), and while Freitas *et al.* (2010) collect speech by means of a service that offers a personalized TTS (as well as through a quiz game), they do so for the purpose of training an ASR system. However, it would be a small step from there to crowdsourcing a large speech corpus specifically for TTS purposes. Interestingly, a huge crowdsourced speech corpus that can be made suitable for TTS training already exists in the form of the LibriVox (http://librivox.org/; accessed 17 October 2012) recordings. The LibriVox project provides free audiobooks from the public domain, recorded entirely by a crowd of volunteers. One of these books provides the data for the current Blizzard Challenge (King and Karaiskos 2012).

Finally, it should be noted that there are even crowdsourced human alternatives to TTS: Services such as http://voicebunny.com/ (accessed 17 October 2012) turn any text into speech by simply using crowdworkers. However, using such services for real-time applications still belongs in the realm of science fiction (as in Stephenson 1995).

References

Ainsley S, Ha L, Jansche M, Kim A and Nanzawa M (2011) A web-based tool for developing multilingual pronunciation lexicons. *12th Annual Conference of the International Speech Communication Association (Interspeech)*, pp. 3331–3332.

Artstein R and Poesio M (2008) Inter-coder agreement for computational linguistics. *Computational Linguistics* **34**(4), 555–596.

Bech S and Zacharov N (2006) *Perceptual Audio Evaluation: Theory, Method and Application*. John Wiley & Sons, Inc.

Bennett CL (2005) Large scale evaluation of corpus-based synthesizers: results and lessons from the Blizzard Challenge 2005. *Proceedings of Interspeech —Eurospeech*.

Bennett CL and Black AW (2006) Blizzard Challenge 2006: results in *Proceedings of Blizzard Challenge*.

Benoit C and Grice M (1996) The SUS test: a method for the assessment of text-to-speech intelligibility using semantically unpredictable sentences. *Speech Communication* **18**, 381–392.

Black AW and Tokuda K (2005) The Blizzard Challenge—2005: evaluating corpus-based speech synthesis on common datasets. *Proceedings of Interspeech—Eurospeech*.

Black AW, Bunnell HT, Dou Y, Muthukumar PK, Metze F, Perry D, Polzehl T, Prahallad K, Steidl S and Vaughn C (2012) Articulatory features for expressive speech synthesis. *Proceedings of ICASSP*. IEEE.

Buchholz S and Latorre J (2011) Crowdsourcing preference tests, and how to detect cheating. *Twelfth Annual Conference of the International Speech Communication Association*.

Chen K, Wu C, Chang Y and Lei C (2009) A crowdsourceable QoE evaluation framework for multimedia content. *Proceedings of the 17th ACM International Conference on Multimedia (MM '09)*, pp. 491–500, ACM.

Chen L, Gales M, Wan V, Latorre J and M. A 2012 Exploring rich expressive information from E-book data using cluster adaptive training. *Proceedings of Interspeech*.

Clark RAJ, Podsiadło M, Fraser M, Mayo C and King S (2007) Statistical analysis of the Blizzard Challenge 2007 listening test results *Proceedings of Blizzard Challenge*.

CrowdFlower 2012 http://crowdflower.com/docs/gold#auto.

Eyben F, Buchholz S, Braunschweiler N, Latorre J, Wan V, Gales MJF and Knill K (2012) Unsupervised clustering of emotion and voice styles for expressive TTS. *Proceedings of ICASSP*. IEEE.

Fedorov A (2012) https://github.com/francois2metz/wavplayer (accessed 5 October 2012).

Fraser M and King S (2007) The Blizzard Challenge 2007. *Proceedings of Blizzard Challenge*.

Freitas J, Calado A, Braga D, Silva P and Sales Dias M (2010) Crowd-sourcing platform for large-scale speech data collection. *FALA 2010 – VI Jornadas en Tecnologa del Habla and II Iberian SLTech Workshop*.

Hashimoto K, Yamagishi J, Byrne W, King S and Tokuda K (2011) An analysis of machine translation and speech synthesis in speech-to-speech translation system. *Acoustics, Speech and Signal Processing (ICASSP), 2011 IEEE International Conference on*, pp. 5108–5111, IEEE.

Hinterleitner F, Möller S, Falk T and Polzeh T (2010) Comparison of approaches for instrumentally predicting the quality of text-to-speech systems: data from Blizzard Challenge 2008 and 2009. *Proceedings of Blizzard Challenge*.

Hinterleitner F, Neitzel G, Möller S and Norrenbrock C (2011) An evaluation protocol for the subjective assessment of text-to-speech in audiobook reading tasks. *Proceedings of Blizzard Challenge*.

Hirth M, Hoßfeld T and Tran-Gia P (2010) Cheat-detection mechanisms for crowdsourcing. *Technical Report*, University of Würzburg, Germany.

House AS, Williams CE, Hecker MHL and Kryter KD (1963) Psychoacoustic speech tests: a modified rhyme test. *Tech. Doc. Rept. ESD-TDR-63-403*, US Air Force Systems Command, Hanscom Field, Electronics Systems Division.

Ipeirotis P, Provost F and Wang J (2010) Quality management on Amazon Mechanical Turk. *Proceedings of the ACM SIGKDD Workshop on Human Computation*, pp. 64–67, ACM.

Ipeirotis PG 2010 Demographics of Mechanical Turk. *CeDER Working Papers*.

Irvine A and Klementiev A (2010) Using Mechanical Turk to annotate lexicons for less commonly used languages. *Proceedings of the NAACL HLT 2010 Workshop on Creating Speech and Language Data with Amazon's Mechanical Turk*, ACL.

ITU (1988a) *Absolute Category Rating (ACR) Method for Subjective Testing of Digital Processes* International Telecommunication Union. Blue Book, Volume V, Annex A to Suppl. 14.

ITU (1988b) *Subjective Perfomance Assessment of Digital Encoders Using the Degradation Category Rating Procedure (DCR)*, Blue Book, Volume V, Annex B to Suppl. 14. International Telecommunication Union.

ITU (1994) *A Method for Subjective Performance Assessment of the Quality of Speech Voice Output Devices,* ITU-T Recommendation P.85. International Telecommunication Union.

Karaiskos V, King S, Clark RAJ and Mayo C (2008) The Blizzard Challenge 2008. *Proceedings of Blizzard Challenge*.

King S and Karaiskos V (2009) The Blizzard Challenge 2009. *Proceedings of Blizzard Challenge*.

King S and Karaiskos V (2010) The Blizzard Challenge 2010. *Proceedings of Blizzard Challenge*.

King S and Karaiskos V (2012) The Blizzard Challenge 2012. *Proceedings of Blizzard Challenge*.

Krstulović S, Buchholz S and Latorre J (2008) Comparing QMT1 and HMMs for the synthesis of American English prosody. *Proceedings of Speech Prosody*.

Latorre J, Gales MJ, Buchholz S, Knill K, Tamura M, Ohtani Y and Akamine M (2011) Continuous f0 in the source-excitation generation for HMM-based TTS: do we need voiced/unvoiced classification? *Proceedings of ICASSP*. IEEE.

Latorre J, Wan V, Gales M, Chen L, Chin K, Knill K and M. A 2012 Speech factorization for HMM-TTS based on cluster adaptive training. *Proceedings of Interspeech*.

Likert R 1932 A technique for the measurement of attitudes. *Archives of Psychology* **140**, 1–55.

Liu F, Weng F, Wang B and Liu Y (2011) Insertion, deletion, or substitution? Normalizing text messages without pre-categorization nor supervision. *Proceedings of the 49th Annual Meeting of the Association for Computational Linguistics: Human Language Technologies*, pp. 71–76. ACL.

Mohammad SM and Turney PD (2011) Crowdsourcing a word-emotion association lexicon. *Computational Intelligence*.

Moore S, Korhonen A and Buchholz S (2009) Number sense disambiguation. *Proceedings of the 12th Conference of the Pacific Association for Computational Linguistics*.

Moore S, Korhonen A and Buchholz S (2010) Annotating the enron email corpus with number senses. *Proceedings of the Seventh Conference on International Language Resources and Evaluation (LREC)*.

MTurk (2012a) https://www.mturk.com/mturk/help?helppage=requester#do_support_outside_us (accessed 8 October 2012).

MTurk (2012b) https://www.mturk.com/mturk/help?helppage=worker#how_paid (accessed 8 October 2012).

MTurk (2012c) https://www.mturk.com/mturk/help?helppage=worker#intl_tax_us_citizen_live_abroad (accessed 8 October 2012).

Novotney S and Callison-Burch C (2010) Cheap, fast and good enough: Automatic speech recognition with non-expert transcription. *Proceedings of the NAACL HLT 2010 Workshop on Creating Speech and Language Data with Amazon's Mechanical Turk*. ACL.

Paolacci G, Chandler J and Ipeirotis PG (2010) Running experiments on Amazon Mechanical Turk. *Judgment and Decision Making*.

Parlikar A and Black AW (2012) Data-driven phrasing for speech synthesis in low-resource languages. *Proceedings of ICASSP*. IEEE.

Pennell DL and Liu Y (2010) Normalization of text messages for text-to-speech. *Proceedings of ICASSP*. IEEE.

Ribeiro F, Florêncio D, Zhang C and Seltzer M (2011) CrowdMOS: an approach for crowdsourcing mean opinion score studies. *Acoustics, Speech and Signal Processing (ICASSP), 2011 IEEE International Conference on*, pp. 2416–2419. IEEE.

Rzeszotarski J and Kittur A (2011) Instrumenting the crowd: using implicit behavioral measures to predict task performance. *Proceedings of the 24th Annual ACM Symposium on User Interface Software and Technology*, pp. 13–22. ACM.

Sheng V, Provost F and Ipeirotis P (2008) Get another label? improving data quality and data mining using multiple, noisy labelers. *Proceeding of the 14th ACM SIGKDD international conference on Knowledge discovery and data mining*, pp. 614–622. ACM.

Sityaev D, Knill K and Burrows T (2006) Comparison of the ITU-T P.85 standard to other methods for the evaluation of text-to-speech systems. *Proceedings of INTERSPEECH—ICSLP*.

Snow R, O'Connor B, Jurafsky D and Ng A (2008) Cheap and fast—but is it good?: evaluating non-expert annotations for natural language tasks. *Proceedings of the Conference on Empirical Methods in Natural Language Processing*, pp. 254–263. Association for Computational Linguistics.

Spitzner L (2002) *Honeypots: Tracking Hackers*. Addison Wesley.

Stephenson N (1995) *The Diamond Age: Or, A Young Ladys Illustrated Primer*. Penguin Books.

Taylor P (2009) *Text-to-Speech Synthesis*. Cambridge University Press.

Vuurens J, de Vries A and Eickhoff C (2011) How much spam can you take? an analysis of crowdsourcing results to increase accuracy. *Proceedings ACM SIGIR Workshop on Crowdsourcing for Information Retrieval (CIR11)*, pp. 21–26.

Wan V, Latorre J, Chin K, Chen, L. Gales MJF, Zen H, Knill K and Akamine M. (2012) Combining multiple high quality corpora for improving HMM-TTS. *Proceedings of Interspeech*.

Watts O, Yamagishi J and King S (2010) Letter-based speech synthesis. *Proceedings of 7th ISCA Speech Synthesis Workshop (SSW7)*. ISCA.

Wherry R (1938) Orders for the presentation of pairs in the method of paired comparison. *Journal of Experimental Psychology* 23, 651–660.

Wolters M, Isaac K and Renals S (2010) Evaluating speech synthesis intelligibility using Amazon Mechanical Turk.

Yamagishi J, Usabaev B, King S, Watts O, Dines J, Tian J, Guan Y, Hu R, Oura K, Wu YJ, Tokuda K, Karhila R and Kurimo M (2010) Thousands of voices for HMM-based speech synthesis-analysis and application of TTS systems built on various ASR corpora. *IEEE Transactions on Audio Speech and Language Processing* 18(5), 984–1004.

Zen H and Gales MJF (2011) Decision tree-based context clustering based on cross validation and hierarchical priors. *Proceedings of ICASSP*. IEEE.

Zen H, Braunschweiler N, Buchholz S, Gales M, Knill K, Krstulović S and Latorre J (2012) Statistical parametric speech synthesis based on speaker and language factorization. *IEEE Transaction on Audio, Speech and Language Processing* 20(6), 1713–1724.

8

Crowdsourcing for Spoken Dialog System Evaluation

Zhaojun Yang[1], Gina-Anne Levow[2] and Helen Meng[3]
[1] University of Southern California, USA
[2] University of Washington, USA
[3] The Chinese University of Hong Kong, China

8.1 Introduction

A spoken dialog system (SDS) is a computer system that supports human-computer conversations in specific knowledge domains. It integrates technologies including speech recognition, natural language understanding, dialog modeling, language generation, and text-to-speech (TTS) synthesis. Advances in speech and language technologies have made SDS an important research area and have brought about systems in a wide variety of applications, such as flight information (Hirschman *et al.* 1993), bus schedule inquiries (Raux *et al.* 2005), stock market information delivery (Meng *et al.* 2004a), tourist information (Wu *et al.* 2006) and student tutoring (Litman *et al.* 2006). To facilitate the development of SDS and compare the performance of different systems, it is necessary to conduct SDS evaluation with appropriate methodologies.

A typical SDS architecture is illustrated in Figure 8.1. This implements a dialog interaction between two interlocutors, which consists of a series of dialog turns. A spoken dialog turn is a process in which one participant *A* utters something to the other *B*, and *B* interprets *A*'s utterance and then responds accordingly.

As illustrated in Figure 8.1, the first step for the system is to recognize the user's speech with automatic speech recognition (ASR) and interpret the underlying meaning with natural language understanding technologies. This involves extracting the user's communicative (and informational) goal and inferring the appropriate follow-up actions and responses. Language understanding involves a variety of methods, such as the use of parsers and grammars (Seneff

Crowdsourcing for Speech Processing: Applications to Data Collection, Transcription and Assessment, First Edition.
Edited by Maxine Eskénazi, Gina-Anne Levow, Helen Meng, Gabriel Parent and David Suendermann.
© 2013 John Wiley & Sons, Ltd. Published 2013 by John Wiley & Sons, Ltd.

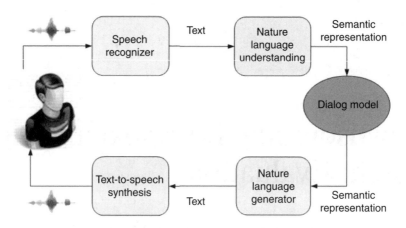

Figure 8.1   A typical spoken dialog turn between the system and the user.

1992; Ward and Issar 1994), belief networks (Meng *et al.* 1999, 2004b), and so on. The dialog model is the principal component of a dialog system that maintains the history of the dialog, decides which action is appropriate based on language understanding, and controls the dialog flow. A dialog model typically incorporates dialog states, state transitions, and a dialog policy. A dialog state represents the results of performing system actions in previous states; state transitions allow dialogs to move forward; and the dialog policy determines how to map dialog states into system actions (McTear 1998). After the dialog model decides upon the most appropriate response, the system needs to convey its response to the user in audio form. This is achieved with language generation (Baptist and Seneff 2000; Cole *et al.* 1997) and TTS synthesis technologies (Taylor 2009).

SDSs are becoming increasingly pervasive in supporting information access by the masses. This calls for sound strategies in evaluating, comparing, and predicting the design and performance of such systems. Therefore, developing principled ways of evaluating an SDS has become a hot research area. Generally, SDS evaluation can be categorized into *component-based* and *holistic* approaches.

The component-based approach covers the performance of individual components such as correctness of speech recognition, ability to understand natural language, appropriateness of response generation, as well as the naturalness of the synthetic speech in conveying the responses. A thorough evaluation of an SDS needs to consider all relevant evaluation metrics covering the functionalities for all the system components (Möller 2005). As a result, different kinds of evaluation metrics have been proposed in previous work, such as query density and concept efficiency for measuring the system's understanding ability (Glass *et al.* 2000). The metrics have also been classified according to their functionalities into five categories: *dialog related*, *metacommunication related*, *cooperativity related*, *task related*, and *speech-input related* (Möller 2005). Component-based evaluation may become challenging as the complexities of systems increase and their components become integrated in more intricate ways. This is especially true when we need to compare two systems with different components, or when some metrics are in conflict with one another. For example, we may have two systems *A* and *B*, where *A* uses explicit confirmations of the user's information, and has dialogs with a longer dialog length but a higher task completion rate. *B* uses implicit confirmations, leading to dialogs with a shorter dialog length but a lower task completion rate. In this situation, we may face a dilemma where *A* has better effectiveness but *B* has better efficiency.

In contrast, holistic evaluation, which assesses not only individual components but also the integrated performance of an SDS, may be more appropriate. It involves the perceived level of system usability, system intelligence, and error recovery capabilities by considering the system in its entirety (McTear 2002). It also needs to cover the wide variety of users' impressions (user judgments) relating to all dimensions of the quality of an SDS (Bonneau-Maynard *et al.* 2000). The ultimate objective of an SDS is to satisfy the demands of real users. Therefore, user satisfaction is considered the most important criterion for system performance among all user impressions (Moller and Skowronek 2003).

Many evaluation methods have been developed in recent years. One popular method of measuring user judgments is to invite subjects to fill out a questionnaire after they interact with an SDS. The questionnaire often involves perceptions of many aspects of the system, such as task completion and user satisfaction. In spite of its popularity, this traditional approach has some disadvantages. First, it is a costly and time-consuming process. Moreover, due to limitations in resources, this approach is often constrained to a small number of evaluators whose feedback may not be statistically representative of the large user population that can access the SDS. Furthermore, in some situations where the system has already been deployed, it is often difficult to ask real users to patiently complete an evaluation survey.

Another popular evaluation method, the PARAdigm for DIalog System Evaluation (PARADISE) framework, has been proposed for automatic inference of overall user satisfaction of unrated dialogs (Walker *et al.* 1997). It assumes that the overall performance of an SDS can be described in terms of a linear regression model of a set of dialog metrics, to maximize this usability-related measure (Walker *et al.* 1998). The trained model can explicitly demonstrate which factors have significant contributions to user satisfaction. Nevertheless, such frameworks still need evaluated dialogs to train the predictive model, whose performance may also be determined by the amount of graded data (Engelbrecht and Möller 2007).

The challenges in SDS evaluation motivated us to explore the use of crowdsourcing as a potentially effective and efficient paradigm. We note that crowdsourcing technologies have been widely applied by many researchers to collect (see Chapters 3 and 5), transcribe (see Chapter 4), and annotate speech and language data in recent years (Snow *et al.* 2008; McGraw *et al.* 2010; Novotney and Callison-Burch 2010). Crowdsourcing refers to outsourcing a task to a crowd of people. Unlike using the traditional method in which data is manually labeled by experts or trained people, tasks can be completed with crowdsourcing in a cost-effective, efficient, and flexible manner. We feel that user judgments for SDS evaluation can also be collected by using crowdsourcing instead of user experiments. In this chapter, we describe our work in developing a crowdsourcing methodology for SDS evaluation through Amazon Mechanical Turk (MTurk) (http://www.mturk.com), a popular crowdsourcing marketplace that makes use of human intelligence online to perform tasks which cannot be completed entirely by computer programs.

This chapter is organized as follows. In Section 8.2, we discuss prior work in crowdsourcing for dialog and speech assessment. In Section 8.3, we provide an overview of related work on SDS evaluation. The remainder of the chapter details a specific approach to and analysis of crowdsourcing for SDS evaluation, to serve as an illustrative example. Section 8.4 presents information about the experimental corpus and how automatic dialog classification is done.

Section 8.5 presents the methodology in the use of crowdsourcing for collecting user judgments on SDSs. We design two types of tasks—the first targets rapid rating of a large number of dialogs with regard to some dimensions of the SDS's performance, and the second aims to assess the reliability of workers in crowdsourcing. To address the particular challenges of crowdsourcing, we structure these tasks to elicit judgments on a representative sample of dialogs, support semiautomatic task validation, and ensure task clarity. To control the quality of ratings from crowdsourcing, we also develop and present a set of approval rules. Section 8.6 presents an analysis of collected results, which demonstrates that the crowdsourcing method for the collection of user judgments is efficient, flexible and cost-effective. The comparison of annotations between experts and workers on SDSs shows a high level of agreement between the two groups, which supports the reliability of crowdsourcing. Finally, this chapter concludes with Section 8.7.

8.2   Prior Work on Crowdsourcing: Dialog and Speech Assessment

8.2.1   Prior Work on Crowdsourcing for Dialog Systems

Some recent approaches have explored the use of crowdsourcing in the development of dialog systems. These approaches have aimed to exploit the crowd to create or enhance text-based dialog interactions. Bessho *et al.* (2012) investigated the automatic generation of replies to tweets on the Twitter microblog. Their approach employed a similarity based technique using a database of tweet-reply pairs. For a new input, the most similar tweet in the database is identified and its corresponding reply is generated as the response. If no sufficiently similar tweet is present, the tweet is sent to a "crowd" in real time, and if a reply is provided within a given time window, that crowdsourced reply is used as the system response.

DePalma *et al.* (2011) provide an example of the use of crowdsourcing to develop models of human–robot dialog (HRI). They describe the use of crowdsourced interactions from a game called "Mars Escape" in which players take the roles of an astronaut and a robot doing a collaborative task. Actions and communication via text-based chat are logged. They also planned a real-world variant with a robot in the Boston Museum of Science. These crowd-sourced interactions provided a much richer set of dialog behaviors than are typically present in hand-coded HRI systems.

8.2.2   Prior Work on Crowdsourcing for Speech Assessment

Crowdsourcing has also been employed for the collection of judgments of speech quality, particularly in the field of speech synthesis. Ribeiro *et al.* (2011) describe a framework for collection of mean opinion scores (MOS) through crowdsourcing, applying the basic approach to assessment of synthesized speech, as well as of image quality and region of interest determination. The tasks required workers to listen to speech samples and provide scores from 1 to 5. They streamlined the tasks through interface design, such as the use of radio buttons, and by dropping prequalification tasks. Instead, they developed strategies for quickly rejecting poor or cheating workers, by rejecting incomplete tasks, tasks completed too quickly, and workers with low correlation coefficients. Known good and bad samples were also included to enable automatic validation. Lastly, they aimed to recruit and retain good workers, by providing bonuses for completing more tasks and maintaining high quality, as measured by correlation coefficient. This approach allowed high throughput at a relatively low cost, important since MOS testing relies on large numbers of subjects for statistical power.

Wolters *et al.* (2010) describe the crowdsourcing of speech synthesis assessment through an intelligibility task. Workers are asked to listen to utterances and transcribe them word-for-word. Since the utterance text itself is available, worker input can be evaluated with respect to that gold standard. Workers with too many errors were classified as cheaters.

Lastly, Buchholz and Latorre (2011) present the crowdsourcing of speech synthesis assessment through a preference task. Workers are asked to listen to pairs of utterances and indicate which they prefer. To validate workers' input, they employed several checks: transitivity of ranking as a pseudo-gold standard, timing of audio play relative to preference entry, and distribution of ranks. They also used CrowdFlower's geolocation service as a nativeness filter. They compared crowdsourced assessments to those of locally recruited researchers. They found that despite the presence of some systematic cheaters, the two groups produced similar preference trends, though the preferences were stronger for the workers, possibly due to differences in nativeness or task wording. Additional details on crowdsourcing and speech synthesis can be found in Chapter 7.

8.3 Prior Work in SDS Evaluation

As introduced earlier, evaluation plays a critically important role in the design and development of SDS. The performance of an SDS can be measured in a multitude of ways, such as task success, the number of dialog turns, speech recognition accuracy, system response delay, naturalness of the output speech, consistency with the user's expectations, and cooperativeness of the system (Möller 2005). These evaluation metrics are usually categorized into *subjective* and *objective* metrics. Subjective metrics, which reflect the users' perceptions of the quality of an SDS, are often obtained from real or test users. Objective metrics, which quantify system behavior during the interaction and the performance of the components of an SDS, can be extracted automatically or labeled manually from the user-system interactions by expert evaluators. The objective metrics are also called interaction metrics in Möller (2005).

8.3.1 Subjective User Judgments

Since subjective metrics mostly rely on user judgments of system quality, distributing questionnaires to users before or after interaction with an SDS is an effective way to collect quantifiable user judgments. Developing a reliable and valid questionnaire for subjective judgment collection involves many design considerations. The subjective assessment of speech system interfaces (SASSI) questionnaire is designed for subjective assessment of speech-based systems (Hone and Graham 2001). SASSI consists of 50 items (or statements), and each is rated by users on a 7-point scale: strongly agree, agree, slightly agree, neutral, slightly disagree, disagree, and strongly disagree. A factor analysis of the collected data from 226 completed questionnaires suggests that six main factors contribute to a user's subjective perceptions of speech-based systems, that is, perceived system response accuracy, likeability, cognitive demand, annoyance, habitability, and speed.

The International Telecommunication Union (ITU) recommendation proposed another list of questions for the evaluation of SDS in telephone services (ITU-T Rec. P.851 2003). Three types of questionnaires are distinguished in the recommendation. Type 1 questionnaires are intended to collect the user's background information and are distributed at the beginning of an evaluation experiment. Type 2 questionnaires are related to the user–system interactions. Type 3 questionnaires are about the user's overall impression of system quality. A list of topics are proposed for each type of questionnaire, and the statements are rated on a 5-point scale.

8.3.2   Interaction Metrics

In contrast to subjective judgments of system performance, interaction metrics can easily quantify the ability of a system or its components to perform the designed functions. Such information is obtained from the log files which record the interactions between the system and its users. Certain metrics that provide an overview of user-system interacions can be automatically extracted from log files, such as dialog duration, recognition confidence scores, etc. Other metrics are related to the content of the interactions and are usually manually labeled by experts or trained annotators, such as the accuracy of understanding, task success, etc.

In recent decades, many metrics have been identified to measure the functionalities of a system and its components. Early metrics are for individual components, such as the speech recognizer and language understanding component. Commonly used metrics are *word accuracy (WA), sentence accuracy (SA), concept accuracy (CA), query density (QD), concept efficiency (CE)*, and so on. Later, metrics for whole systems have been developed, including *Task Success (TS)* to measure the extent to which the system achieves the task, *number of dialog turns* for measuring the dialog cost, or *contextual appropriateness (CA)* for measuring the degree to which the system provides an appropriate response (McTear 1998; Glass *et al.* 2000).

Based on the literature of interaction metrics, Möller *et al.* summarize a set of metrics for SDSs evaluation and classify them into five categories:

1. **Dialog- and communication-related category**: Metrics about the overall dialog, such as overall dialog duration, dialog turns, or average number of words per system turn during the dialog.
2. **Metacommunication-related category**: Metrics describing the recognition and understanding capabilities, such as the number of help requests and the number of barge-in attempts from the user.
3. **Cooperativity-related category**: Metrics about the cooperativity of system actions (responses). The contextual appropriateness of system responses directly measures cooperativity, which is often judged by experts based on Grice's maxims.
4. **Task-related category**: Successful task completion is a key requirement for task-oriented systems. Möller defined seven levels of task success: success by providing a completely right answer; success with constraints from users, or from systems, or from both users and systems; success by determining that no solutions exist; failure resulting from the user's noncooperative behavior or the system's inappropriate response.
5. **Speech-input-related category**: Metrics about the capability of systems to recognize the input speech and to understand the meaning of inputs. Commonly used metrics are *WA, SA*, or *CA* as introduced above.

This categorization and the metrics in each category have been incorporated in the ITU recommendation (ITU-T Suppl. 24 to P-Series Rec. 2005).

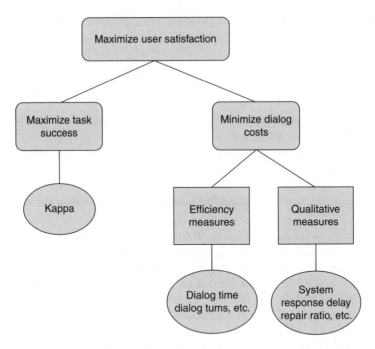

Figure 8.2   The PARADISE structure of objectives for spoken dialog performance (Walker *et al.* 1997).

8.3.3   PARADISE Framework

PARADISE is a general framework for evaluating and comparing the performance of SDSs. It quantifies the contributions of different system properties to system usability and supports the development of predictive models of system performance (Walker *et al.* 1997). The PARADISE framework uses decision-theoretic methods to relate a collection of dialog metrics to the system's overall performance and determine the significant contributors. The PARADISE performance model is shown in Figure 8.2. The overall performance is correlated with user satisfaction, and the primary objective of a system is to maximize user satisfaction. This objective can be further decoupled into two subobjectives: maximizing task success and minimizing dialog cost, based on the assumption that task success and dialog cost are two main types of contributors to user satisfaction. In the original PARADISE framework, task success is measured with the use of the kappa coefficient and attribute-value matrix (AVM). Dialog costs can be categorized into two types: dialog efficiency and quality. Dialog efficiency is represented by the number of dialog turns or the dialog duration, while dialog quality is measured in terms of the appropriateness of system response, system repair ratio, and so on.

The PARADISE framework posits that the objective structure in Figure 8.2 can be realized by building a performance model through multivariate linear regression with user satisfaction as the target and the dialog metrics of task success, dialog efficiency and quality as predictors. Building the performance model requires a dialog corpus be collected through controlled user experiments during which users subjectively rate their satisfaction. Moreover, the predictors of the model, that is, the dialog metrics, can be either automatically extracted from dialog log files or manually labeled by experts. Based on these illustrations, the performance model of an SDS is defined below:

$$S_u = (\alpha * N(\kappa)) - \sum_{i=1}^{n} w_i * N(c_i), \tag{8.1}$$

where S_u is system performance correlated with user satisfaction, κ is a measure for task success, c_i is a measure for dialog cost, α is a weight on κ, w_i is a weight on c_i, and $N(\cdot)$ is a z-score normalization function (Cohen 1995). Both κ and c_i can be represented as dialog measures m, and Equation (8.1) is transformed into a simpler one:

$$S_u = \sum w_i * N(m_i). \tag{8.2}$$

Since the dialog measures have been normalized into the same scale by $N(\cdot)$, the weight w_i reflects the relative contribution of the corresponding measure m_i to user satisfaction.

By applying the performance model, values of user satisfaction for SDSs are directly predicted from a suite of dialog metrics that are simply extracted from dialogs, without the need to conduct user experiments to assess user satisfaction. In addition, system developers can directly figure out which system components have a greater impact on user satisfaction by observing the coefficients of dialog metrics in the performance model, so that they can focus on improving the performance of those "important" components. In this way, an efficient system design process is ensured, that is, focusing on high service quality for the end user.

PARADISE has been widely applied in evaluating many SDSs, such as the ITSPOKE tutoring system (Forbes-Riley and Litman 2006) and the DARPA Communicator (Walker et al. 2001). It is applied to test different dialog models with two systems: ELVIS and TOOT (Walker et al. 1998). The ELVIS experiments tested different initiative dialog models, while the TOOT experiments tested models with different information presentation policies. Regression functions are derived for each model to analyze their performance. Kamm et al. discuss the generalization ability of PARADISE across three different systems (Kamm et al. 1999). Walker et al. study its generalization across different user populations and find that it does not generalize well from novice users to expert users (Walker et al. 2000). Additionally, researchers have extended PARADISE to evaluate multimodal systems (Beringer et al. 2002; Dybkjaer et al. 2004).

8.3.4   Alternative Approach to Crowdsourcing for SDS Evaluation

Aside from the approach presented later in this chapter, Jurcicek et al. (2011) present the only other approach to date for evaluating spoken dialog systems themselves through crowdsourcing. In their case, the task requires workers to interact with an SDS to perform a prespecified task and complete a brief questionnaire focused on task completion, asking if they obtained the information sought. Due to technical difficulties with enabling the SDS interaction through the MTurk interface itself, workers interacted with the SDS over the telephone and were given a code to enter in the crowdsourcing interface as part of task verification. Task performance was validated by comparison with the goals specified in the assigned task and those inferred from the interaction logs. The crowdsourced results were compared with those from locally recruited users across different dialog system models and found to be largely consistent.

Table 8.1   An example of the Let's Go! dialogs.

S: *Welcome to the CMU Let's Go bus information system.*
S: *What can I do for you?*
U: *When is the next 64A from Forbes and Murray to East Liberty?*
S: *The 64A. Did I get that right?*
U: *Yes.*
S: *Leaving from Forbes and Murray. Correct?*
U: *Yes.*
S: *Going to East Liberty. Am I right?*
U: *Yes.*
S: *I think you want the next bus.*
U: *Yes.*
S: *Hold on. I'll look that up.*
S: *The next 64A leaves Murray Avenue at Forbes at 10:32 p.m. and arrives at Highland Avenue at Penn Avenue at 10:55 p.m.*

8.4   Experimental Corpus and Automatic Dialog Classification

The corpus used in this case study contains about 50,000 dialogs drawn from the log files collected through the Let's Go! dialog system developed by Carnegie Mellon University (Raux *et al.* 2005). This system provides bus schedule information for Pittsburgh residents when the customer service hotline is not staffed. Input utterances from users are transcribed automatically by ASR. Table 8.1 shows an example of the Let's Go! dialogs.

We develop a simple characterization of the dialogs in the corpus. We survey the dialog logs and identify several cue phrases that enable us to develop a set of heuristics to automatically classify the dialog logs into five categories in terms of task success. The dialog logs are tracked from the end toward the beginning with a set of ordered rules:

1. Dialogs with no more than three turns are classified as "TooShort."
2. Dialogs with the cue phrases, such as "Start a new query" and "New query," in the user turns are classified as "MultiTask."
3. Dialogs with the cue phrase "... result[1] ..." in the system turns are classified as "TaskComplete."
4. Dialogs with the cue phrase "I'm sorry. ...." in the system turns are classified as "OutofScope."
5. If none of the above rules apply, the dialog is classified as "TaskIncomplete."

Example dialogs in each of these five categories are shown in Table 8.2. In TooShort dialogs, the user usually hung up after only a single utterance, indicating that the user did not really want to use the system. Therefore, such dialogs are separated from the TaskIncomplete ones. The distribution of the classification results is shown in Figure 8.3. About 35% of the dialogs are in the TaskComplete category, while the remaining categories share comparable percentages. As will be described later, this rough classification is used to determine the proportion of the different types of dialogs that are uploaded to MTurk for crowdsourcing. The

---

[1]This cue phrase is observed in the grammar tags of the Let's Go! System's dialog logs.

Table 8.2   Samples of the Let's Go! dialogs in the different categories of automatic classification.

Case 1: Task complete
S: *Welcome to the CMU Let's Go bus information.*
S: *What can I do for you?*
U: *When is the next 64A from Forbes and Murray to East Liberty?*
S: *The 64A. Did I get that right?*
U: *Yes.*
S: *Leaving from Forbes and Murray. Correct?*
U: *Yes.*
S: *Going to East Liberty. Am I right?*
U: *Yes.*
S: *I think you want the next bus.*
U: *Yes.*
S: *Hold on. I'll look that up.*
S: *The next 64A leaves Murray Avenue at Forbes at 10:32 p.m. and arrives at Highland Avenue at Penn Avenue at 10:55 p.m.*

Case 2: TooShort
S: *Welcome to the CMU Let's Go bus information*
S: *What can I do for you?*
U: *OAKLAND . . . ANY. . . FIND. . . SOUTH SIDE. WANT.*

Case 3: MultiTask
. . . . . .
S: *There is a 28X leaving THE AIRPORT. . .*
U: *NEW QUERY*
S: *Okay, let's start from the beginning.*
. . . . . .

Case 4: OutofScope
. . . . . .
S: *I'm sorry, I don't know any routes.*

Case 5: TaskIncomplete
. . . . . .
S: *Leave at 8 a.m. Is this correct?*
U: *YES*
S: *Hold on, I'll look that up.*

rough classification also provides a means by which we can verify whether the human raters on the crowdsourcing platform have put effort into rating our dialogs, as opposed to providing random (and unreliable) inputs. We will elaborate on these points in the following sections.

8.5   Collecting User Judgments on Spoken Dialogs with Crowdsourcing

As introduced in Section 8.3, the subjective user judgments on SDSs are considered the most important criterion for SDS performance. However, user judgments are usually measured by asking subjects to fill out a questionnaire after they interact with an SDS, which is a

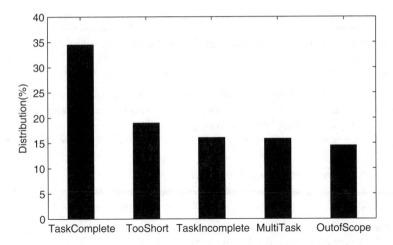

Figure 8.3 Distribution of the five dialog categories based on the automatic, heuristic-based classification.

costly, tedious, and time-consuming process. This section presents an approach to the use of crowdsourcing for collection of user judgments on SDSs (Yang *et al.* 2010). The objective of collecting user judgments with crowdsourcing is to have a large number of dialogs evaluated by a large (hence more statistically representative) group of people in an efficient and cost-effective manner, which is difficult to achieve using traditional methods. In the experiments detailed below, the crowdsourcing approach is implemented through MTurk. MTurk was chosen based on its well-established status at the time of the experiments; however, the basic approach could be deployed on alternative platforms (see Chapter 9 for other candidates). The MTurk platform organizes work in the form of human intelligence tasks, which we will refer to here as "tasks" consistent with the terminology in Chapter 1. A task is designed by the "Requester" (i.e., the research team) and is completed by many "Workers" (i.e., anyone who is interested in the task) over the Internet. We describe a design methodology for two types of tasks—the first targets rapid rating of a large number of dialogs on several dimensions of SDS performance, and the second aims to assess the reliability of workers on the crowdsourcing platform through an assessment of interannotator agreement in ratings among different workers and between workers and experts. We structure the tasks to elicit ratings of a representative set of dialogs, to enable semiautomatic validation of workers' submissions, and to achieve clear tasks. In addition, we develop and apply a set of approval rules, required to exclude submissions with nonsensical ratings, which would affect the overall quality of the ratings obtained. We describe the tasks and the approval process below.

8.5.1   Tasks for Dialog Evaluation

This type of task is designed to outsource the assessment of the SDS to workers. The assessment focuses on selected dimensions of performance regarding the SDS, based on a large number of dialogs selected from the logs. To achieve this goal, we have authored a set of questions that constitute the task in Table 8.3.

Table 8.3   Questions constituting the task on dialog evaluation (Q: question; opt: options).

Q1	**Do you think you understand from the dialog what the user wanted?**
Opt	(1) No clue (2) A little bit (3) Somewhat (4) Mostly (5) Entirely
Aim	*Elicit the worker's confidence in his or her ratings.*
Q2	**Do you think the system is successful in providing the information that the user wanted?**
Opt	(1) Entirely unsuccessful (2) Mostly unsuccessful (3) Half successful/unsuccessful (4) Mostly successful (5) Entirely successful
Aim	*Elicit the worker's perception of whether the dialog has fulfilled the informational goal of the user.*
Q3	**Does the system work the way you expect it?**
Opt	(1) Not at all (2) Barely (3) Somewhat (4) Almost (5) Completely
Aim	*Elicit the worker's impression of whether the dialog flow suits general expectations.*
Q4	**Overall, do you think that this is a good system?**
Opt	(1) Very poor (2) Poor (3) Fair (4) Good (5) Very good
Aim	*Elicit the worker's overall impression of the SDS.*
Q5	**What category do you think the dialog belongs to?**
Opt	(1) TS:Fu—Failed because of the user behavior, due to noncooperative user behavior
	(2) TS:Fs—Failed because of the system behavior, due to system inadequacies
	(3) TS:SN—Succeeded in spotting that no solution exists
	(4) TS:CsCu—Succeeded with constraint relaxation both from the system and from the user
	(5) TS:Cu—Succeeded with constraint relaxation by the user
	(6) TS:Cs—Succeeded with constraint relaxation by the system
	(7) TS:S—Succeeded
Aim	*Elicit the worker's impression of whether the dialog reflects task completion.*

Note: The questionnaire covers *the user's confidence, the perceived task completion, the expected behavior, the overall performance* and *the categorization of task success.*

In Table 8.3, we include the explanation of the aim for each question, but this is not shown to the workers. These questions cover *the user's confidence, perceived task completion, expected behavior, overall performance,* and *categorization of task success.* We chose these aspects for evaluation because they represent a range of important factors in system quality and because our data included only the textual transcription of dialogs, which made evaluation of other aspects of the systems, such as speech output quality, impractical. This final questionnaire was developed in the course of pilot experiments, in which questions that caused confusion among workers were revised and clarified. For example, the definition of task completion was particularly problematic and led us to the options in Q5 based on the ITU Recommendation (ITU-T Suppl. 24 to P-Series Rec. 2005). We have purposely designed the questions in such a way that they can cross-validate each other (Q2 and Q5 both aim to assess task completion), which will be used for approval of ratings from crowdsourcing later.

Each task contains the text transcription of one dialog and the questionnaire in Table 8.3 for assessment by the workers, who are paid US$0.05 for each task completed. We have uploaded 11,000 dialogs in total, including samples from the three major dialog categories and in proportions that follow the percentages obtained from the automatic classification, that is, TaskComplete (55%), TaskIncomplete (27%), and OutofScope (18%). TooShort and MultiTask dialogs are excluded from the task. The former are easily detectable as unsuccessful. The latter can be easily segmented into monotask dialogs, which can then follow the three-way categorization (TaskComplete/TaskIncomplete/OutofScope) directly.

8.5.2   Tasks for Interannotator Agreement

This type of task is an extension of those in Section 8.5.1 and is designed to assess the reliability of workers in crowdsourcing through interannotator agreement. Each task includes the text transcriptions of 30 selected Let's Go! dialogs drawn from log files (10 dialogs from the categories of TaskComplete, TaskIncomplete and OutofScope, respectively). Each dialog is associated with the questionnaire in Table 8.3. Workers are paid US$1.5 for each task completed. Altogether, we have three groups of workers (each with 16 individuals) rating two sets of dialogs (each with 30 dialogs). Groups 1 and 2 evaluate the first set of dialogs, while Group 3 evaluates the second set. In this way, we can assess whether the interannotator agreement varies across different raters and different dialogs.

8.5.3   Approval of Ratings

It is important to verify the quality of inputs from workers. Since the quality of the ratings directly impacts the credibility of the SDS evaluation, some basic rules have to be set to ensure that the workers are devoting effort to their tasks and to guarantee the reliability of ratings, in addition to the qualification requirement preset for the workers. We have developed an approval mechanism as follows:

R1: We reject tasks for which the working time is less than 15 seconds since we feel that careful (and thus high quality) ratings cannot be completed within such a short period.

R2: If a worker completes a large number of tasks (e.g., over 20) but provides identical answers for all of them, his or her work will be rejected.

R3: Approval requires consistency between the answers to related questions (Q2 and Q5). Consistency is based on four main heuristics:
1. Answers to Q2 being "Entirely successful" or "Mostly successful" can go with answers to Q5 being TS:S, TS:CS, TS:Cu, or TS:CsCu.
2. Answers to Q2 being "Entirely unsuccessful" or "Mostly unsuccessful" can go with answers to Q5 being TS:Fs or TS:Fu.
3. The answer to Q2 being "Half unsuccessful / successful" can go with any answer in Question 5.
4. The answer to Q5 being TS:SN can go with any answer to Q2.

R4: Approval requires consistency between the answers to Q5 and the automatic classification of the dialogs (see Section 8.4). In particular, the heuristics are:
1. TaskComplete can match with TS:S, TS:Cs, TS:Cu and TS:CsCu.
2. TaskIncomplete can match with TS:Fs and TS:Fu.
3. OutofScope can match with TS:SN.

R5: If these above heuristics are not satisfied, the dialog will be checked carefully. Random (incorrect) ratings are rejected. However, we have approved some ambiguous cases, explained in Section 8.6.2.

8.6   Collected Data and Analysis

This section describes the collected data, as well as presents the approval rates and the feedback from the workers. We also present an analysis of the collected data in terms of consistency among ratings and interannotator agreement.

8.6.1   Approval Rates and Comments from Workers

A total of 11,000 dialogs were rated by about 700 online workers in 45 days. Three persons in our team completed the verification of the rated tasks and approved 8394 of them. The total expenditure paid to the workers is US$350. Approval rates for each dialog category, that is, TaskComplete, TaskIncomplete, and OutofScope, are shown in Figure 8.4. OutofScope is the highest because some workers consider a task to be successful if they think that the absence of the information is due to the database but not the ability of the system. Others consider such cases to be failures since the system does not provide the requested information for the users. We approve either decision from the workers.

Rejected dialogs led to some controversies. Some apologized for their errors and others complained about the rejections. We received feedback from the workers concerned, many of

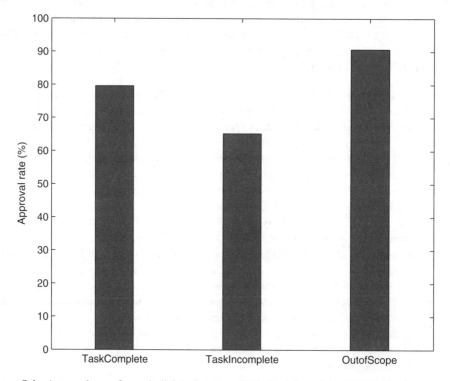

Figure 8.4   Approval rates for each dialog category, that is, TaskComplete, TaskIncomplete, and OutofScope. OutofScope has the highest approval rate.

which are useful to help enhance our understanding of SDS evaluations. Here, we list some typical comments associated with their implications as follows:

- The system does not provide the exact information that the user wanted although it provides some related results. (*Retrieval result from the database is a vital aspect of SDS performance.*)
- The system's understanding ability is very important, so good understanding may lead workers to choose task success even if the system does not provide any information to the user. (*Good language understanding ability plays an important role in improving user satisfaction.*)
- The system succeeds in providing a result based on the user's initial input but fails to follow up with the user's updated information. (*Timely updates of the dialog state impact users' perception of SDS performance positively and greatly.*)

8.6.2 Consistency between Automatic Dialog Classification and Manual Ratings

To assess the quality of ratings from crowdsourcing, we investigate the consistency between automatic dialog classification (see Section 8.4) and the manual ratings from workers (with respect to Q2 and Q5) based on the approved tasks of about 8000 dialogs.

According to the approval rules, the tasks whose ratings of Q5 are not consistent with the automatic classification will be rejected (see R4 in Section 8.5.3). However, as mentioned in Section 8.5.3, we still approve some ambiguous dialogs for which it is difficult to determine task success. The manual ratings of Q5 for the ambiguous dialogs do not agree with automatic classification. Table 8.4 shows an example where the dialog is terminated midway. Some

Table 8.4   An ambiguous dialog from TaskIncomplete category.

S: *Welcome to the CMU Let's Go bus information system.*
S: *To get help at any time, just say "Help" or press zero.*
S: *What can I do for you?*
U: *TWENTY. .EIGHT. .AT. .IN. .DOWNTOWN .FIFTY. AIRPORT.*
S: *Leave from AIRPORT. Is this correct?*
U: *FROM DOWNTOWN.*
S: *Leave from DOWNTOWN. Is this correct?*
U: *YES*
S: *Where are you going to?*
U: *TO AIRPORT*
S: *You want to go to AIRPORT. Is this correct?*
U: *YES*
S: *What time are you leaving?*
U: *EIGHT AM*
S: *leave at 8 a.m. Is this correct?*
U: *YES*
S: *Alright*
S: *Hold on. I'll look that up.*

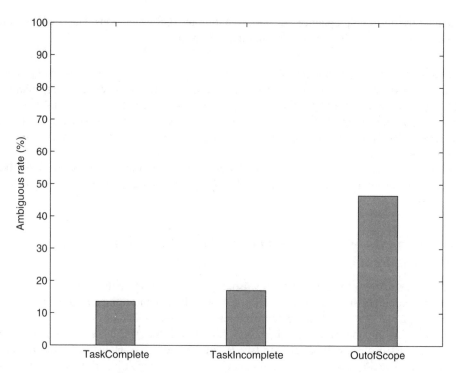

Figure 8.5   Percentages of ambiguous dialogs over the total approved ones in each category of TaskComplete, TaskIncomplete, and OutofScope. A higher percentage of ambiguous dialogs are approved for the OutofScope category.

workers regard the dialog as successful (in Q5) because they feel that the system would have completed the task had the user not hung up. Nevertheless, since the dialog does not have the necessary cue phrase (i.e., "... result. ... ") in the log, it is automatically classified as TaskIncomplete. Percentages of ambiguous dialogs over the total approved ones in each category of TaskComplete, TaskIncomplete, and OutofScope are shown in Figure 8.5. Note that a higher percentage of ambiguous dialogs are approved for the OutofScope category, mainly due to workers' diverse understandings of such dialogs, as we have discussed in Section 8.6.1.

Moreover, Figure 8.6 plots the mean answer scores of the two questions for approved dialogs in each of the three categories, where a higher score maps to a higher level of success:

- Answer scores to Q2 range from 0 for "Entirely unsuccessful" to 1 for "Entirely successful."
- Answer scores to Q5 range from 0 for "TS:Fu" to 1 for "TS:S."

The scores to Q2 and Q5 have been normalized to the same range from 0 to 1.

Generally, although some ambiguous dialogs are approved, we still observe reasonable agreement; that is, the dialogs automatically classified as TaskComplete receive high scores from the workers, those automatically classified as TaskIncomplete receive low scores, and those in the OutofScope category receive middling scores. Such consistency verifies the reliability of the approved ratings from crowdsourcing to some extent.

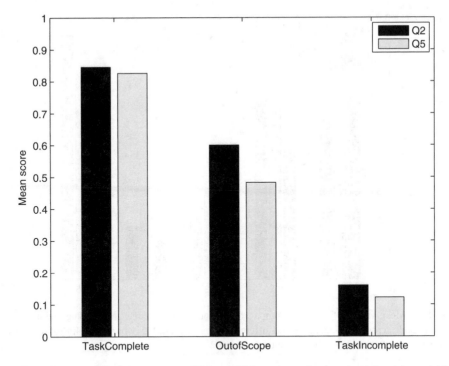

Figure 8.6 The normalized mean scores of Q2 and Q5 for approved ratings in each category. A higher score maps to a higher level of task success.

8.6.3 Interannotator Agreement among Workers

As mentioned earlier, the second type of task (see Section 8.5.2) is designed to assess the level of interannotator agreement (ITA) among the workers. We adopt Cohen's weighted kappa measure, which is often applied to ordinal categories,

$$\text{Weighted kappa} = \frac{\sum_{i=1}^{c} \sum_{j=1}^{c} w_{ij}(n_{ij}/N - n_{i.}n_{.j}/N^2)}{1 - \sum_{i=1}^{c} \sum_{j=1}^{c} w_{ij}n_{i.}n_{.j}/N^2}, \tag{8.3}$$

where c is the number of categories (i.e., answer options for each question here, $c = 5$ for Q1–Q4 and $c = 7$ for Q5), $w_{ij} = 1 - \frac{(i-j)^2}{(1-c)^2}$, n_{ij} is the element in the observed matrix, $n_{i.} = \sum_j n_{ij}$, and $n_{.j} = \sum_i n_{ij}$. Details can be found in Shoukri (2004). A higher weighted kappa value indicates a higher interannotator agreement.

Recall that we have three groups of workers rating two sets of dialogs. These ratings are accepted directly and do not undergo the approval process. For any pair of workers in each group, we compute the weighted kappa value for each question. We then compute the mean weighted kappa value for each question over the entire group. Results are shown in Figure 8.7. Despite the fact that groups 1 and 2 evaluated the same dialog set, while group 3 evaluated a different dialog set, the three plots remain close, which illustrates that the

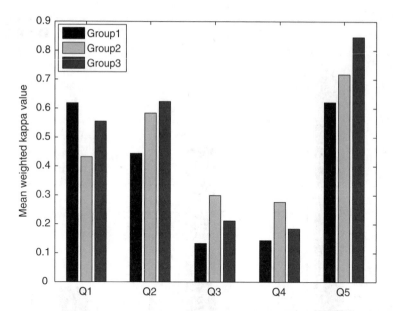

Figure 8.7 Mean values of weighted kappa of five questions for three groups. The weighted kappa values are stable across different raters and different dialogs. Q2 and Q5, which can gain "official" ratings from reliable raters, have high levels of interannotator agreement.

interannotator agreement for each question remains stable across different raters and different dialogs. In particular, Q5 (categorization of task success) achieves mean weighted kappa values above 0.6 and Q2 (perceived task completion) achieves reasonable values above 0.4, which is indicative of a moderate level of agreement (Landis and Koch 1977). Q2 and Q5 are related to task success and thus elicit relatively objective ratings from reliable raters, so the moderate and stable agreement partially shows the reliability of workers and provides support for the utilization of crowdsourcing as a judgment collection mechanism. On the other hand, Q3 (expected behavior) and Q4 (overall impression of system performance) have low values below 0.3, which is indicative of a lack of agreement. This suggests that evaluation based on overall user satisfaction may be quite subjective and that it may be necessary to further refine questionnaires to better elicit such assessments. The low agreement in user satisfaction may lead to the low prediction accuracy for the evaluation model, which has been analyzed in (Engelbrech *et al.* 2009).

Figure 8.7 also shows an interesting observation that for Q2 and Q5 the relative value gap between groups 1 and 2 is larger than that between groups 2 and 3. Intuitively, the values of interannotator agreement in groups 1 and 2 should be closer, since workers in the two groups evaluate the same set of dialogs. We investigate the annotations of workers in each group. Results show that three workers in group 1 are obviously unreliable. For example, they rate the TaskComplete dialog in Table 8.5 as TaskIncomplete, which contradicts the ratings of most group members, while there are no such inconsistent inputs in the other two groups. These inconsistent annotations from the few unreliable workers lead to the lower interannotator agreement in group 1. Despite this difference, the level of interannotator agreement over the

Table 8.5 A TaskComplete dialog from the dialog set in the task for interannotator agreement.

S: *Welcome to the CMU Let's Go bus information system.*
S: *To get help at any time, just say "Help" or press zero.*
S: *What can I do for you?*
U: *I'D LIKE TO GO FROM FORBES AND BIGELOW TO SQUIRREL HILL*
S: *Leaving from FORBES AND BIGELOW. Did I get that right?*
U: *YES*
S: *Right. Going to SQUIRREL HILL. Did I get that right?*
U: *YES*
S: *Okay. When do you want to travel?*
U: *NEXT AVAILABLE*
S: *When would you like to travel?*
U: *RIGHT NOW*
S: *You want the next bus. Is this correct?*
U: *YES*
S: *Just a minute. I'll look that up.*
S: *The next 61A leaves FORBES AVENUE AT BIGELOW SCHENLEY DRIVE at 10 40 p.m. and arrives at FORBES AVENUE AT MURRAY at 10 49 p.m.*

---

two groups is still the same. If we exclude the ratings from the three unreliable workers, the interannotator agreement of Q2 will increase from 0.444 to 0.542, and that of Q5 will increase from 0.621 to 0.698.

We conceive of several factors that cause low interannotator agreement in overall user satisfaction. First, different users may emphasize different aspects of system performance, ranging from the system's intelligence, task completion, dialog efficiency, and so on. Second, raters with different levels of domain knowledge may have different expectations of the system. It may not be meaningful to compute an overall average score of user satisfaction across a diversity of users. Instead, we may consider the following possibilities:

- Evaluate SDSs along different dimensions individually.
- Evaluate SDSs in terms of different types of user queries, targeting different system functionalities.
- Evaluate SDSs based on different user groups with different levels of domain knowledge.

8.6.4 Interannotator Agreement on the Let's Go! System

Section 8.6 has assessed the reliability of the SDS evaluation from workers using crowdsourcing by comparing the crowdsourced annotations to automatic classification and analyzing the interannotator agreement among workers. Nonetheless, there remains a key issue not resolved: whether the SDS evaluation by workers is as reliable as that by experts, in this case, researchers in the area of SDSs. In this section, we will investigate the level of agreement between the nonexpert and expert annotations through two case studies.

The second type of task in Section 8.5.2 contains a set of 30 dialogs from the Let's Go! dialog corpus, of which 10 dialogs each are from the categories of TaskComplete, TaskIncomplete, and OutofScope. We have 4 experts who evaluate the set of dialogs and compare their results

Table 8.6   Expert–workers and expert–expert agreement.

	E_1–W	E_2–W	E_3–W	E_4–W	E–E
Q1	0.5625	0.6107	0.5508	0.5525	0.6937
Q2	**0.6744**	**0.7037**	**0.7160**	**0.7434**	**0.8709**
Q3	0.2838	0.4368	0.2441	0.4214	0.5288
Q4	0.3686	0.4295	0.3188	0.4416	0.6240
Q5	**0.7984**	**0.8242**	**0.8209**	**0.81**	**0.9435**

Note: There is a high level of agreement between experts and workers for Q2 and Q5, which are about classification of task completion, while the other questions about the overall user satisfaction have a lower agreement.

with those of one group of 16 workers who evaluate the same set of dialogs in Section 8.5.2. The weighted kappa in Equation (8.3) is again employed to measure the interannotator agreement (ITA) (Shoukri 2004).

For any pair of experts, we first compute the ITA value for each question. We then calculate the mean expert–expert (E–E) ITA for each question over all the pairs. Given an expert E, the ITA value of any pair of E and workers is calculated, and then the E-workers (E–W) ITA is obtained by averaging the ITA values over all the E-worker pairs. Intuitively, reliable workers are expected to have a high level of agreement with experts.

Table 8.6 shows E–W and E–E ITA values for each question. There is a high level of agreement between experts and workers for Q2 and Q5 (the weighted kappa values are around 0.7) which are about classification of task completion and represent relatively objective ratings (Landis and Koch 1977), and the E–W ITA values approach those of E–E ITA. This agreement indicates that workers perform well on these measures and the crowdsourced data could be considered reliable. On the other hand, similar to the results of ITA among workers in Section 8.6.3, Q3 (expected behavior) and Q4 (overall impression on system performance) have lower ITA values for both E–W and E–E, which is indicative of lower agreement. Although agreement is slightly higher among the experts, it remains lower overall and suggests that evaluation based on overall user satisfaction may be quite subjective.

8.6.5   Consistency between Expert and Nonexpert Annotations

The 2010 Spoken Dialog Challenge (SDC) was organized by the Dialog Research Center at CMU. The aim of this challenge was to realistically compare different SDSs on the same dialog task and make the collected dialogs from the systems available for the development of evaluation techniques.

In the 2010 SDC, there are four participating systems (the three competitors and the CMU reference system). All these systems provide Pittsburgh bus schedule information. Four dialog corpora were collected through the SDC-controlled test, which was conducted by asking subjects to call all four systems. The organizers manually transcribed all the dialogs for each system and manually labeled task success for each dialog. A dialog is labeled as successful if one piece of acceptable information is provided (Black *et al.* 2010). We submit the text transcriptions of all the dialogs on MTurk using the task template in Section 8.5.1. Each dialog

Table 8.7 Comparison of labels from CMU experts and workers.

	Sys1	Sys2	Sys3	Sys4
Total dialogs	91	61	75	83
SR (workers)	56.1%	36.1%	86.7%	78.3%
SR (CMU)	64.8%	37.3%	89.3%	74.7%
CR	70.3%	73.7%	85.3%	86.7%
CSR	88.1%	80%	85.9%	88.5%

is evaluated by three workers. Task completion of a dialog is determined by the answers to Q5 with the use of majority vote. For example, if two or more workers choose one of the options of TS:S, TS:SCs, TS:SCu, and TS:SCsCu, the dialog will be tagged as success.

We compare the task completion labels from workers with those from CMU experts. Several measures are defined for comparison. Given a dialog corpus $D = \{d_1, d_2, \cdots, d_N\}$, the crowd-sourced labels of the corpus from workers are denoted as $L^W = \{l_1^W, l_2^W, \cdots, l_N^W\}$, and labels from CMU experts are denoted as $L^E = \{l_1^E, l_2^E, \cdots, l_N^E\}$, where $l_i^W, l_i^E \in \{success, failure\}$. The first measure called success rate (SR) is defined in the following way:

$$SR = \frac{\sum_{i=1}^{N} I(l_i = success)}{N}, \tag{8.4}$$

where $I(\cdot)$ is

$$I(\omega) = \begin{cases} 1 & \text{if } \omega = true \\ 0 & \text{otherwise.} \end{cases} \tag{8.5}$$

The second measure of consistency rate (CR) is defined as below:

$$CR = \frac{\sum_{i=1}^{N} I(l_i^W = l_i^E)}{N}. \tag{8.6}$$

The third measure is consistent success rate (CSR):

$$CSR = \frac{\sum_{i=1}^{N} I((l_i^W = success)\&(l_i^E = success))}{\sum_{i=1}^{N} I((l_i^W = success) \parallel (l_i^E = success))}. \tag{8.7}$$

Comparison results are shown in Table 8.7. The SR per system by the workers is quite close to that by the CMU experts. The CRs are all around 70%, even above 80% for systems 3 and 4. The CSRs (around 80%) indicate that the "Success" annotations from experts and nonexperts have a large overlap. Table 8.7 demonstrates that worker judgments are generally consistent with expert judgments. This consistency also provides support for the utilization of crowdsourcing for collecting user judgments on dialogs.

Next, we investigate the cases where the workers differ from the experts. We find that experts care more about the accuracy of the bus information that the system provides when they label the dialogs. In some dialogs, the systems provide the bus information that cannot exactly meet the user's request, for example, wrong bus arrival time though correct bus route. The experts regard such dialogs as failure, while workers often consider them as success. On the other hand, workers seem to focus on the user's final intent. In some cases where the user changes his or her mind and triggers a new query after the system provides acceptable information, the experts tag them as success, while workers often label them as failure since they feel that the system does not accomplish the user's final goal. Although the classification of task completion is more objective than other ratings (like user satisfaction), there are still some differences of opinion which may cause inconsistency.

8.7   Conclusions and Future Work

This chapter presented our strategy for the use of crowdsourcing in collection of user judgments on SDSs. We describe a design methodology for two types of tasks—the first for rapid, efficient collection of ratings of a large number of dialogs and the second for assessment of the reliability of these ratings through interannotator agreement among workers. These tasks were designed specifically to overcome several challenges associated with crowdsourcing of SDS evaluation: creating a representative sample of dialogs for evaluation, providing sufficiently clear, simple tasks for workers, and establishing a framework for semiautomatic validation of crowdsourced results. A set of approval rules ensured the quality of ratings from crowdsourcing.

Compared with the traditional method of inviting subjects to fill out a questionnaire after interaction, the results we achieved show that the crowdsourcing method is more efficient, flexible and inexpensive, and could access a more statistically representative population. At the same time, the quality of ratings can also be controlled. Reliable ratings for 8,394 dialogs by around 700 online workers are approved. Approval rates for each dialog category, that is, TaskComplete, TaskIncomplete, and OutofScope, are 79.59%, 65.23%, and 90.65%, respectively. Reasonable consistency between the manual crowdsourced ratings and the automatic dialog categorization in terms of task success is an indicator of the reliability of the approved ratings from crowdsourcing. The moderate level of interannotator agreement among workers for ratings in task completion partially verifies the reliability of such workers.

This chapter has also provided support for the use of crowdsourcing for evaluation of SDSs by investigating the agreement between workers and experts through two case studies. Experimental results showed a high level of interannotator agreement (around 0.7) between experts and workers in terms of task completion when we compared their annotations on the Let's Go! dialog corpus. There was also a high degree of consistency between expert and nonexpert labels of task success for the dialogs from the four SDC systems.

8.8   Acknowledgments

This work was done as part of a project related to the SDC in 2010 (http://dialrc.org/sdc), which was organized by Professor Maxine Eskénazi and Professor Alan Black of the CMU Dialog Research Center. This work was conducted when Zhaojun Yang was a master's student and Professor Gina-Anne Levow was a visiting scholar at The Chinese University of Hong Kong. The project team also included Professor Irwin King, Baichuan Li, and Yi Zhu from The Chinese University of Hong Kong. The work is partially supported by a grant from the HKSAR Government Research Grants Council (Project No. 415609).

References

Baptist L and Seneff S (2000) Genesis-II: a versatile system for language generation in conversational system applications. *Proceedings of ICSLP*.

Beringer N, Kartal U, Louka K, Schiel F, Türk U (2002) Promise: a procedure for multimodal interactive system evaluation. *Proceedings of the LREC Workshop on Multimodal Resources and Multimodal Systems Evaluation*, pp. 90–95.

Bessho F, Harada T and Kuniyoshi Y (2012) Dialog system using real-time crowdsourcing and Twitter large-scale corpus. *Proceedings of SIGDial 2012*.

Black A, Burger S, Langner B, Parent G and Eskenazi M (2010) Spoken dialog challenge 2010. *Proceedings of IEEE Workshop on Spoken Language Technology*.

Bonneau-Maynard H, Devillers L and Rosset S (2000) Predictive performance of dialog systems. *Proceedings of Language Resources and Evaluation Conference (LREC)*.

Buchholz S and Latorre J (2011) Crowdsourcing preference tests and how to detect cheating. *Proceedings of Interspeech 2011*.

Cohen P (1995) *Empirical Methods for Artificial Intelligence*, vol. 55. MIT press.

Cole RA, Mariani J, Uszkoreit H, Zaenen A, Zue V (1997) *Survey of the State of the Art in Human Language Technology*. Citeseer.

DePalma N, Chernova S and Breazeal C (2011) Leveraging online virtual agents to crowdsource human-robot interaction. *Proceedings of CHI Workshop on Crowdsourcing and Human Computation*.

Dybkjaer L, Bernsen N and Minker W (2004) Evaluation and usability of multimodal spoken language dialogue systems. *Speech Communication* **43**(1–2), 33–54.

Engelbrecht K, Gödde F, Hartard F, Ketabdar H and Möller S (2009) Modeling user satisfaction with Hidden Markov Model. *Proceedings of SIGDIAL Workshop on Discourse and Dialogue*, pp. 170–177.

Engelbrecht K and Möller S (2007) Pragmatic usage of linear regression models for the prediction of user judgments. *Proceedings of SIGdial Workshop on Discourse and Dialogue*.

Forbes-Riley K and Litman D (2006) Modelling user satisfaction and student learning in a spoken dialogue tutoring system with generic, tutoring, and user affect parameters. *Proceedings of Association of Computational Linguistics (ACL)*, pp. 264–271.

Glass J, Polifroni J, Seneff S and Zue V (2000) Data collection and performance evaluation of spoken dialogue systems: the MIT experience. *Proceedings of International Conference on Spoken Language Processing (ICSLP)*.

Hirschman L, Bates M, Dahl D, Fisher W, Garofolo J, Pallett D, Hunicke-Smith K, Price P, Rudnicky A and Tzoukermann E (1993) Multi-site data collection and evaluation in spoken language understanding. *Proceedings of the Workshop on Human Language Technology*, pp. 19–24.

Hone K and Graham R (2001) Subjective assessment of speech-system interface usability. *Seventh European Conference on Speech Communication and Technology*.

ITU-T Rec. P.851 (2003) Subjective quality evaluation of telephone services based on spoken dialogue systems. *International Telecommunication Union, Geneva*.

ITU-T Suppl. 24 to P-Series Rec. (2005) Parameters Describing the Interaction with Spoken Dialogue Systems. *ITU, Geneva*.

Jurcicek F, Keizer S, Gasic M, Mairesse F, Thomson B, Yu K and Young S (2011) Real user evaluation of spoken dialogue systems using Amazon Mechanical Turk. *Proceedings of Interspeech 2011*.

Kamm C, Walker M and Litman D (1999) Evaluating spoken language systems. *Proceedings of AVIOS*.

Landis J and Koch G (1977) The measurement of observer agreement for categorical data. *Biometrics* **33**(1), 159–174.

Litman D, Rosé C, Forbes-Riley K, VanLehn K, Bhembe D and Silliman S (2006) Spoken versus typed human and computer dialogue tutoring. *International Journal of Artificial Intelligence in Education* **16**(2), 145–170.

McGraw I, Lee C-y, Hetherington L and Glass J (2010) Collecting voices from the crowd. *Proceedings of Language Resources and Evaluation Conference (LREC)*.

McTear M (1998) Modelling spoken dialogues with state transition diagrams: experiences with the CSLU toolkit. *Proceedings of ICSLP*, pp. 1223–1226.

McTear M (2002) Spoken dialogue technology: enabling the conversational user interface. *ACM Computing Surveys (CSUR)* **34**(1), 90–169.

Meng H, Ching P, Chan S, Wong Y and Chan C (2004a) ISIS: an adaptive, trilingual conversational system with interleaving interaction and delegation dialogs. *ACM Transactions on Computer-Human Interaction (TOCHI)* **11**(3), 268–299.

Meng H, Lam W and Wai C (1999) To believe is to understand. *Proceedings of European Conference on Speech Communication and Technology.*

Meng H, Wai C and Pieraccini R (2004b) The use of belief networks for mixed-initiative dialog modeling. *IEEE Transactions on Speech and Audio Processing* **11**(6), 757–773.

Möller S (2005) Parameters for quantifying the interaction with spoken dialogue telephone services. *Proceedings of SIGdial Workshop on Discourse and Dialogue*, pp. 166–177.

Moller S and Skowronek J (2003) Quantifying the impact of system characteristics on perceived quality dimensions of a spoken dialogue service. *Eighth European Conference on Speech Communication and Technology.*

Novotney S and Callison-Burch C (2010) Cheap, fast and good enough: Automatic speech recognition with non-expert transcription. *Proceedings of Association for Computational Linguistics* (ACL), pp. 207–215.

Raux A, Langner B, Bohus D, Black A and Eskenazi M (2005) Let's go public! taking a spoken dialog system to the real world. *Proceedings of Interspeech.*

Ribeiro F, Florencio D, Zhang C and Seltzer M (2011) CrowdMOS: an approach for crowdsourcing Mean Opinion Score studies. *Proceedings of ICASSP 2011.*

Seneff S (1992) TINA: A natural language system for spoken language applications. *Computational Linguistics* **18**(1), 61–86.

Shoukri M (2004) *Measures of Interobserver Agreement.* CRC Press.

Snow R, O'Connor B, Jurafsky D and Ng A (2008) Cheap and fast—but is it good?: evaluating non-expert annotations for natural language tasks. *Proceedings of the Conference on Empirical Methods in Natural Language Processing*, pp. 254–263.

Taylor P (2009) *Text-to-Speech Synthesis.* Cambridge University Press.

Walker M, Kamm C and Litman D (2000) Towards developing general models of usability with PARADISE. *Natural Language Engineering* **6**(3&4), 363–377.

Walker M, Litman D, Kamm C and Abella A (1997) PARADISE: a framework for evaluating spoken dialogue agents. *Proceedings of Association for Computational Linguistics (ACL).*

Walker M, Passonneau R and Boland J (2001) Quantitative and qualitative evaluation of DARPA Communicator spoken dialogue systems. *Proceedings of Association for Computational Linguistics (ACL)*, pp. 515–522.

Walker MA, Litman DJ, Kamm CA and Abella A (1998) Evaluating spoken dialogue agents with PARADISE: two case studies. *Computer Speech & Language* **12**(4), 317–347.

Ward W and Issar S (1994) Recent improvements in the CMU spoken language understanding system. *Proceedings of the Workshop on Human Language Technology*, pp. 213–216.

Wolters M, Isaac K and Renals S (2010) Evaluating speech synthesis intelligibility using Amazon Mechanical Turk. *Proceedings 7th Speech Synthesis Workshop (SSW7).*

Wu Z, Meng H, Ning H and Tse S (2006) A corpus-based approach for cooperative response generation in a dialog system. *Chinese Spoken Language Processing*, pp. 614–626.

Yang Z, Li B, Zhu Y, King I, Levow G and Meng H (2010) Collection of user judgments on spoken dialog system with crowdsourcing. *Proceedings of IEEE Workshop on Spoken Language Technology (SLT).*

9

Interfaces for Crowdsourcing Platforms

Christoph Draxler

Ludwig-Maximilian University, Germany

9.1 Introduction

This chapter on interfaces for crowdsourcing platforms and applications is divided into three sections. The first section gives a brief outline of the current web technology and terminology. It introduces network architectures, communication protocols, markup languages, programming paradigms, and technology for media playback. The terms are needed to understand the technological principles of crowdsourcing and the properties—including the limitations—of the interfaces provided by crowdsourcing platforms to their services.

The second section presents three commercial crowdsourcing platforms (CrowdFlower, Amazon Mechanical Turk (MTurk), and Clickworker) as typical examples of such platforms. This section also covers WikiSpeech, a system for speech recording, annotation and online perception experiments that offer many features typical of crowdsourcing platforms. Its origins date back to the year 1997, when WWWTranscribe, the first web-based transcription tool, was proposed (Draxler 1997). WikiSpeech is targeted at academic applications, and has been used in several speech database projects in Germany and other countries.

Finally, the third section describes in detail how tasks are implemented using the interfaces of crowdsourcing platforms. Three main options are available: implementing tasks manually using a graphical user interface, using command-line tools that generate tasks from configuration files, or writing application programs that interact directly with the crowdsourcing platform programming interfaces. This section also outlines how recording, annotation, and experiment tasks are defined in WikiSpeech.

Crowdsourcing for Speech Processing: Applications to Data Collection, Transcription and Assessment, First Edition.
Edited by Maxine Eskénazi, Gina-Anne Levow, Helen Meng, Gabriel Parent and David Suendermann.
© 2013 John Wiley & Sons, Ltd. Published 2013 by John Wiley & Sons, Ltd.

In this chapter, a crowdsourcing *job* corresponds to a set of *tasks*. A *unit task* is a single task out of this set. It is common practice to give the same unit task to several workers to improve the quality of the result returned—an *assignment* is the pairing of a unit task and a worker. The corresponding MTurk terms are *human intelligence task (HIT) group*, *HIT*, and *assignment*.

9.2 Technology

This section introduces the technology underlying the World Wide Web and crowdsourcing. To illustrate the technology, a small website with a typical crowdsourcing set of tasks will be used as an example.

9.2.1   TinyTask Web Page

The TinyTask web page implements a very basic but typical crowdsourcing job. This job is to determine the language of a given utterance. It consists of a number of unit tasks, for example, a list of audio samples that workers must categorize. Each unit task should be processed by more than one worker, that is, each task should have multiple assignments.

The website consists of a single web page with a headline, a short list of instructions, an interactive icon to start audio playback, and a set of labeled input buttons (see Figure 9.4).

A worker clicks on the icon to listen to the audio and selects the button corresponding to his or her judgment. The page then displays a message whether the decision was correct or not.

On the computer, the website consists of a dedicated directory named `tinytask/`, which resides in the root directory of the web server. This directory contains the HTML source code in the file `tinytask.html`, the style sheet file `tinytask.css,` and a number of subdirectories, such as `audio/` for the audio files used by the set of tasks, `icons/` for logos and other icons, and `scripts/` for JavaScript source code.

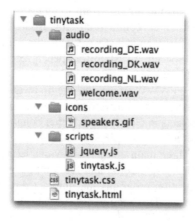

9.2.2   World Wide Web

A *client–server architecture* consists of computers connected by a network; in this network, computers may have different roles: *clients* request a response or a service from *servers* via a standardized communication protocol.

Figure 9.1   Client–server architecture of the World Wide Web.

The World Wide Web is a client–server architecture that uses the global Internet as a network and the hypertext transport protocol to exchange requests and responses between clients and servers.

Servers and services they provide are accessed using a *uniform resource locator* (URL). A URL is a unique address of a resource on the Internet. It *identifies* a resource, that is, it gives it a name; it *locates* the resource, that is it provides an address, for example, a server name, and it *specifies* how the resource can be accessed, that is, which protocol to use. A *uniform resource identifier* (URI) is more general in that it uniquely identifies a resource, but does not necessarily locate it.

A *user agent* is a software application that:

- Runs on the client.
- Sends a *request* for a service to the server.
- Receives a *response* that is used by the client, for example, displayed on the screen, saved to a file, or played via the audio output (see Figure 9.1).

Typically, a user agent is either a web browser with a graphical user interface, for example, Mozilla Firefox, Internet Explorer, Google Chrome, Safari, Opera or other, or a command-line tool such as `curl`.

9.2.3   Hypertext Transfer Protocol

The hypertext transfer protocol (*HTTP*) is a "protocol with the lightness and speed necessary for distributed, collaborative, hypermedia information systems. It is a generic, stateless, object-oriented protocol" (Berners-Lee *et al.* 1996, p. 1). For a detailed description of the standard, see (Gourley *et al.* 2002).

In HTTP, a method is applied to a resource that is identified and located by a URL. Allowed method names are GET, POST, PUT, DELETE, and so on. An HTTP request consists of a method name, the resource URL, and the version string of the protocol. Parameters are optional; they may be provided with the URL as with the GET method, or separately as with the POST method.

Example   The following HTTP request

```
GET http://www.w3.org/pub/WWW/TheProject.html HTTP/1.1
```

sends the GET method to the server www.w3.org to locate the resource `TheProject.html` in the file directory `pub/WWW/` using version 1.1 of the protocol. The server response consists of a status message, and if the request succeeds, the resource itself, for example, a web page in HTML.

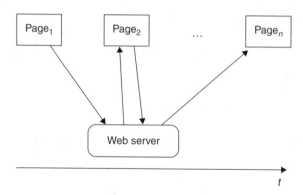

Figure 9.2 In the WWW, each HTTP request sent to a server returns a response that replaces the original page in the browser. As a consequence, session information cannot be passed from one page to the next. Instead, it has to be stored on the client, for example, in cookies, or included in every request–response pair exchanged between the server and the client, for example, as a unique session identifier.

HTTP is a *stateless* protocol. This means that no information can be carried over from one request to the next. In a browser, this means that each request–response pair leads to a new page being displayed (see Figure 9.2). For an alternative, see Section 9.2.9 on AJAX.

The secure version of the hypertext transfer protocol is HTTPS where the communication is encrypted. HTTPS is typically used where confidentiality is required, for example, payment transactions or access to personal account settings.

9.2.4   Hypertext Markup Language

HTML, proposed by Tim Berners-Lee in 1990, is a markup language describing the structure and content of web pages. Markup tags are written in angled brackets, for example, <head>, and most markup tags come in pairs, for example, <body> and </body>. The opening tag may contain attributes with optional argument values, for example, id="p1".

An HTML document consists of a directive, an <html> root element, a header and a body part organized in a hierarchical structure. The directive specifies the HTML version and compliance levels; the header contains the title of the web page plus optional information on character encoding, links to style sheets and scripts. The body contains the visible contents of the web page, that is, elements such as headings, text, lists, forms, input elements, and images All elements may be identified by an id attribute whose value must be unique within a page.

The *document object model* (DOM) is a data structure that describes an HTML page. It can be processed by scripts, for example, to hide and show elements, or to modify, add or delete page elements.

DOM is the technology underlying modern Web 2.0 applications because it allows dynamic modifications of web pages depending on the state of an application program.

Crowdsourcing web pages are typically independent web pages contained within surrounding web pages of the crowdsourcing platforms. In HTML this can be achieved by using the <iframe> tag in the surrounding page. The effect of this tag is that a two-dimensional space is spared out in the web page; this space is then filled with the web page for the worker. Using this technique, a worker may access both information from the crowdsourcing platform, for example, his or her account data and payments, and the task relevant data in one browser window.

HTML is standardized by the World Wide Web consortium. The current standard is HTML 4.01, and HTML5 is currently being developed (Ragett *et al.* 1998; Hickson 2012). Numerous books and websites dedicated to HTML5 are available, for example, w3schools.com/html5/.

HTML5 offers many interesting new features; in the context of crowdsourcing for speech processing, certainly the most important are audio and video output. Although HTML5 has at the time of writing not been finalized, many browsers already support it not only on traditional computers, but also on new devices such as tablets and smartphones. This allows recruiting new worker populations for crowdsourcing, for example, commuters.

Example   The HTML5 code for the TinyTask web page is given in Figure 9.3. The display of the raw HTML is shown in Figure 9.4a.

```html
<!DOCTYPE HTML>
<html>
  <head>
    <title>Tiny Task</title>
    <meta http-equiv="content-type" content="text/html;charset=utf-8" />
    <!-- this is a comment ---
      <script src="script/tinytask.js"></script>
      <script src="scripts/jquery.js"></script>
      <link rel="stylesheet" href="tinytask.css" type="text/css" />
    -->
  </head>

  <body>
    <h1>Determine the language</h1>
    <audio id="expaudio" src="audio/welcome.wav">
    </audio>
    <ol class="left">
      <li>Click on the loudspeaker icon to listen to the audio</li>
      <li>Select the button corresponding to the language you heard</li>
    </ol>
    <div class="centered">
      <img src="icons/speakers.gif" onclick="playaudio()"; />
      <form method="post">
        <input type="radio" name="userinput" onclick="submitvalue('DE');" />German
        <input type="radio" name="userinput" onclick="submitvalue('NL');" />Dutch
        <input type="radio" name="userinput" onclick="submitvalue('DK');" />Danish
        <input type="radio" name="userinput" onclick="submitvalue('__');" />other
      </form>
    </div>
  </body>
</html>
```

Figure 9.3   TinyTask web page in HTML. Note that the two `<script>` tags and the `<link>` tag in the header are inside a comment. Hence, they will not be loaded. Note also that the HTML5 audio tag is used. This element is not visible on the screen.

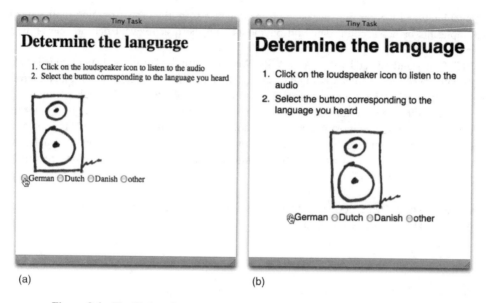

(a)   (b)

Figure 9.4 TinyTask web page raw display (a) and with a style sheet applied (b).

9.2.5 Cascading Style Sheets

The presentation of HTML files on output media, for example, large computer screens or small handheld devices, is defined using styles in the cascading style sheets (CSS) language. CSS code can either be embedded in a web page, or held in an external text file.

A browser generally comes with a predefined set of styles for the different types of text and content in a web page. However, web page providers may define their own styles that overwrite the predefined settings to ensure that a web page looks the same on any browser.

Example The following CSS code changes the appearance of the TinyTask web page (see Figure 9.4). It specifies a larger sans-serif font and centers the headline, the icon and the input options on the page.

```
.centered { text-align : center; }
.left { text-align : left; }

body { font-size : large; font-family : sans-serif; }
ol li { margin : 5px auto; }
form { width : 300px; margin : 10px auto; }
```

To apply the style sheet, the comment tags were removed from the header of the TinyTask HTML page.

9.2.6 JavaScript

JavaScript is a "prototype-based scripting language that is dynamic, weakly typed and has first-class functions. It is a multiparadigm language, supporting object-oriented, imperative, and functional programming styles" (see `http://en.wikipedia.org/wiki/JavaScript` for a description).

Originally, JavaScript was primarily used to validate form input, for example, to check whether the string entered by a user conforms to a valid e-mail address. With the specification of the DOM, JavaScript is increasingly being used to dynamically modify web pages. HTML elements can be created, modified, deleted, shown, or hidden using JavaScript commands. The event mechanism in JavaScript allows the implementation of highly dynamic interactive web pages. Furthermore, computers have become so fast that even complex computations can be performed using interpreted languages such as JavaScript.

In a browser, a runtime engine executes JavaScript programs; it strictly limits access to the host computer and thus prevents programs loaded from the web from harming the host computer.

For a detailed description of JavaScript, see (Flanagan 2011) or the official standard (ECMA International 2009), and (MacCaw 2011) or (Stefanov 2010) for the development of complex software using JavaScript.

To support and simplify JavaScript programming, a number of tools/frameworks have been developed. JSLint is a powerful code verifier for JavaScript (Crockford 2012). A particularly popular framework is jQuery. It combines the selectors used in CSS with functional language elements to provide compact and powerful commands to manipulate the structure and display of web pages. An introduction to jQuery as well as documentation and tutorials can be found at `jquery.com`.

Example In the TinyTask web page, JavaScript is used to detect mouse clicks by the worker. Three functions are defined:

```
var languages = ["DE", "NL", "DK"];

selectlanguage = function () {
  var index = Math.floor(Math.random() * languages.length);
  return languages[index];
}

playaudio = function () {
  var url, player;
  url = "audio/recording_" + selectlanguage() + ".wav";
  player = document.getElementById("expaudio");
  player.setAttribute("src", url);
  player.play();
}

submitvalue = function (value) {
  var src;
  src = $("#expaudio").attr("src");
  if (src.indexOf(value) > 0)
    alert("Congratulations. You're right!");
  else
    alert("Oops. You're wrong.");
}
```

When the worker clicks on the loudspeaker, an event is fired that triggers the execution of the `playaudio()` function. This function randomly selects one of the language codes held in the array `languages`, computes a relative URL from the language code, assigns the URL to the source attribute of the `<audio>` tag and starts playback of the audio signal.

When the worker clicks on a radio button, the function `submitvalue()` is executed which checks whether the language code associated with the selected radio button is part of the name of the audio URL. If so, then the worker's input is correct.

The script is stored in an external text file named `tinytask.js`. Note that the `playaudio()` function uses pure JavaScript only, whereas the `submitvalue()` function uses jQuery commands. In pure JavaScript, the `getElementById("expaudio")` function accesses the element with an ID of `expaudio` in the DOM, and the function `setAttribute` sets its value. In jQuery, a single line of code is sufficient: `$('#expaudio').attr("src")` accesses the same element and gets the `src` attribute value via the jQuery function `attr()`.

To use this script, the links to the script files `tinytask.js` and `jquery.js` must be moved out of the comment in the header part of the HTML file in Figure 9.3.

9.2.7   JavaScript Object Notation

JavaScript object notation (JSON) is a simple text-based data format. It is compact, human-readable, and easy to parse, and a JSON-formatted string can directly be executed by the JavaScript runtime engine. JSON "is used primarily to transmit data between a server and a web application, serving as an alternative to XML" (see `en.wikipedia.org/wiki/JSON` for more information).

For many programming languages, libraries exist to read, create, and process JSON strings (see `json.org` for details).

Example   In the TinyTask script, the list of languages is given in JSON format:

```
["DE", "NL", "DK"]
```

9.2.8   Extensible Markup Language

Extensible markup language (XML) is a markup language originally proposed by (Bray *et al.* 1998) and standardized by the W3 consortium (Bray *et al.* 2005). An XML document is strictly hierarchical; that is, it consists of a single root element and possibly many nested elements. XML documents can be linked via hyperlinks using a *standoff* notation. With this stand-off notation, nonhierarchical structures can be modeled using XML—such nonhierarchical structures are typically found in linguistic annotations.

In the WWW, XML is commonly used as a transfer format to exchange data between the server and the client.

XML Syntax

An XML document consists of a directive and the XML document proper. Elements are written in tags in angled brackets and they usually come in pairs, for example, `<customer>` and

</customer>. The content of elements is written between the opening and closing tags; elements without content may be written as a single tag with the character sequence /> to mark the end of the tag.

Opening tags may contain attributes with optional values. The values must be enclosed in quotes, for example, id="expaudio".

The nesting of elements is strict; that is, outer tag pairs may not be closed as long as inner tag pairs are still open.

XML tags may belong to a namespace, indicated in the tag by an identifier followed by a colon, for example, <s:Envelope>. This namespace serves to disambiguate tag names.

Well-Formed and Valid XML

Two types of XML conformance can be checked: an XML document is *well formed* if it complies with XML syntax, and it is *valid* if it complies with a given document-type definition (DTD).

A DTD specifies the elements, their attributes and hierarchical relationships in an XML document. DTDs come in three formats: plain text DTD format, and Relax-NG or XML Schema, which are both written in XML. The DTD format is very simple and does not allow the specification of data types other than string (#PCDATA for parsable character data). XML Schema and Relax-NG allow the definition of complex data types, a prerequisite for data transfer via the web or the definition of web services.

The location of a DTD file for a given XML document is specified in the XML directive of the document, that is, the first line of the document. This location can be a local file, or it can be a link to a resource on a website.

XML-Based Technologies

Building on XML, a number of technologies have been developed, for example, XPath to access elements of an XML document via navigational directives, XQuery to formulate search queries on XML documents, XPointer to refer to other XML documents, XSLT to transform XML documents to other text formats, SOAP for data transfer and remote procedure calls via a network, and many others.

All programming languages have extensive libraries to handle XML documents and technology.

9.2.9 Asynchronous JavaScript and XML

Many applications in the WWW require that session information be available even if web pages change. For example, in a typical crowdsourcing session, the worker logs in, works on one assignment after the other, and when completed receives his or her payment. The crowdsourcing platform must make sure that information about which worker performed which assignments must be maintained during the entire session during which the worker is logged in.

As described in Section 9.2.3, HTTP is a stateless protocol; that is, a request does not know anything about earlier requests. To store information between HTTP connections, it is either

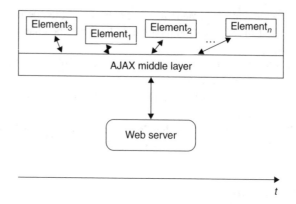

Figure 9.5 A single web page with dynamic page elements accessing a server via a middle-layer program. Over time, this AJAX middle layer sends requests to the server and updates the page elements with data extracted from the responses without reloading the web page.

stored on the client machine, for example, in cookies, or included in the request–response pairs exchanged between the server and the client.

An alternative approach to storing information between sessions is to delegate sending requests to and retrieving responses from the server to a middle layer in the client software. In this approach, a single web page is on display in the browser on the client machine. This web page holds all information relevant to the given task, including session data. Some elements of the web page are visible, while others are hidden, depending on the state of the program.

User actions, for example, input in a text field, trigger events that are interpreted by the middle-layer program and sent to the server; the server's response is received by the middle-layer program, which then modifies parts of the web page. This is achieved by using a technique known as asynchronous JavaScript and XML (AJAX) (see Figure 9.5). The advantage of this approach is that pages are not reloaded, but simply updated. Session information can be stored as part of the page, for example, in a hidden input field or as part of the text, and it is available as long as the page lives in the browser. Furthermore, since only parts of the page are updated, users will experience a smooth and highly responsive interactive interface.

JavaScript frameworks usually have easy-to-use functions that implement AJAX functionality. For example, the jQuery `ajax()` function may be used with the URL to which the request is sent, and a list of settings. These settings are, among others, the data format, encoding, callback function to execute when the response is received from the server, and so on.

9.2.10 Flash

Adobe Flash is a proprietary software to create and display multimedia content. Flash processes raster and vector images, audio, and video and allows access to the host computer, for example, to capture keyboard input or record audio.

Flash content is created either with Adobe's commercial Flash Builder or using open-source software development kits such as Apache Flex (`incubator.apache.org/flex/`) or Haxe (`haxe.org`). To display Flash content in a web browser, the free Flash plug-in is needed. It is available for all major browsers on traditional computers, but not on all tablets or smartphones, and Adobe will not publish new Flash versions for mobile devices (Levine 2011). The multimedia content is embedded into a web page using the <embed> tag; however, this tag is not part of the HTML standard.

In the context of crowdsourcing, Flash is commonly used for audio playback, and it is the preferred means of recording audio via an HTML page; for an example, see Chapter 3 or (McGraw *et al.* 2011). As a substitute for Flash audio and video playback, HTML5 is slowly gaining acceptance, especially on mobile devices. However, it does not (yet) support audio or video recording.

9.2.11 SOAP and REST

In web applications, users manipulate data objects and resources on the client, but these objects and resources effectively reside on the server. To standardize the communication between the server and the client on a more abstract level than HTTP, the protocol SOAP and, more recently, the paradigm REST have been developed. These technologies greatly facilitate the development of complex web applications.

SOAP is a network protocol to exchange data and execute remote procedure calls on computers in a network (Box *et al.* 2000). SOAP uses XML to package data exchanged between the server and the client. A SOAP message is expressed as an XML element named Envelope in namespace s:. This element contains an optional header and a mandatory body. The header contains metadata for the message, for example, routing or encryption information, the body contains the message itself. This message is again an XML element with a tag name from an application-specific namespace, in general a programming language data object or method.

Example   The following XML code (taken from de.wikipedia.org/wiki/SOAP) sends a request for a title search with the search words *"DOM," "SAX," "SOAP"* to a server. The <Envelope> tag is validated against the soap-envelope DTD at w3.org, the TitleInDatabase tag against an application specific DTD at lecture-db.de/soap.

```
<?xml version="1.0"?>
<s:Envelope xmlns:s="http://www.w3.org/2003/05/soap-envelope">
  <s:Body>
    <m:TitleInDatabase xmlns:m="http://www.lecture-db.de/soap">
      DOM, SAX and SOAP
    </m:TitleInDatabase>
  </s:Body>
</s:Envelope>
```

The response to this request is again a SOAP-envelope.

In contrast to SOAP, which is a protocol, REST is a paradigm for addressing resources on a server via URLs and HTTP:

> SOAP architectures are highly focused on transparent mapping of programming language objects. [...] however, aspects such as distribution and scalability are reduced to playing a second role. [...] the major driver of the REST architecture is distribution and scalability.
>
> (Roth 2009)

In the REST paradigm, only the methods defined in HTTP may be used to access and modify resources. GET, PUT, POST, and DELETE typically are basically interpreted as read, create, update, and delete commands applied to resources on the server. The resource itself is addressed by the file name part of the URL. Arguments to the commands are sent as key=value pairs in GET commands, or separately as in, for example, the POST command.

URLs containing REST commands may be treated just like any other URL; that is, they may be stored as bookmarks in a browser, be used in the curl command, or issued by an application program, for example, to automatically feed tasks into a crowdsourcing platform.

A REST service on a given server must implement the following four principles (from en.wikipedia.org/wiki/REST):

(i) Identification of resources via URI and retrieval of resources in HTML, JSON, or XML representation.
(ii) Manipulation of resources through these representations on the client.
(iii) Self-descriptive messages describe how to process the message.
(iv) Hypermedia as the engine of application state.

Many crowdsourcing platforms, for example, MTurk or Clickworker, offer RESTful services, that is, web services implemented using HTTP and following the four principles. Such interfaces are also called *web API* or simply *API* (application programming interface).

Example The request

```
GET https://sandbox.clickworker.com/api/marketplace/v2/customer/
```

will retrieve information about a given customer from the protected sandbox environment on the server sandbox.clickworker.com in XML format:

```
<?xml version="1.0"?>
<customer_response>
  <request_status>
    <id>+rSF+zat1qe7LCV5eJH5d7fwTeo=</id>
    <valid>true</valid>
    <status_code>200</status_code>
    <status_text>OK</status_text>
  </request_status>
  <customer>
    <link href="https://sandbox.clickworker.com/en/api/marketplace/v2/customer"
     rel="self" type="application/xml"/>
    <balance_amount>0.00</balance_amount>
    <reserved_amount>0.00</reserved_amount>
    <credit_limit>200.00</credit_limit>
    <currency_code>EUR</currency_code>
  </customer>
</customer_response>
```

9.2.12 Section Summary

This section has introduced the technology of the World Wide Web. The focus has been on the technology and terms relevant to crowdsourcing in general, and crowdsourcing for speech processing in particular. This technology and the underlying principles are applied to crowdsourcing for speech in Section 9.3 where crowdsourcing platforms are described in more detail.

9.3   Crowdsourcing Platforms

A crowdsourcing platform is an entity, usually a company or research institute, that provides the software and administrative infrastructure necessary to run a crowdsourcing service, and that has access to a pool of workers who select and work on tasks.

A *requester* is a person or institution who wants specific jobs to be performed by crowdsourcing. Using a crowdsourcing platform, the requester breaks the job down into *tasks* small enough to be performed by *workers*. The requester specifies the minimum qualifications that workers must provide to work on the given task. An *assignment* associates a worker with a task; very often, a unit task is processed by more than one worker to improve the reliability of the results obtained, for example, by selecting the results of the majority of workers. In general, a worker receives a *reward* for his or her work if the work is accepted by the requester.

The software implementing a crowdsourcing platform is a web application that provides the following:

- Administration of requesters and workers.
- **For requesters**: Definition, test, and deployment of tasks.
- **For workers**: Selection from classified lists of available tasks, and processing of assignments.
- Quality control.
- Processing of payment.
- Forums for workers and requesters.

The pool of workers is a set of registered users who are willing to work on tasks in exchange for some reward, for example, payment or purchase vouchers.

9.3.1   Crowdsourcing Platform Workflow

The typical workflows on a crowdsourcing platform are shown in Figure 9.6.

Both requesters and workers must register with a crowdsourcing platform. The requester workflow *define-test-deploy* submits tasks into the task database. The worker workflow *select-process-submit* reads assignments from and updates the corresponding tasks in the database. Finally, the requester workflow *review-retrieve* exports completed assignments from the task database to the requester.

Registration and Administration

The administration of requesters consists of registering the necessary business and legal information from the requester: company or individual's name, address, legal status, bank account data, and so on. Once sufficient credentials have been provided and a money deposit has been made, the requester is given an account on the platform. Many platforms allow testing the platform in a so-called *sandbox* environment without registration so that requesters can check whether their task can be run on a given platform.

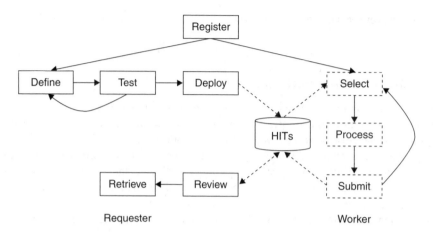

Figure 9.6 Typical requester (solid boxes) and worker (dashed boxes) workflows on a crowdsourcing platform. The database of tasks is accessed both by requesters and workers.

Workers must also register with a given crowdsourcing platform. In general, they must provide some means of identification, bank details and information about their qualification, for example, languages spoken, reading or writing skills, and hearing impairments. Any additional qualification, possibly achieved as a worker on this platform, is also stored. Depending on the platform, anonymous participation is also possible, for example, in online surveys or experiments.

Definition, Test, and Deployment of Tasks

Requesters define the tasks according to the technical infrastructure provided by the crowdsourcing platform. In general, there are two ways of defining a task:

1. *Manually* via a graphical user interface (GUI).
2. *Programmatically* via an application programming interface (API).

In the *manual* approach, a software wizard guides the requester through the definition of a task in a graphical editor running within the setup pages on the platform's website. The editor provides the interface elements to be used in the task page, for example, input fields for the text to be displayed, or templates for input elements. These templates are optimized for a given type of input, for example, a group of radio buttons for mutually exclusive input options.

This approach is suited for prototyping tasks, and if the number of different task types is small.

In the *programming* approach, the requester defines the contents and appearance of tasks using a software development kit (SDK) exposing the API of a crowdsourcing platform. The set of available elements to define a task is determined by the crowdsourcing platform—in general, the requester downloads an SDK to a local machine and opens it in an integrated development environment (IDE) such as Eclipse or some other editors.

This approach is more flexible in that it allows the requester to make use of the full set of API features. Furthermore, tasks can be generated automatically from source data via scripts.

Once editing or generating the task is done, it is uploaded to a *sandbox* on the crowdsourcing platform's server. The sandbox is a protected area on the server, which offers all features of the final deployment platform, but the task is not visible to the workers. The requester tests the task—does it look right, is the text readable, are the instructions clear, do the input options work as expected, how long does it take workers to perform a task, and so on?

When the requester is satisfied with the performance, display and stability of the task, it is submitted to the database of available tasks and assignments in the public area of the server.

Selecting and Processing Tasks

A worker views the available tasks from a classified task list and selects an assignment. Some platforms distribute tasks automatically to workers based on their skills. Most crowdsourcing platforms offer a search function to allow workers to quickly find suitable tasks.

The page associated with the task opens in the browser and the worker processes the assignment. When the worker submits his or her work, it is transferred to the server together with the information about the worker having processed this assignment.

Quality Control

The requester reviews the result of the processed tasks and decides about the payment. Different platforms have different policies: the requester may retrieve all results or only a subset of the results without any payment, or the requester must pay to retrieve the results. Some platforms have auto-accept policies by which a worker must be paid if the requester does not reject his or her work within a given time span (in general, a day or two).

Typically, there are two reasons why the results provided by the worker are rejected by the requester:

1. A worker may not have followed the instructions, or is not capable of doing the task, or simply because of typos or incorrect input.
2. A requester may have provided ambiguous instructions, or the task is too difficult for the workers (who are often unqualified), or the time constraints are too strict, or some other reason.

Some platforms allow the requester to exclude workers from specific tasks if they do not achieve satisfactory results. Some platforms give workers the opportunity to redo their work if the requester did not accept the first version. Some platforms provide a mediation service in the case that workers and requesters disagree about the payment. Some platforms provide forums for workers and requesters to exchange information, and others try to prevent any direct exchange between workers and requesters or even between workers.

In linguistic crowdsourcing tasks, quality control is often implemented by having several workers perform the same task in different assignments. When all assignments of a task are done, the results are compared, for example, for interannotator agreement. If a given threshold of agreement is reached, the corresponding assignments are accepted and the worker can be paid.

Quality control is one of the most critical issues in crowdsourcing. Several studies have been published (see Parent and Eskénazi 2011 for an overview), but the results are still inconclusive and are not necessarily valid for speech-related tasks. (Aker *et al.* 2012) find higher quality for higher payment for simple objective tasks, (Mason and Watts 2010) only more but not better results for object detection in images. In domains with subjective results, for example, translation, transcription or rating, quality control is particularly difficult (Kunchukuttan *et al.* 2012). There is a trade-off between speed, quality, and cost—if checking worker input for quality cannot be automated or at least performed very efficiently, then crowdsourcing this particular task may not be a good idea (see Chapter 4 for details).

Example   Audhkhasi and a colleague from SAIL laboratories at the University of Southern California have used MTurk for linguistic tasks (K. Audhkhasi, personal communication, 2012). According to them, the most important shortcoming of MTurk is the "limited quality control."

In their MTurk projects, they submitted tasks to more than one worker. The rationale behind this was that if several workers produce the same input for a given task independently of each other, then this input must be correct. However, they experienced serious problems with this approach: requesters must quickly decide whether to pay a worker or not; if the delay between two workers processing the same task is too long, then the requester cannot compare the workers' input and hence not decide which input is correct.

Furthermore, many linguistic tasks, for example, quality or acceptability ratings, or speech recordings, do not have correct solutions.

Processing of Payment

When the requester accepts the result of an assignment, the platform transfers the amount to be paid from the requester's deposit to the worker. In general, the platform keeps a part (commonly 10–50%) of the payment for itself.

Retrieving the Task Results

In general, the requester is given access to the full results of the tasks after the work has been paid. (However, some platforms, notably MTurk, give requesters access to the results without payment.) Depending on the type of result data, the data is mailed to the requester, or made available via download using REST or some other data delivery service, for example, DropBox or Amazon S3. The data format is usually CSV or XML for text or other well-structured data, and raw formats for media data.

9.3.2   Amazon Mechanical Turk

MTurk is one of Amazon's web services (aws.amazon.com/). It is "a web service that provides an on-demand, scalable, human workforce to complete jobs that humans can do better than computers" (MTurk API reference, p. 1, Amazon Web Services LLC 2011). MTurk was one of the first crowdsourcing platforms on the market, and it has served as a model for many similar platforms.

At the time of writing, MTurk accepts only institutions with a US billing address and a "US ACH-enabled bank account," or individuals with a valid US driver's license number as requesters. Workers can be from almost any country, but they must be residents of a rather small list of countries, for example, United States, Canada, and India, to receive payment. Workers from other countries are paid with shopping vouchers.

In order to use MTurk, a requester has to set up a requester account and provide either a bank account, or credit or debit card information. After setting up the account, the requester may upload tasks either to the sandbox for testing, or to the production system so that workers can select and work on assignments.

Speech-Related Projects Performed with MTurk

MTurk is being used quite a lot in speech-related research and commercial applications.

A search performed via Google through the conference proceedings of Interspeech, LREC and ICPhS for the years 2008–2012 has resulted in 330 hits for MTurk, and 249 for "crowd-sourcing" (last checked: 2012-07-06)[1].

On March 22, 2012, a total of 726 tasks containing the words "audio transcription" were found in the MTurk list of available tasks, and 430 for "speech transcription." For "audio recording" or "speech recording," 91 resp. 95 tasks were found (for an alternative to using MTurk for speech recordings, see Section 9.3.5).

The recording projects simply ask the worker to record an utterance via some external audio recorder such as WaveSurfer and then upload the audio file; others provide a visible Flash element with a recording button. The transcription tasks ask the worker either to transcribe a given utterance according to some transcription guidelines (which, e.g., specify special markup for noise events and mispronunciations), or to simply write down the words of an utterance to result in a readable text (see Chapters 3 and 4 for details on crowdsourcing for audio recording and transcription and Figure 9.7 for sample screens of speech related tasks in MTurk).

Discussion of MTurk

A number of studies have analyzed the type of tasks performed via MTurk. This is facilitated by analysis tools provided by the Amazon Web Services—see the introduction to this book for details.

Most MTurk tasks are useful. However, studies have found that MTurk is also used to generate spam or promotional content for social networks, discussion forums, and customer feedback texts in commercial websites.

---

[1]The search query was `allintext: LREC OR Interspeech OR ICPhS "Mechanical Turk" site:.org filetype:pdf`

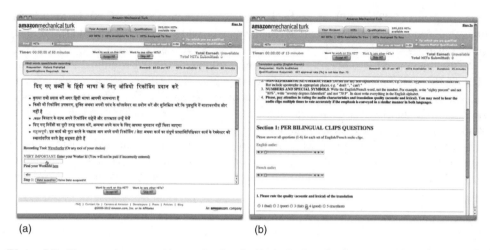

(a)   (b)

Figure 9.7 Two sample pages with speech-related tasks in the list of HITs of MTurk. Note that the task itself is inside an internal frame embedded within the MTurk page. (a) This is a recording task: workers are asked to record an utterance using an external audio recorder such as WaveSurfer and to upload the audio file to MTurk. (b)This is a rating task where workers are asked to rate the quality of a translation from English to French.

Example   In his blog contribution "Mechanical Turk: Now with 40.92% spam." (Ipeirotis 2010b) analyzed how many MTurk tasks are "spammy," that is, tasks that generate ad clicks, create social network followers and friends, and so on. He arrives at the quite disturbing conclusion that approximately 41% of all tasks actually fall into the category "spammy." Interestingly, he was able to compute this number by using the gold-standard technique of CrowdFlower (see Section 9.3.3 for details).

In a second contribution, "Be a Top Mechanical Turk Worker: You Need $5 and 5 Minutes," he describes how workers can achieve a high number of completed tasks with a 100% approval rate within a few minutes (Ipeirotis 2010a). Both the number of completed tasks and the approval rate are commonly used as qualification criteria for workers.

Another issue is the reward paid to the workers, which may be ridiculously low for quite qualified work. Of course, what is considered low in one country may be acceptable in others—however, common sense must be used.

Transcribing audio is a time-consuming task. An experienced and quick typist will need two to five times the duration of the audio file for transcription, depending on the comprehensibility of the audio file and the familiarity of the worker with the domain of the lecture. In general, workers are required to use correct punctuation and spelling. To correct their spelling, they are encouraged to use the autocorrection feature of Google, which slows down transcription considerably.

In this particular case, MTurk workers have warned each other about this requester in forums dedicated to MTurk.

Example The transcription of a 10-minute audio file of a gynecology lecture pays $0.30. The requester states that he knows the pay rate is low, but he claims to be a student who cannot afford to pay more. There is no way a worker can verify this claim.

For an in-depth discussion, see Chapter 11 on ethical issues.

9.3.3 CrowdFlower

CrowdFlower (`crowdflower.com`) is a commercial crowdsourcing platform. It differs from other crowdsourcing platforms in that it does not have its own pool of workers. Instead, Crowd-Flower submits its jobs to other crowdsourcing platforms, for example, MTurk, InboxDollars, or Swagbucks.

CrowdFlower allows requesters to set up tasks in the sandbox environment with a simple user registration; no bank details are needed.

CrowdFlower encourages the requester to define so-called *gold-standard* data which is used to determine the quality of the results returned from the workers. Gold-standard data specifies the correct and accepted input to given tasks. This data is provided by the requester.

With these gold-standard data, CrowdFlower can automatically determine a worker's performance, and it can give workers feedback. These gold-standard tasks are inserted inconspicuously in the list of available tasks. A worker's performance is measured via his or her agreement with the gold standard. Note that if the quality of work delivered by a worker matches the gold standard, it has to be accepted by the requester.

Finally, after the data has been uploaded and the task has been defined, the requester must calibrate the price and choose a worker channel, that is, a crowdsourcing platform that provides the workers for this task. According to CrowdFlower, it normally takes a few iterations until a price is found that is both attractive to workers and acceptable to the requester. Clearly, higher paid tasks will be completed quicker than lower paid ones—but not necessarily with better results. Results are retrieved from CrowdFlower either again via a CSV text file, or via JSON data objects.

9.3.4 Clickworker

Clickworker is a German company with associated offices in the United States and Switzerland. It focuses on three main areas:

1. Text generation and translation
2. Data categorization and tagging
3. Web search and surveys.

Text generation is divided into item descriptions, search-engine-optimized texts and translations. Data categorization and tagging refers to media, usually video content. Web search and survey tasks request workers to search for specific information in the web, for example, opening times of local small businesses, or to take part in surveys.

It has standard tariffs for text tasks, and individual tariffs for all other tasks. All text generation results are checked for plagiarism or copyright violations via the web service copyscape.com. Furthermore, Clickworker claims that it makes sure that its workers receive fair payment, claiming that its workers should be able to reach a payment of €9 per hour (Clickworker, personal communication, 2012).

The target audience for Clickworker is large and medium-sized companies in the areas of media, Internet, e-commerce, and directory services.

Example In the download section of their website, Clickworker presents some descriptions of successful use cases. For example, a set of tasks was to categorize moves in US football with a fixed vocabulary for Swink.tv (Humangrid GmbH 2010). Clickworkers qualified in US football were asked to tag 4000 videoclips, with a total of 6500 moves. Clickworker delivered 200–300 tagged moves per day, at a cost of €0.13 per move.

9.3.5 WikiSpeech

Since 1997, when WWWTranscribe was presented (Draxler 1997), the Bavarian Archive for Speech Signals (BAS, phonetik.uni-muenchen.de) hosted by the Institute of Phonetics and Speech Processing at the University of Munich has been developing web-based speech tools. Originally, the motivation for the development of web-based tools was to become independent of the operating system and proprietary annotation software and hence be able to use all available computers in the laboratory. A second motivation was to distribute the annotation work over as many people as there were available. This latter aspect became ever more important as the size of the speech databases grew dramatically.

WikiSpeech (webapp.phonetik.uni-muenchen.de/wikispeech) is the current web application for speech recording and annotation via the web at the BAS (Draxler and Jänsch 2008). Its components are SpeechRecorder (Draxler and Jänsch 2004) for web-based scripted speech recordings, and WebTranscribe, a flexible web-based multilingual annotation editor (Draxler 2005). Recently Percy, a framework for web-based perception experiments was developed (Draxler 2011), and it is currently being integrated into WikiSpeech. A similar system, targeted at web-based speech recordings for the training of speech recognition systems, is SPICE at CMU (Schultz *et al.* 2007).

WikiSpeech qualifies as a crowdsourcing platform in that:

- It distinguishes requesters and workers.
- Workers select and perform tasks from a list of projects.

In WikiSpeech, the requesters are project administrators; they are responsible for defining the contents of a recording, for retrieving the audio data collected, for defining annotation tasks, or for the design of online experiments. Workers are speakers, annotators, or experiment participants – workers known to the system can be assigned to projects, and new workers may at any time sign up and join WikiSpeech.

WikiSpeech is different from other crowdsourcing platforms in the following respects:

- The use of WikiSpeech is free for academic projects and institutions.
- Workers do not get paid.

- The basic layout and screen presentation of tasks is fixed.
- The initial installation of a project must be performed by the system administrator of WikiSpeech.

Two large speech database creation projects show how WikiSpeech can be used for crowdsourcing; in the Ph@ttSessionz project, more than 1100 pupils aged 12–20 were recorded via the web in 41 public secondary schools in Germany (Draxler and Steffen 2005). In the VOYS project, 251 pupils were recorded in schools in 9 metropolitan areas of Scotland (Dickie *et al.* 2009). Both speech databases contain read and nonscripted speech; the read material consists of digits, numbers, money amounts, date and time expressions, proper names and phonetically rich sentences. VOYS also contains a 300-word long story in the form of a dialog ("Dog and Duck" story). The number of recordings in Ph@ttsessionz exceeds 110,000, and 25,000 in VOYS. The annotations adhere to the SpeechDat transcription guidelines (Senia and van Velden 1997), and they were performed using the WebTranscribe component of WikiSpeech.

9.4 Interfaces to Crowdsourcing Platforms

The creation of a task on a crowdsourcing platform comprises two basic steps, *task design* and *task implementation*. The task design relates to domain expertise, the task implementation to technological expertise. The domain expert knows what the task is about, what qualification is necessary to process the task, how it can be presented to a worker, what instructions and auxiliary information are necessary, what input is expected, and what constitutes an acceptable result. The technology expert knows how to implement a task given the specific technology of a crowdsourcing platform.

Domain and technology experts can be the same person, or they can be different people in a team. The technology provided by crowdsourcing platforms for the definition of tasks ranges from easy-to-use graphical editors to large programming language-specific SDKs and APIs. Thus, the complexity of a task, the available domain and technology expertise, and the technology provided by the platform determine how this task can be implemented.

This section describes four different approaches in detail to the implementation of tasks:

(i) Manual implementation using a graphical editor within a GUI.
(ii) Command-line tools using an SDK and the platform's API.
(iii) A RESTful web service.
(iv) XML configuration files.

They are presented with examples for the crowdsourcing platforms CrowdFlower, MTurk, and Clickworker, respectively. The implementation of tasks for WikiSpeech can be achieved via a graphical editor, or via uploading configuration files in XML.

Most crowdsourcing platforms provide at least one of these approaches to creating tasks. These approaches differ from each other in terms of ease-of-use, functional capabilities or expressive power, and support and documentation.

Novice users generally start with a manual approach using a graphical editor provided by the crowdsourcing platform. This approach is suitable for tasks with few variable elements, a simple layout and quite a small number of tasks. To design a task, the requester opens a graphical editor on the crowdsourcing platform's website. Generally, the platform already provides a number of predefined designs. The requester selects a design, adapts the screen layout to his or her requirements, and connects the screen elements with the underlying data. Most graphical editors are limited in their functional capabilities. In general, they provide only a subset of available input elements, have only standard input validation filters, and cannot be customized or extended.

Expert users will prefer using an SDK in conjunction with the platform's API or a RESTful web service. This requires programming skills in one of the programming languages supported by the crowdsourcing platform or libraries built around the crowdsourcing API (such as the Python Boto library, `github.com/boto`, built around the MTurk API), and an established programming workflow or IDE such as Eclipse (`eclipse.org`). Such an integrated programming environment loads the SDK and exposes the API of the platform to the programmer. In general, using an SDK gives the requester full access to the capabilities of a platform. It allows implementing customized input validation filters, dynamic adaptation of screen layout, and a flexible access to the underlying data.

9.4.1   Implementing Tasks Using a GUI on the CrowdFlower Platform

The CrowdFlower platform provides both a GUI and an API to implement tasks. Using the GUI, implementing a task consists of three steps:

   (i) Upload the data using the data panel.
  (ii) Create the worker form and instructions in the edit panel.
 (iii) Create gold-standard data in the gold panel.

Initially, the CrowdFlower website displays the so-called *Jobs dashboard*. The dashboard is divided into the panels: overview, data, edit, gold, analytics, and reports. Overview gives a bird's eye view of the current requester and the status of the tasks submitted by this requester.

Data Panel

The data panel lists the data rows that underlie the open tasks; each data row corresponds to a task. It carries a unique identifier and consists of instructions to the worker and the data proper, which is displayed in the task web page.

The requester prepares the data in a spreadsheet file on the local computer and then uploads the file to the CrowdFlower platform via a standard HTTP file upload. CrowdFlower recommends using Unicode in UTF-8 encoding, and it assumes that the first row in the data table contains the name of the data item.

Edit Panel

The edit panel contains the CrowdFlower's task editor. This editor has three modes: a graphical mode, a CML mode, and a preview mode.

In the graphical mode, the left side of the editor window contains a list of available web page input elements with editable labels. The right side of the editor window contains the web page elements of the task.

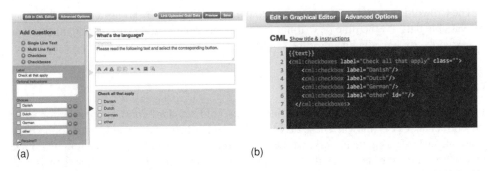

(a)   (b)

Figure 9.8   Details of the CrowdFlower editors. (a) In the graphical editor, data items on the left side are graphically linked to visible elements in the task web page. (b) The CML editor shows the CML commands with syntax highlighting.

By default, the CrowdFlower graphical editor divides a task page into four parts: title, instructions, data and input elements. The title and instructions are simply text fields with the text set by the requester. The data element takes its visible contents from the data items uploaded to the platform, and the input elements are the visible graphical elements which the worker uses to enter his or her input. In the graphical editor, arrows between the left and right sides of the window show which text or data goes into which visible page elements (see Figure 9.8a).

CML (CrowdFlower markup language) is a markup language optimized for defining crowdsourcing tasks. All CML tags belong to the namespace `cml:`; they require less typing than their HTML counterparts and allow restrictions for input validation, for example, `validates="required numeric"`.

Example   A typical CML tag for a rating element is shown below. Note that in order to produce this in HTML and CSS, quite a lot of codes would have been necessary, for example, to evenly distribute the buttons or to place the text above the buttons (compare with Figure 9.3)

```
<cml:ratings label="Rate me" points="4" />
```

In CML mode, the editor displays the CML code that implements the task web page (see Figure 9.8b).

The advanced options are used for complex and application-specific input widgets, for example, hours of operation per weekday, or a taxonomy, that is, a hierarchical tree structure. Additionally, CML provides logical checks via attributes such as `only-if` which are checked when the worker enters data and either proceeds to the next input element or submits the entire form.

The CrowdFlower GUI editor is limited to the capabilities of CML plus a few reserved tags for variable data. At the time of writing, it was not possible to add extra markup to the code, for example, for a media player or the `<audio>` tag. For this, the API must be used.

Gold Panel

In the next step, a gold standard is defined. This gold standard consists of tasks for which the correct answer is known. These tasks are defined by the requester and added to the list of available tasks. During their work, workers will process gold-standard tasks without knowing this. The results of these tasks will be compared to the expected input. If a worker consistently produces correct results, then his or her work is considered as trusted. This helps both the requester, who now has workers he or she can rely on, and the worker who may thus gain extra qualifications that might be a prerequisite for other tasks.

Submitting a Task

When a task has been defined and tested and is ready to be worked on, it must be submitted to the CrowdFlower platform as an available task. The requester specifies the amount to be paid for each task, and whether any restrictions, for example, minimum worker qualifications, apply. When these details have been completed, the task is submitted.

While the job is active, that is, available for workers to select, the dashboard gives Crowd-Flower customers an up-to-date overview of the available jobs, allows editing these jobs and the definition of gold-standard answers. Finally, the dashboard also contains an analytics package to monitor the performance of workers, and a report generator.

9.4.2 Implementing Tasks Using the Command-Line Interface in MTurk

MTurk offers various interfaces to requesters: the requester user interface, a command-line interface, and the API accessed from SDKs. These interfaces not only support the implementation of tasks, but also allow the requester to monitor work progress, accept or reject the workers' results, and pay workers or grant them extra qualifications.

In MTurk terminology, a task is called a *human intelligence task (HIT)*. In order to implement a task, a requester has to have a valid requester account with Amazon. The identity of a requester is determined by the RequesterID. The authentication of most AWS APIs (including MTurk API) uses the access key and a secret key.

This section presents both the requester user interface and the command-line interface of MTurk. However, since the requester user interface is very similar to that of CrowdFlower, the emphasis is on the command-line interface and the SDK. Note that in the following listings some details of the code have been removed for legibility. The code has been tested in the sandbox environment of MTurk.

Implementing a Task via the Command-Line Interface

MTurk Getting Started guide recommends the command-line interface "to have a hands-on approach [...] and a relatively small number of assignments and results" (MTurk Getting Started Guide Amazon Web Services LLC 2011, p. 28).

Basically, the definition of a task in MTurk consists of three files:

(i) Input file.
(ii) Template or question file.
(iii) Properties file.

If an external HTML file is used, this file must of course be accessible via a URL.

Input File

The *input file* is a tab-delimited plain text file with the standard extension `.input`. It contains the questions the worker has to answer. Note that in this list, `[...]` is a placeholder for a valid server address:

```
audio_url
http://[...]/audio/recording_DE.wav
http://[...]/audio/recording_DK.wav
http://[...]/audio/recording_NL.wav
```

Template or Question File

The *template* or *question file* is an XML file with the standard extension `.question`. MTurk defines two types of question files: a *regular* question file and an *external* question file. The regular question file defines task web pages that will reside on the MTurk server; these regular question files come in two formats: either a pure XML format that may contain only a limited set of elements, or XML with embedded HTML. The external question file refers to an HTML-formatted task page file on an external server. This external task file is embedded in the final task web page within an `<iframe>` element.

There is a trade-off between flexibility and ease of use: regular question files restrict requesters in the choice of available page elements, but automate the display of data and the submission of worker input. External question files give requesters full control over the appearance of their task web page and allow them to use their preferred page editors or even automatic page generators, but require that they take care of the data display and input submission implementation themselves.

The question file defines a template for the task and conforms to the `QuestionForm`, the `HTMLQuestion` or the `ExternalQuestion` schema defined in the MTurk API Reference (Amazon Web Services LLC 2012). This template lists all elements that appear on the task web page and the type of input expected, and it is populated with data from the input file. Placeholders for data are marked using the $ sign and the variable name in curly braces, which must correspond to a name in the input file.

Template File According to the `QuestionForm` Schema

A question file according to the `QuestionForm` schema is a pure XML file. Elements such as `<List>` or `<AnswerSpecification>` define the visual input elements of the task web page (see Figure 9.9).

```xml
<?xml version="1.0"?>
<QuestionForm xmlns="http://[...]/QuestionForm.xsd">
<Overview>
 <Title>Determine the language</Title>

 <List>
  <ListItem>Click on the playback button to listen to the audio.</ListItem>
  <ListItem>Select the button corresponding to the language you heard</ListItem>
 </List>

 <EmbeddedBinary>
  <EmbeddedMimeType>
   <Type>audio</Type><SubType>wav</SubType>
  </EmbeddedMimeType>
  <DataURL>${audio_url}</DataURL>
  <AltText> </AltText>
  <Width>400</Width><Height>20</Height>
 </EmbeddedBinary>
</Overview>

<Question>
 <QuestionIdentifier>1</QuestionIdentifier>
 <QuestionContent><Text>Determine the language</Text></QuestionContent>
 <AnswerSpecification>
  <SelectionAnswer>
   <StyleSuggestion>radiobutton</StyleSuggestion>
    <Selections>
     <Selection>
      <SelectionIdentifier>DE</SelectionIdentifier>
      <Text>German</Text>
     </Selection>
     <Selection>
      <SelectionIdentifier>NL</SelectionIdentifier>
      <Text>Dutch</Text>
     </Selection>
     <Selection>
      <SelectionIdentifier>DK</SelectionIdentifier>
      <Text>Danish</Text>
     </Selection>
     <Selection>
      <SelectionIdentifier>__</SelectionIdentifier>
      <Text>other</Text>
     </Selection>
    </Selections>
  </SelectionAnswer>
 </AnswerSpecification>
</Question>

</QuestionForm>
```

Figure 9.9 Sample MTurk question file in XML format with `<QuestionForm>` root element. Note how many lines of code are needed for four radiobuttons, and the use of `<EmbeddedBinary>` to allow the browser to display the preferred media player controller on the screen

The template file features tags for media content, which can be of type image, audio, or video. Media content can be displayed in the task web page either via Flash or Java applets (see the MTurk API Reference, Amazon Web Services LLC 2012, pp. 187 for details).

The MTurk platform generates the web page from the question file according to MTurk's predefined styles (for an example, see Figure 9.11a).

Template File According to the ExternalQuestion Schema

If external files are to be used, then two files are needed: the question file and the HTML file it refers to on the external server. The question file is a pure XML file with an <ExternalQuestion> element:

```
<ExternalQuestion xmlns="http://[...]/ExternalQuestion.xsd">
  <ExternalURL>
  http://[...]/tinytask.html?url=${helper.urlencode($audio_url)}
  </ExternalURL>
  <FrameHeight>450</FrameHeight>
</ExternalQuestion>
```

The corresponding HTML file may be any HTML file. However, this file must use the <form> element for submitting input data to the MTurk server, and it must provide a number of predefined form fields and JavaScript functions to process data, for example, tasks contents or assignment IDs (see Figure 9.10).

The task page appears exactly as defined in the external HTML and style pages. Thus, the requester may apply his or her own style sheets to the page (see Figure 9.11b for an example).

Properties File

The *properties file* is a plain text file with *key:value* pairs. It has the extension .properties and contains the administrative data on the task, for example, title of the task, short description, reward, and timing information:

```
#####################################
## Basic HIT Properties
#####################################

title:Determine the language
description:Please listen to the audio contained in this HIT
keywords:audio, language, v2
reward:0.03
assignments:1
annotation:${audio_url}
```

```
<!-- This file needs to be hosted on an external server.  -->
<html>
 <head>
  <title>Tiny Task</title>
   <meta http-equiv="content-type" content="text/html;charset=utf-8" />
  <link rel="stylesheet" href="tinytask.css" type="text/css" />
 </head>
 <script language="Javascript">

 function gup( name ) {
  var regexS = "[\\?&]"+name+"=([&#]*)";
  var regex = new RegExp( regexS );
  var tmpURL = window.location.href;
  var results = regex.exec( tmpURL );
  if( results == null )
   return "";
  else
   return results[1];
 }

 function decode(strToDecode) {
  var encoded = strToDecode;
  return unescape(encoded.replace(/\+/g,  " "));
 }

 submitvalue = function (value) {
  document.getElementById('selection').value = value;
  document.getElementById('mturk_form').submit();
 }
</script>

<body>
 <form id="mturk_form" method="POST"
   action="http://www.mturk.com/mturk/externalSubmit">
  <input type="hidden" id="assignmentId" name="assignmentId" value="">
  <input type="hidden" id="selection" name="selection" value="">

  <div class="centered">
   <h1>Determine the language</h1>
   <audio id="expaudio" src="audio/welcome.wav">
   </audio>
   <ol class="left">
    <li>Click on the loudspeaker icon to listen to the audio</li>
    <li>Select the button corresponding to the language you heard</li>
   </ol>
   <img src="icons/speakers.gif"
    onclick="document.getElementById('expaudio').play()"; />
   <br />
   <input type="radio" name="userinput" onclick="submitvalue('DE');" />German
   <input type="radio" name="userinput" onclick="submitvalue('NL');" />Dutch
   <input type="radio" name="userinput" onclick="submitvalue('DK');" />Danish
   <input type="radio" name="userinput" onclick="submitvalue('__');" />other
  </div>
 </form>
 <script language="Javascript">
  document.getElementById('assignmentId').value = gup('assignmentId');
  document.getElementById('expaudio').src = decode(gup('url'));
 </script>
</body>
</html>
```

Figure 9.10 HTML code for the TinyTask page embedded in an MTurk task page. Note the use of the required <form> element with fixed action and method attributes, and predefined form fields and JavaScript functions.

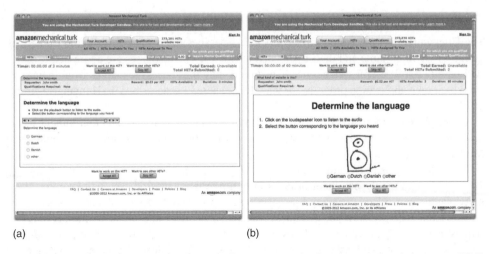

(a)   (b)

Figure 9.11   The task page in the MTurk sandbox, generated using the `<QuestionForm>` XML document (a), and via an external HTML page (b). Note how the HTML page style is modified by the CSS style definitions.

```
#####################################
## HIT Timing Properties
#####################################

# this Assignment Duration value is 60 * 3 = 3 minutes
assignmentduration:180

# this HIT Lifetime value is 60*60*24*7 = 1 week
hitlifetime:604800

# this Auto Approval period is 60*60*24*3 = 3 days
autoapprovaldelay:259200
```

Once these three files have been created, they are uploaded to the MTurk developer sandbox for testing, or submitted to the production area via the `loadHITs` command-line tool.

```
loadHITs -input tinytask.input -question tinytask.question
    -properties tinytask.properties -sandbox
```

The `loadHITs` command checks whether the requester has provided the necessary access credentials. If so, then it validates the template file against the schema and finally submits the task to the sandbox or the production environment.

9.4.3   Implementing a Task Using a RESTful Web Service in Clickworker

The Clickworker API is called *Marketplace*. It describes the resources available to the programmer via their name and their associated attributes. The resources are addressed using an API based on a RESTful web service. Customers may thus access resources from their tasks from their own application programs or via URLs. The API is described in (Humangrid GmbH 2012).

Marketplace distinguishes three runtime environments: the production environment where tasks are worked on, a sandbox to test tasks, and a sandbox beta where upcoming versions of Marketplace can be used by developers.

The API supports localization via language tags, with English being the default language. This allows a quick development of tasks in multiple languages; in fact, Clickworker offers native language workers for many European languages, including Turkish and Russian.

Using the REST paradigm, a request transfers data from the client to the server. This request may carry parameters to address the appropriate resources on the server, for example, customers, accounts, and tasks. A response is the data returned from the server to the client. In Marketplace, it is formatted in XML or JSON. The HTTP verb GET retrieves resources, POST creates a new resource on the server, PUT updates a known resource and DELETE removes it from the server. (Note that this deviates from the usual operational semantics of the HTTP verbs as described in Section 9.2.11.) For security reasons, communication with the server is encrypted.

Figure 9.12 shows the REST plug-in of Mozilla Firefox connected to the Clickworker Marketplace web service with a request–response pair of data on the current customer. Note that the customer has identified himself or herself by logging into the Marketplace. Hence, customer ID or name need not be transferred via a request—they can be inferred from the login and session key.

On the Clickworker platform, once a customer has received an account, he or she creates a task template resource, task resources, and so on. All resources are defined using the RESTful web service. This can be done manually, which is quite time-consuming, or via an application program that generates the appropriate REST statements from an internal database of task data.

The Marketplace API documentation (Humangrid GmbH 2012) lists all accessible resources and gives examples for their retrieval and creation. The API supports the standard user input elements such as single-line text fields and multi-line text areas, numeric and date fields, URL data and e-mail addresses, plus a media player for audio and video playback (at the time of writing, only a proprietary player is supported). Hence, this API provides the basic requirements for speech-related tasks.

9.4.4   Defining Tasks via Configuration Files in WikiSpeech

WikiSpeech is a crowdsourcing platform for the speech-related tasks such as recording, annotation, and online experiments. The use of WikiSpeech for academic projects and institutions is free.

Because in WikiSpeech the visual appearance is predefined, only the contents of the task need to be specified. For this, CSV or XML configuration files are used.

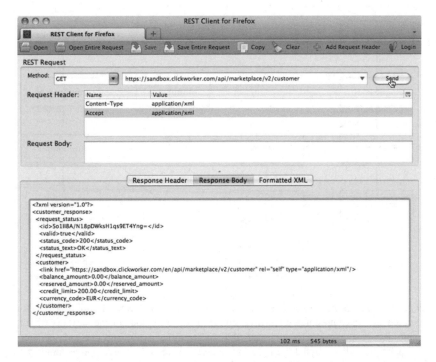

Figure 9.12   REST plug-in for Mozilla Firefox connected to the Clickworker server. the request asked for information about the current customer, and the response body contains the server's response in XML format.

In order to use WikiSpeech, requesters apply for a project administrator account. With the project administrator account, they can set up recording, annotation, or online experiment projects. A project assigns workers, that is, speakers, transcribers, and experiment participants, to tasks defined by requesters. Such tasks typically are recording speech, transcribing recorded audio files, or responding to experiment stimuli.

Technically, WikiSpeech is based on a global corpus data model covering the entire workflow in speech database creation, annotation and distribution.

Figure 9.13 shows a simplified representation of the data model relevant for speech recordings and annotations: a recording project consists of sessions that contain recording scripts which in turn are grouped into sections arranged in sequence; a section contains one or more recordings in sequential or random order. A recording prompts the speaker with some media item, which may be text, an image or audio. The result of a recording is a signal file.

Once a signal file has been recorded, it may be transcribed. An annotation project consists of sessions which in turn contain individual annotations. A annotation may consist of several annotation tiers that contain time-aligned or symbolic segments, depending on the type of tier.

Figure 9.14 shows the data model for online experiments in WikiSpeech. The requester defines an experiment project that may contain several experiments. Experiments consist of one or more experiment scripts, which in turn are divided into sections. A section contains the experiment items which will be presented to the participant. Sections are ordered sequentially;

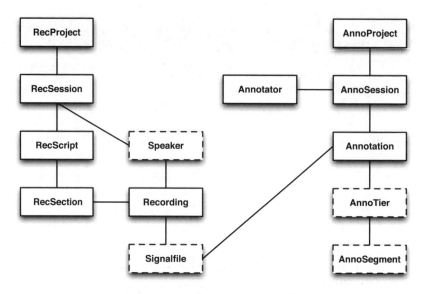

Figure 9.13 The recording and annotation task data model in WikiSpeech. Solid boxes denote data objects defined by the requester for the task, and dashed boxes contain the workers' input, that is, demographic data on the speaker and signal files for recording tasks, and tier and segment data for annotation tasks. Note that a signal file is the outcome of the recording task, and becomes the input or task data for an annotation task.

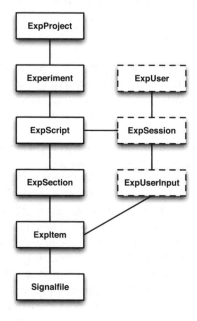

Figure 9.14 An online experiment data model in WikiSpeech. Note that the signal file may reside on the WikiSpeech server, or be any publicly accessible media file in the WWW.

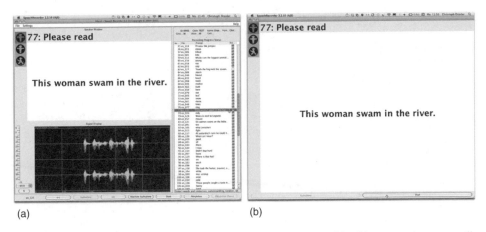

(a)   (b)

Figure 9.15   WikiSpeech supervisor screen (a) and speaker screen (b) with prompt item, recording indicator and control buttons

items within sections may be ordered or in random sequence. The worker input to an experiment is data on the participant and the current experiment session, and the actual response to an experiment item.

Defining a Recording Task

Recordings in WikiSpeech are performed using the embedded Java Web Start version of the SpeechRecorder software (Draxler and Jänsch 2004). SpeechRecorder is an application for scripted audio recordings; its main features are platform independence, full support for UTF-8, HTML pages, audio and image data.

SpeechRecorder offers two views of individual recording sessions. The supervisor window contains the prompt item, a signal display, various control buttons for playback, creation and deletion of elements, and the full list of items to record. The speaker view contains only the prompt item (text, image, audio), a small traffic-light symbol to tell the speaker when to speak, and some control buttons (see Figure 9.15).

A recording project definition in WikiSpeech requires setting some technical specifications such as number of channels, sample rate, lossless compression (e.g., FLAC), or PCM wav-files. These data are commonly entered via a graphical user interface.

The GUI is convenient to enter isolated data items into the system. However, a recording script typically contains dozens if not hundreds of prompt items. For such bulk data, recording scripts in XML format can be imported. These scripts can be created off-line in the stand-alone version of SpeechRecorder, in a text or XML editor, or generated by programming scripts or application programs. For a sample script, see Figure 9.16.

The DTD of recording script files (`SpeechRecPrompts.dtd`) is available on the WikiSpeech server (`webapp.phonetik.uni-muenchen.de/wikispeech/`).

To record speech with WikiSpeech, speakers are invited to participate, for example, by sending them the speaker login information via mail. After login, a speaker creates a new

```
<?xml version="1.0" encoding="UTF-8" standalone="no" ?>
<!DOCTYPE session SYSTEM "SpeechRecPrompts.dtd">

<session id="IE_2000">
  <metadata>
    <key>
      DatabaseName
    </key>
    <value>
      VOYS Scottish English Corpus Recordings
    </value>
    <key>
      ScriptAuthor
    </key>
    <value>
      Chr. Draxler, LMU University of Munich
    </value>
  </metadata>

 <recordingscript>
  <section name="1_1" order="random" speakerdisplay="yes"
  mode="autorecording" promptphase="idle">
    ...
   <recording prerecdelay="500" recduration="8000" postrecdelay="500" itemcode="S3">
     <recinstructions mimetype="text/plain">
      Please read
     </recinstructions>
     <recprompt>
      <mediaitem mimetype="text/plain">
      This woman swam in the river.
      </mediaitem>
     </recprompt>
   </recording>

   <recording  prerecdelay="500" recduration="8000" postrecdelay="500" itemcode="S4">
     <recinstructions mimetype="text/plain">
      Please read
     </recinstructions>
     <recprompt>
      <mediaitem mimetype="text/plain">
      Those people caught a tame horse.
      </mediaitem>
     </recprompt>
   </recording>

   <recording  prerecdelay="500" recduration="8000" postrecdelay="500" itemcode="S5">
     <recinstructions mimetype="text/plain">
      Please read
     </recinstructions>
     <recprompt>
      <mediaitem mimetype="text/plain">
      Touch the frog with the spoon.
      </mediaitem>
     </recprompt>
   </recording>
    ...
  </section>
 </recordingscript>
</session>
```

Figure 9.16 Fragment of a WikiSpeech recording script definition in XML format. The top part contains metadata about the recording project, and the lower part is divided into sections, which in turn contain recording items.

recording session and enters his or her personal data into a standard form. Then he or she starts the recording procedure. The recording software is downloaded to the local machine and started. The worker starts the recording and from then on is guided by the recording software.

One item after the other is presented on the screen. When a recording is done, the signal file is transferred to the server in a background process, and the next item is presented.

The project administrator may check the quality of the incoming audio as soon as it is available on the server. After a session has been recorded, the project administrator may export eitherall session data or the entire project data via the WikiSpeech GUI. The data is then packed and compressed into a zip file and made available for download.

Defining an Annotation Task

The definition of an annotation task by a requester is quite straightforward. All that is necessary is a text file in CSV format with four fields in each line: the name of the annotation project, an identifier for a session that groups together one or more annotations, a tier identifier, and the annotation text to display in the annotation editor. This annotation text will in general be a normalized version of the text the speaker had to read, but it may also be empty, for example, in the case of spontaneous speech or free responses to questions.

The worker input for annotation tasks is the annotation text for a given annotation tier, as well as timestamps when required.

In general, the WikiSpeech annotation editor is suitable for simple, orthographic transcripts of speech. By default, the SpeechDat guidelines for the transcription of speech are implemented by the editor (Senia and van Velden 1997). The editor thus features keyboard shortcuts and editing buttons for typical tasks, and it performs a syntax check before data is saved to the server (see Figure 9.17). Alternative editors are available.

Note that because WebTranscribe is a Java web start application, it does not run on smartphones or most tablets.

Defining Online Experiments

Online experiments in WikiSpeech are performed using the Percy framework (Draxler 2011). Percy was designed to run on any type of web-capable device, including smartphones and tablets. The software is implemented using HTML5 for audio playback, and the JavaScript framework jQuery to dynamically change the screen display and for the data exchange with the server via AJAX. The Percy framework may be used independently of WikiSpeech and is distributed as open-source software.

The client software basically consists of five different screens: experiment selection, welcome, registration, experiment input, and thank you screens. These screens are implemented as <div> elements in one large HTML document, and they are displayed and hidden dynamically according to the experiment phase. Communication with the server is handled via AJAX.

Typically, when a user registers for participation, his or her data is sent to the server and the experiment items for an entire experiment session are returned. They are processed one after the other, with every user input being sent to the server immediately. Finally, when all experiment items are done, a thank you screen is displayed, which may, depending on the experiment, provide some feedback to the participant.

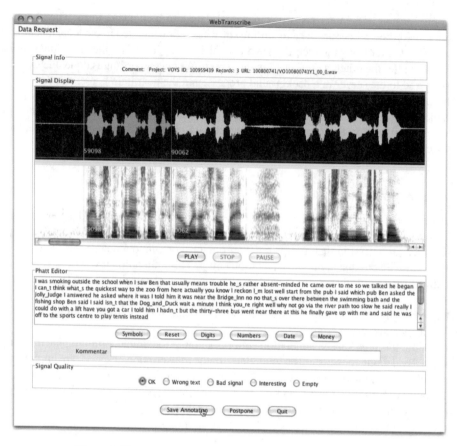

Figure 9.17   WebTranscribe editor with the default editing panel.

A requester defines an online experiment via an XML file. The DTD of this file is available from the WikiSpeech server `webapp.phonetik.uni-muenchen.de/wikispeech`.

Worker input to online experiments consists of the automatically logged data such as item presentation and input timestamps, counters for the number of playback repetitions, and the actual input made by the worker, for example, the selected button.

Requesters may monitor the progress of experiments online and access the workers' input data as soon as it is saved on the server.

Example   To answer the question whether digits spoken in isolation are sufficient to determine the regional origin of a speaker, an experiment was carried out. In this experiment, participants were asked to judge whether they heard a particular phoneme in a specific position in a given digit utterance. For example, participants were asked whether the first sound of the word "drei" (*three*) sounded more like the "d" in the word "Dritten" (*third*) or the "t" in the word "Tritten" (*kicks*).

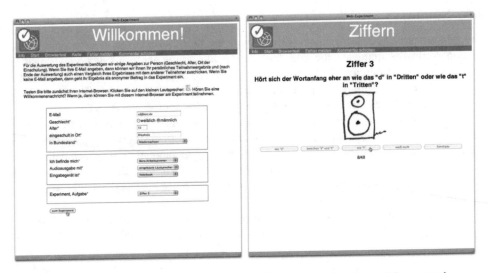

Figure 9.18 Registration and experiment page for the regional variants of German task.

An experiment session consisted of three test items, 45 production items, and a final assessment item.

At the end of each experiment session, an interactive map of Germany with color-coded participant's judgments was shown. The participant could then click on a data point to listen to the utterance again (see Figures 9.18 and 9.19).

Figure 9.19 Result presentation page for the regional variants of German task. The final page shows an interactive map.

9.5 Summary

This chapter presented the technology underlying the World Wide Web. It gave a brief introduction to crowdsourcing and the workflow that crowdsourcing platforms support. Using three commercial crowdsourcing platforms as examples, the definition of tasks was described in some detail: manual definition via graphical editors or command-line interfaces, programming task pages using SDKs, web-based modifications of task resources via REST.

As an alternative to commercial crowdsourcing platforms, WikiSpeech was presented. WikiSpeech is *not* a typical crowdsourcing platform, but it offers similar services. The underlying principle of collaborative work for the recording and annotation of speech databases is very close to that of crowdsourcing, but it is less general. WikiSpeech is focused at speech-related tasks. It gives a requester less flexibility in defining his or her tasks than other platforms, but thus simplifies setup and running tasks.

For academic institutions, WikiSpeech is a viable alternative to commercial crowdsourcing platforms because it is a free service, technologically mature, and stable.

References

Aker A, El-Haj M, Albakour M and Kruschwitz U (2012) Assessing crowdsourcing quality through objective tasks *Proceedings of LREC*, pp. 1456–1461, Istanbul.

Amazon Web Services LLC (2011) Amazon Mechanical Turk Getting Started Guide.

Amazon Web Services LLC (2012) Amazon Mechanical Turk API Reference.

Berners-Lee T, Fielding R and Nielsen HF (1996) Hypertext Transfer Protocol. http/1.0 www.w3.org/Protocols/HTTP/1.0/spec (accessed 18 October 2012).

Box D, Ehnebuske D, Kakivay G, Leyman A, Mendelsohn N, Nielsen HF, Thatte S and Winer D (2000) Simple Object Access Protocol (SOAP) 1.1. www.w3.org/TR/soap (accessed 18 October 2012).

Bray T, Paoli J and Sperberg-McQueen C (1998) Extensible Markup Language (XML) 1.0. http://www.w3.org/TR/1998/REC-xml-19980210 (accessed 18 October 2012).

Bray T, Paoli J, Sperberg-McQueen CM, Maler E and Yergeau F (2005) Extensible Markup Language (XML) 1.0, 5th edn. http://www.w3.org/TR/xml/ (accessed 18 October 2012).

Crockford D (2012) JSLint the JavaScript Code Quality Tool. http://www.JSLint.com/ (accessed 18 October 2012).

Dickie C, Schaeffler F, Draxler C and Jänsch K (2009) Speech recordings via the internet: an overview of the VOYS project in Scotland. *Proceedings of Interspeech*, pp. 1807–1810, Brighton.

Draxler C (1997) WWWTranscribe—a modular transcription system based on the World Wide Web. *Proceedings of Eurospeech*, pp. 1691–1694, Rhodes.

Draxler C (2005) Webtranscribe—an extensible web-based speech annotation framework. *Proceedings of TSD*, pp. 61–68, Karlsbad, Czech Republic.

Draxler C (2011) Percy—an HTML5 framework for media rich web experiments on mobile devices. *Proceedings of Interspeech*, pp. 3339–3340, Florence, Italy.

Draxler C and Jänsch K (2004) SpeechRecorder—a universal platform independent multichannel audio recording software. *Proceedings of LREC*, pp. 559–562, Lisbon.

Draxler C and Jänsch K (2008) Wikispeech—a content management system for speech databases. *Proceedings of Interspeech*, pp. 1646–1649, Brisbane.

Draxler C and Steffen A (2005) Ph@ttSessionz: Recording 1000 adolescent speakers in schools in Germany. *Proceedings in Interspeech*, pp. 1597–1600, Lisbon.

ECMA International (2009) ECMAScript Language Specification.

Flanagan D (2011) *JavaScript—The Definitive Guide*, 6th edn. O'Reilly & Associates, Sebastopol.

Gourley D, Totty B, Sayer M, Aggarwal A and Reddy S (2002) *HTTP: The Definitive Guide*. O'Reilly & Associates, Sebastopol.

Hickson I (2012) HTML5 a vocabulary and associated APIs for HTML and XHTML. www.w3.org/TR/html5/.

Humangrid GmbH (2010) On-Demand Data Categorization—Case Study—How clickworker.com Indexed and Tagged Thousands of Sports Videos for Swink.tv. `www.clickworker.com/en/2010/10/28/swink-tv` (accessed 18 October 2012).

Humangrid GmbH (2012) ClickWorker Marketplace API.

Ipeirotis P (2010a) Be a Top Mechanical Worker: You Need $5 and 5 Minutes. `http://www.behind-the-enemy-lines.com/2010/10/be-top-mechanical-turk-worker-you-need.html` (accessed 18 October 2012).

Ipeirotis P (2010b) Mechanical Turk: Now with 40.92% Spam. `http://www.behind-the-enemy-lines.com/2010/12/mechanical-turk-now-with-4092-spam.html` (accessed 18 October 2012).

Kunchukuttan A, Roy S, Patel P, Ladha K, Gupta S, Khapra M and Bhattacharyya P (2012) Experiences in resource generation for machine translation through crowdsourcing. *Proceedings of LREC*, pp. 384–391, Istanbul.

Levine B (2011) Adobe drops flash for mobile devices news.yahoo.com/adobe-drops-flash-mobile-devices-193154292.html.

MacCaw A (2011) *JavaScript Web Applications*. O'Reilly & Associates, Sebastopol.

Mason W and Watts DJ (2010) Financial incentives and the "performance of crowds" *SIGKDD Explor. Newsl.*, vol. 11, pp. 100–108.

McGraw I, Glass J and Seneff S (2011) Growing a spoken language interface on Amazon Mechanical Turk. *Proceedings of Interspeech*, pp. 3057–3060, Florence, Italy.

Parent G and Eskenazi M (2011) Speaking to the crowd: Looking at past achievements in using crowdsourcing for speech and predicting future challenges. *Proceedings of Interspeech*, pp. 3037–3040, Florence, Italy.

Ragett D, Hors AL and Jacobs I (1998) HTML 4.0 Specification `www.w3.org/TR/REC-html40/`.

Roth G (2009) RESTful HTTP in practice. `www.infoq.com/articles/designing-restful-http-apps-roth` (accessed 29 October 2012).

Schultz T, Black A, Badaskar S, Hornyak M and Kominek J (2007) SPICE: web-based tools for rapid language adaptation in speech processing systems. *Proceedings of Interspeech*, pp. 2125–2128.

Senia F and van Velden J (1997) Specification of orthographic transcription and lexicon conventions. Technical Report SD1.3.2, SpeechDat-II LE-4001.

Stefanov S (2010) *JavaScript Patterns*. O'Reilly & Associates, Sebastopol.

10

Crowdsourcing for Industrial Spoken Dialog Systems

David Suendermann[1,2,3] and Roberto Pieraccini[3]

[1] *DHBW Stuttgart, Germany*
[2] *Synchronoss, USA*
[3] *ICSI, USA*

10.1 Introduction

The history of speech technology has shown a number of times how industrial settings behave differently from academic environments. Speech recognizers may use handcrafted rule-based language models (Pieraccini and Huerta 2005) rather than a 1M word vocabulary statistical model trained on hundreds of billions of tokens (Schalkwyk *et al.* 2010). Speech synthesizers may use diphones with a footprint of 500 kB (Hoffmann *et al.* 2003) rather than a highly tuned unit selection voice based on 10 hours 96 kHz 24 bit studio recordings (Fernandez *et al.* 2006). Dialog managers used in spoken dialog systems may rely on manually designed decision trees (aka call flows) (Minker and Bennacef 2004) rather than statistically optimized and adaptable statistical engines such as POMDP (Young 2002). The speech output of dialog systems may be realized based on tens of thousands of well-formulated and prerecorded prompts (Pieraccini *et al.* 2009) rather than on dynamic language generation and synthesis (Oh and Rudnicky 2000).

10.1.1 Industry's Willful Ignorance

Why does industry sometimes seem to ignore the respectable accomplishments academic research has brought up in the many years of speech science, thereby seemingly reinventing

Crowdsourcing for Speech Processing: Applications to Data Collection, Transcription and Assessment, First Edition.
Edited by Maxine Eskénazi, Gina-Anne Levow, Helen Meng, Gabriel Parent and David Suendermann.
© 2013 John Wiley & Sons, Ltd. Published 2013 by John Wiley & Sons, Ltd.

the wheel? Here are a number of reasons that are not intended to raise too much dust since the answer to this question is not the subject of the text at hand:

1. **Industry needs quick development**: Building and tuning context-specific language models for months is just not feasible.
2. **Industry needs predictability**: Stakeholders of applications need to know what the application is going to do in all possible situations. For example, statistical techniques for dialog management are treated with skepticism since the application can do all sorts of crazy things depending on the millions of parameters of the involved probability models.
3. **Industry needs a low footprint**: Some interactive voice response systems process thousands of concurrent calls. Usually, a single server handles several dozens of calls at once strongly limiting the available resources per instance.
4. **Industry needs robustness**: Speech recognizers and synthesizers, spoken language understanding modules, dialog managers, and voice browsers of some industrial systems process more than one million calls per day. Hence, a minor bug in one of the involved components can affect a huge number of callers. A typical industrial service availability of such systems is above 99.99% (SpeechCycle 2008a).
5. **Industry protects its intellectual property and customer privacy**: Commercial applications are mostly distributed under strong license restrictions and fees; source code almost never gets opened; technical details rarely get published, and if so, mostly as patents; training data is kept proprietary and may contain confidential contents.
6. **Industry has money**: Software licenses, telephony ports, extensive hardware, and man hours can be treated rather flexibly when required by a profitable speech application. For example, it is common to hire a professional voice talent to record tens of thousands of voice prompts to achieve a higher speech quality than a speech synthesizer.

10.1.2   Crowdsourcing in Industrial Speech Applications

How does the role industry is playing in the arena of speech science affect the way crowdsourcing is used in industrial speech applications? To give an answer and prepare this chapter's outline, we want to look at three examples of speech applications featuring both *commercial impact* (i.e., they are of specific interest to industry) and components that could greatly benefit from *typical crowdsourcing tasks*:

1. **Speech transcription**: A popular example is the automatic generation of medical transcriptions, one of the most profitable domains of speech processing these days (Mohr *et al.* 2003). Naturally, medical reports often contain personal information not permitted to be shared with the public under penalty of law. Consequently, sourcing an anonymous crowd of workers as is done in most publications on crowdsourcing for transcription (see Chapter 4) is not an option since privacy of the data is not ensured.
2. **Spoken language annotation**: Semantic annotation of spoken utterance transcriptions is a (semi-)manual task (Tur *et al.* 2003) that can be distributed among a group of workers (Basile *et al.* 2012) in a crowdsourcing framework. This task does not necessarily require workers to listen to speech utterances but can be effectively done by *reading* utterance transcriptions generated by a transcription step carried out beforehand. Again, the transcriptions

to be annotated as well as the semantic annotations returned may contain confidential information. For example, an Internet troubleshooting application the authors have been working on can collect e-mail addresses, account passwords, and credit card numbers, which can by no means be shared with anonymous workers in a public crowdsourcing community.

3. **Measuring performance of spoken dialog systems via subjective scores**: Even though there are a number of performance measures for spoken dialog systems based on solely objective criteria (such as automation rate, average handling time, or retry rate), very often, industrial stakeholders request extensive subjective evaluation of these systems. This is to complement shortcomings of objective measures that have no way of evaluating things like
   (a) caller experience;
   (b) the number of missed speech inputs;
   (c) whether the application contains logical flaws.
   A popular way to measure these is to have subjects (experts or nonexperts) listen to call recordings of human/computer dialogs and have them give their rating according to certain scales or provide free-form text feedback on their observations. If no explicit expert knowledge of the system or the underlying technology is required, one could set up a crowdsourcing task. Obviously, opening call recordings to an anonymous crowd is a no-go in most industrial settings. Not only the privacy of the caller has to be respected, but confidential information may be provided, for example, when an account balance or a credit card number is mentioned during the call.

These three examples are mainly associated with item 5 (industry protects its intellectual property) of the willful ignorance list and seem to discourage the use of crowdsourcing in industrial speech processing. So, is there anything we can do, or is this the end of the present chapter?

10.1.3   Public versus Private Crowd

It seems as if item 6 (industry has money) of the list found in Section 10.1.1 may be able to help us out. One of the major reasons for hiring anonymous workers in the *public crowd* is doubtless that they are inexpensive. For instance, the price for a transcribed utterance via Amazon's Mechanical Turk has been reported to be as low as 0.13 US cents (Marge *et al.* 2010) whereas professional transcription services may charge one US dollar or more per utterance. A trade-off between these two extremes is the use of what we call a *private crowd*, that is, a crowd of workers willing to disclose their identity in exchange for a significantly higher payment. Engaging a private crowd comes with a number of advantages compared with the conventional public one:

- Workers can be asked to sign a nondisclosure agreement protecting the contents of the processed data. They can be explicitly required to work on dedicated hardware making sure that data does not reside anywhere outside the requestor's control. This solves the main drawback of using crowdsourcing for industrial speech applications.
- Workers can be held responsible for the quality of their work. Due to the ability to associate a task with the worker, it is now possible to give appropriate feedback, perform individual training, or isolate underperforming individuals. In turn, workers are more likely to provide a higher quality standard to begin with.

- Workers' throughput can be controlled more easily since the individual workload and availability can be agreed upon prior to starting a task.
- Workers can access resources and tools beyond those that can be implemented in conventional crowdsourcing platforms over the public Internet. For example, a Virtual Private Network (VPN) tunnel can be established between the requester's and the workers' infrastructures for the latter to access internal data sources (on file servers, databases, or the intranet). Special crowdsourcing software (such as the one displayed in Figure 10.6) can be distributed among the workers and installed on their private hardware possibly requiring them to sign a licensing agreement.
- The physical location of workers can be controlled to cope with legal or taxation requirements. In a public crowd, there is no reliable method to tell where a worker is physically located. IP addresses are not trustworthy (due to IP tunneling or server remote connections).
- Workers usually maintain a mid- to long-term relationship with the requester (as opposed to public crowdsourcing). This fact allows for
  ○ enduring and sustainable quality improvement mechanisms;
  ○ submitting substantially more complex tasks to be crowdsourced (it is OK when workers have to be trained for a couple of days on how to work with a complicated tool or on a new domain since they may be working on this task for a long period of time).

Even though, from the standpoint of the speech industry, advantages are predominant, compared to the public crowd, the use of a private crowd comes with a number of disadvantages, too:

- As aforementioned, workers disclosing their identity are usually more expensive than an anonymous crowd (a fact playing a less significant role in industrial settings).
- Engaging workers in a private crowd is not as dynamic as in the public crowd (e.g., the public crowd is spread across multiple continents and time zones resulting in workers being available usually around the clock; in a private crowd setting this would require dedicated personnel at multiple remote locations, which may not be feasible).

Having set the stage and delivered strong arguments for why commercial vendors primarily use private crowds, the following sections of this chapter are going into further details on typical industrial use cases. Section 10.2 describes an architecture accommodating all the components and data connections required for the examples in this chapter. Sections 10.3 and 10.4 are reporting on how private crowds are used to transcribe and semantically annotate spoken utterances. The third example for industrial crowdsourcing is the subjective evaluation of spoken dialog systems as covered in Section 10.5. Conclusions are drawn in Section 10.6.

10.2 Architecture

For the mentioned industrial applications, the sheer amount of processed data can be considerable. For example in Nuance (2009), a large US physician group is cited to have achieved a transcription volume of nearly 50 million lines in 2008. In the same year, tuning services for spoken dialog systems installed at a large North American cable provider used over two million caller utterances transcribed within 3 months to update language models and semantic classifiers (Suendermann *et al.* 2009b). To be able to keep up with a transcription volume of this magnitude and assure quality of service at the same time, a complex architecture is required, an example of which is described in the following.

Depending on the kind of task at hand, the crowdsourcing architecture may combine multiple capabilities at once. Considering the three examples discussed in this chapter (transcription, semantic annotation, and subjective evaluation of spoken dialog systems), we will present an architecture supporting all of these crowdsourcing services.

Apparently, the type of data required depends on the specific task: for transcription, speech data is necessary, semantic annotation relies on transcription and log data describing the recognition context, and for the subjective evaluation of spoken dialog systems, full-duplex call recordings and additional call information retrieved from call logs are necessary (details below). Accordingly, the architecture must be designed to accommodate the diversity of different data sources and types.

The components of industrial spoken dialog systems are mostly distributed across different servers, which, in turn, often reside in different data centers. It is common that these data centers are thousands of miles apart adding a level of local redundancy encouraged in industrial spoken dialog systems. Given the speed of today's Internet connections, data centers located on the East and West Coasts of the United States may communicate with each other with round-trip times below 20 ms. This ensures that spoken dialog systems operate without perceivable latency caused by network delays. Particularly, the application servers hosting dialog managers and knowledge bases of the spoken dialog systems are often located far away from voice browsers and speech recognizers. As shown in Figure 10.1, these two components of our infrastructure communicate using the VXML standard (McGlashan *et al.* 2004). On the application servers, web applications are hosted that dynamically produce one VXML page per user transaction. This page is sent to the voice browser which posts information back to the application server, for example, recognition results from the speech recognizer.

The speech recognizers, in turn, are often hosted in the same facilities as the voice browsers since the data being transmitted (audio) is more voluminous than that between application servers and browsers (VXML pages). Hence, it is recommendable to have browsers and speech recognizers share the same local area network to avoid latency or signal quality issues due to dropped packets. Communication between these two components is most commonly based on the Media Resource Control Protocol (Burke 2007).

Application servers, voice browsers, and speech recognizers are common components of industrial spoken dialog systems. In Figure 10.1, they are combined in the application layer of the infrastructure. The log layer is dedicated to logging events happening in all the components of the application layer including the recording of audio data. Application servers write multiple logs including web server logs and, most importantly, application logs which are, preferably, stored in the form of relational database tables on dedicated database servers. These databases have to store information including

- high-level call information (application version, product, and customer, call start and end time, duration, call categorization)—this information is used to extract performance parameters such as automation rate, opt-out or hang-up rates, and average handling time;

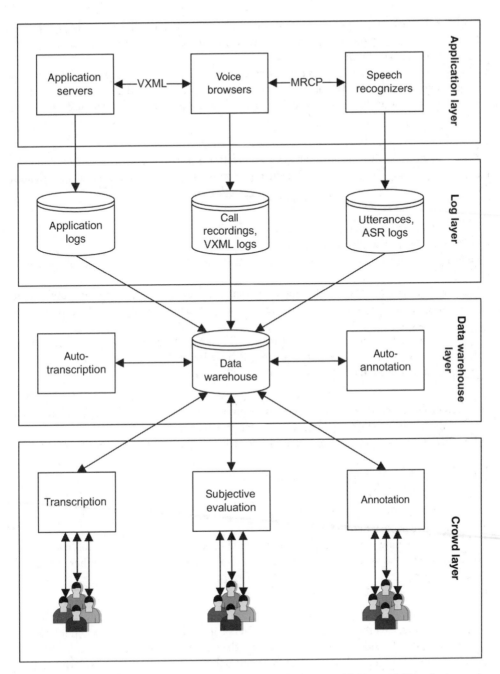

Figure 10.1 Example of an industrial crowdsourcing architecture combining capabilities for transcription, semantic annotation, and subjective evaluation of spoken dialog systems.

- workflow logs (for every call flow activity engaged in the specific call, name, time stamp, outbound transition, and possible exceptions are logged including information from telephony components, backend requests, and so on);
- call parameters (here, hundreds of call-specific properties are stored, for example, the caller's phone number, customer ID, and equipment type, whether the caller had called recently, whether and how many times the caller had asked for an operator or produced a speech recognition error).

Voice browsers serve as the link between application servers, speech recognizers, and the telephony network. They store their own logs (browser logs), which contain information about:

- establishment of a telephony session;
- URLs requested by application servers;
- language models, classifiers, prompts, and other files loaded from application or media servers;
- local decisions (e.g., often end-point detection, confidence thresholding, and touch tone processing are performed directly by the browser without engaging speech recognizer or application server);
- which recognition contexts are engaged (often, contexts such as confirmation, reprompt, time-out, no-match, and so on are handled directly by the browser without engaging the application server), and so on.

As voice browsers handle call traffic coming from and going to the telephony network, they are also able to store call recordings of the entire processed call.

For every processed recognition event, the speech recognizer can log information about:

- the active language models/semantic classifiers;
- call flow activity;
- the n best recognition hypotheses;
- the respective acoustic confidence scores;
- the m best semantic hypotheses (depending on whether semantic classification/parsing can be executed by the speech recognizer directly as supported by several commercial speech recognizers);
- the respective semantic confidence scores.

Similarly to the voice browsers, speech recognizers can store all processed speech utterances.

As estimated in Suendermann and Pieraccini (2011), the amount of data to be stored totals approximately 3MB per call on average in Synchronoss's[1] applications. Considering a gross number of 50 million processed calls per year (SpeechCycle 2008b), this amounts to over

[1] Synchronoss Technologies, Inc. (NASDAQ: SNCR) is a provider of on-demand multichannel transaction management solutions to the communications services sector. Through the merger with SpeechCycle, Inc. it added a speech solutions vertical to its product portfolio including large-scale spoken dialog systems.

140TB of data that would need to be stored. The vastness of data accumulated results in nontrivial problems when it comes to:

- random-accessing information from certain calls;
- deriving statistics across subsets of calls fulfilling certain filter criteria;
- joining information from heterogeneous sources (such as different log types).

All the above operations can be performed by a central data warehouse providing an external interface between the log and audio data coming from production applications on the one hand and crowdsourcing applications for transcription, annotation, and subjective evaluation, on the other hand. Essentially, the data warehouse layer consists of a database cluster as well as a file server optimized for high-speed access to all required pieces of information. It also contains some data processing services to automatically transcribe and semantically annotate certain portions of data as discussed in Sections 10.3 and 10.4.

The lowest layer of the presented hierarchy is the crowd layer. It features a number of software pieces tailored to the specific crowdsourcing task at hand (in our example, transcription, semantic annotation, and subjective evaluation of spoken dialog systems).

10.3   Transcription

Speech transcription can be considered the first (Draxler 1997) and predominant application of crowdsourcing to speech processing. Transcriptions of spoken data belong to one of the major revenue drivers of the speech industry. For example, an individual industry branch focuses on *medical transcription* with multiple specialized companies and study programs at numerous colleges (Nuance 2009). Furthermore, transcriptions can be used to train acoustic and language models of speech recognizers (Evermann *et al.* 2005) or as input for natural language processing tasks such as spoken language understanding (Tur and de Mori 2011), machine translation (Och 2006), or speech data retrieval (Rose 1991).

Most of the aforementioned cases of speech transcription are based on end-pointed utterances originating from the speech recognizer as described in Section 10.2. Considering the amount of data accumulated in the data warehouse layer, even using a well-organized crowdsourcing setting it is not feasible to transcribe the set of available utterances in its entirety[2]. That is, subsets of utterances have to be selected according to certain criteria forming tasks of reasonable size, which are then submitted for transcription to the workers. Utilizing log information provided by the speech recognizer (see a list of examples in Section 10.2), these subsets can focus on utterances recorded in certain

- call flow or application versions;
- date or time ranges;
- call flow activities;
- recognition contexts with certain language models/semantic classifiers active;

to provide some typical examples. While such utterance filters can be defined manually, in an industrial setting, several transcription tasks involving tens of thousands of utterances have to

---

[2]If it was feasible, the use of ASR would become questionable.

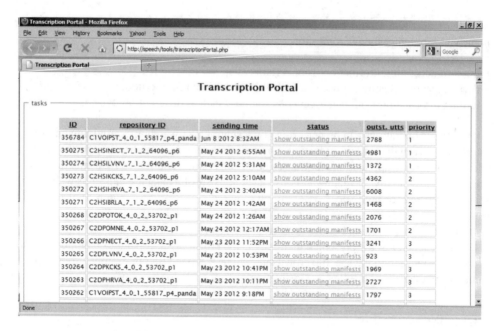

Figure 10.2   Example of a transcription portal.

be issued on a daily basis. In doing so, a broad range of transcriptions covering all applications, call flows, and customers has to be accommodated. Rather than having a transcription manager define tasks manually, an automatic transcription task service can take over the day-to-day responsibility, making sure that

- tasks cover all current applications, call flows, and products;
- tasks involve a minimum number of calls to be transcribed (e.g., 1000 calls, i.e., between 2000 and 20,000 utterances depending on the application);
- workers always have enough but not too much workload.

An industrial transcription task service can also accommodate different priority levels (in case emergency transcription is required). An example web interface of a transcription task service is shown in Figure 10.2. In this table, every row represents an individual task with

- a unique ID;
- a repository ID (identifying customer, application, data center location, and particular call flow version);
- the time the task was issued;
- a link to available subtasks (since tasks can involve tens of thousands of utterances, typically, they are subdivided into smaller chunks, e.g., of 1000 utterances each);
- the number of outstanding utterances to be transcribed;
- a priority level.

Workers now access (and lock) individual subtasks in the web portal and process them by means of a transcription client[3]. This client can make use of a number of data points retrieved from the data warehouse to render transcription as effective as possible. In particular, the client can display

- IDs of individual calls and utterances;
- call flow activity names;
- names of language models and semantic classifiers;
- the system prompt preceding the utterance;
- the first best recognition hypothesis;
- the first best acoustic confidence;
- the history of all the above items.

The availability of the first-best hypothesis and the respective confidence score have a significant impact on transcription speed. This is because, in the majority of cases, the speech recognizer correctly recognized the input utterance, in which case the utterance can be accepted, for example, by a single key stroke of the worker. This can increase transcription throughput to about 1000 utterances per worker and hour (Suendermann *et al.* 2010a). This way, we have been able to collect a corpus containing about 22 million transcribed utterances from almost all recognition contexts of a multitude of spoken dialog systems operated by Synchronoss.

In fact, the concept of speech-recognition-confidence-based transcription assistance can be extended by having cloud-based speech recognizers reprocess utterances, significantly decreasing the overall word error rate. After training a language-model-specific rejection threshold to make sure that the word error rate of accepted utterances is on par with that of human transcribers, a significant number of utterances will not even be included in tasks anymore but automatically transcribed (Suendermann *et al.* 2010b). Autotranscription is carried out in the data warehouse layer as shown in Figure 10.1.

In order to estimate the word error rate associated with human transcription as used for autotranscription as well as for the ongoing quality assurance of manual transcription, partially overlapping tasks can be assigned to multiple transcribers by means of an automated process. Transcriptions of identical material assigned to multiple workers are then compared by calculating the word-level Levenshtein distance for each possible pair of workers. The average of this distance can be considered a measure of the disagreement among transcribers on the same body of data and, hence, is an estimate of the word error rate of human transcription. By comparing the average individual word-level Levenshtein distance to other workers with the word error rate of human transcription, seemingly underperforming workers can be isolated. As a consequence, transcriptions of such workers can be initially discarded or reworked. Furthermore, as discussed in Section 10.1.3, a clear advantage of the private crowd is that workers are not anonymous. Consequently, it is possible to engage workers whose work does not satisfy common quality standards in individual training sessions. See Chapter 4 for more details on quality assurance of crowdsourced transcription.

---

[3] In the authors' case, a proprietary Java-based client.

10.4   Semantic Annotation

Many spoken language processing tasks require not only transcriptions of spoken utterances but also a semantic interpretation of said utterances. Examples for disciplines involving semantics of spoken language include

- topic classification (Rochery *et al.* 2001);
- emotion recognition (Lee and Pieraccini 2002);
- language/accent identification (Zissman 1993);
- semantic parsing/spoken language understanding (Tur and de Mori 2011).

In simple scenarios, the association between spoken utterance and meaning can be expressed in a manually designed rule-based fashion. There are a number of W3C standards supporting rule-based semantic parsers, which are often used as *grammars* in spoken dialog systems (Java Speech Grammar Format (Sun Microsystems 1998), Speech Recognition Grammar Specification (SRGS) (Hunt and McGlashan 2004), Semantic Interpretation for Speech Recognition (van Tichelen and Burke 2007)). The standardization of speech recognition grammars made them very popular starting around 2004. They found their way into almost all industrial spoken dialog systems since major vendors of speech processing software adhere to the above-mentioned standards preventing vendor lock-in and, more importantly, they are easy to build and maintain. Figure 10.3 shows an example of a speech recognition grammar to parse responses to the prompt

> *Which of these services did the outage affect: Internet or cable?*

The example grammar covers responses such as

> *Internet,*
> *It was TV,*
> *I guess cable,*

and returns a semantic class $out \in \{\texttt{Internet}, \texttt{Video}\}$ depending on the input utterance. At runtime, the speech recognizer of a deployed spoken dialog system will return the semantic class (hypothesis) to the voice browser, which, in turn, will forward it to the application server (all of this happens in the application layer, see Figure 10.1). The application server (more precisely, the *dialog manager* portion of the application server (Pieraccini *et al.* 2001)) only cares about the semantic hypothesis (e.g., $\texttt{Video}$) rather than the exact wording pronounced by the caller (e.g., $\texttt{I guess TV}$).

Despite the good arguments for using rule-based speech recognition grammars, there is a severe counterargument that concerns the performance of this approach. Recent publications (e.g., Suendermann *et al.* 2009b) suggest that rule-based speech recognition grammars will almost always be significantly outperformed by an architecture composed of statistical language models and semantic classifiers. The reason for this is that statistical language models and classifiers can be trained and optimized on large bodies of data whereas rule-based grammars are limited in their complexity due to the fact that they are manually updated. Moreover, generating statistical language models and classifiers is an automatic procedure (given training

```xml
<?xml version="1.0" encoding="utf-8" ?>
<grammar xml:lang="en-US" version="1.0" xmlns="http://www.w3.org/
2001/06/grammar" root="selection">
  <rule id="selection" scope="public">
      <item repeat="0-1"><ruleref uri="#pre_phrases" /></item>
      <one-of>
        <item><ruleref uri="#rule_internet"/><tag>out='Internet';
        </tag></item>
        <item><ruleref uri="#rule_video"/><tag>out='Video';
        </tag></item>
      </one-of>
  </rule>
  <rule id="pre_phrases">
      <one-of>
        <item>it was</item>
        <item>i guess</item>
      </one-of>
  </rule>
  <rule id="rule_internet">
      <one-of>
        <item>internet</item>
      </one-of>
  </rule>
  <rule id="rule_video">
      <one-of>
        <item>video</item>
        <item>t_v</item>
        <item>cable</item>
      </one-of>
  </rule>
</grammar>
```

Figure 10.3 Example of a speech recognition grammar.

data), whereas tuning rule-based grammar requires the attention of experts ("speech scientists"). As an example, Figure 10.4 shows how semantic accuracy (in terms of True Total, see Suendermann *et al.* 2009a) changed over time as the initial rule-based grammar (September 2008) was replaced by multiple versions of statistical language models and classifiers. The reason why performance does not seem to be on a constant rise is that test sets kept changing over time. We recommend to always use the newest batch of utterances for testing, since they best represent the current situation of the application.

The training data of semantic classifiers (see an overview in Evanini *et al.* 2007) consists of the speech utterances' transcription as well as their semantic classes. Transcriptions are produced as described in Section 10.3. Semantic annotation is generally a manual task as well, requiring annotators to assign one out of a set of possible semantic classes to a given utterance.

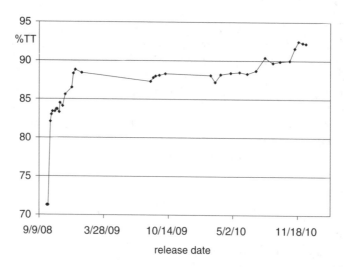

Figure 10.4 Improvement of semantic accuracy when replacing rule-based grammars by statistical language models and classifiers. Each dot represents a different version of a statistical language model and classifier.

In order to perform this task, a number of workers residing in the crowd layer (see Figure 10.1) make use of an annotation software providing functionality such as

- a field displaying the transcription, application version, proposed semantic class, confidence, and so on for all processed utterances;
- a hierarchical structure of the available semantic classes (this structure has to be established in accordance with the call flow the classifier is used in; a change to the call flow may provoke the association between utterances and classes to be affected or the set of classes to be altered);
- selection and filtering capabilities for the utterances to be processed (including the exact context in which the current classifier is being used; the application version, client, product, data center location, time the utterance was recorded; the person who transcribed/annotated the utterance, the transcription/annotation time, and so on);
- audio playback for revising the utterance;
- acceleration features allowing for annotations to be performed by a single click on a semantic class. Furthermore, all utterances with identical transcriptions can be annotated at the same time. Depending on the recognition context chosen and the amount of preexisting annotations, this simple feature can make a significant difference. In contexts of low variability (such as in voice menus or at yes/no questions), more than 95% of the utterances can be automatically annotated. Figure 10.5 shows how the ratio of automatically annotated utterances depends on the amount of preexisting data for two typical contexts: a yes/no context and an open prompt distinguishing about 80 semantic classes. The resulting throughput can exceed 5000 utterances per annotator and hour and is mainly limited by the throughput of transcription (see Section 10.3). In fact, more than 98% of all the utterances transcribed in the course of the last 3 years have been semantically annotated.

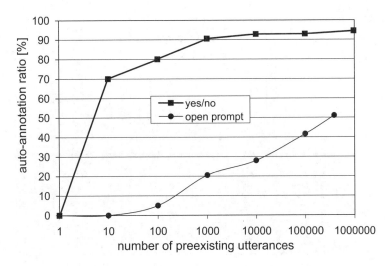

Figure 10.5 Dependency of the ratio of automatically annotated utterances on the number of preexisting annotations for two different contexts.

When a worker selects a portion of data to work on (in Figure 10.6, 100 utterances), the data is locked for annotation by this worker in the data warehouse. This ensures that workers do not interfere on the same data at any given time. Upon completing a task, semantic annotations are written to the data warehouse along with a number of tags including the time of annotation, the IP address/machine the annotation software was running on, and the user

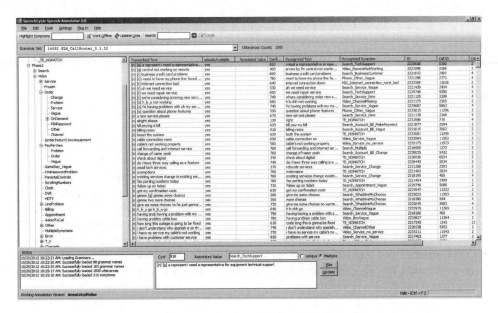

Figure 10.6 Example of a semantic annotation software.

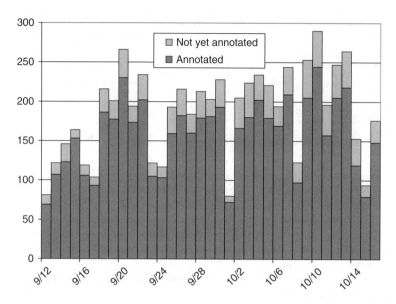

Figure 10.7   Ratio of annotated utterances of a selected application for a time interval of 35 days.

ID of the worker. This allows for performance analysis of individual workers w.r.t. quality and throughput.

Especially in industrial settings, quality and consistency of semantic annotation is a critical prerequisite for deploying semantic classifiers trained on this data. For example, it will never happen that a classifier returns a semantic class not accounted for in the call flow. It is also important to have an estimate of the classifier's performance before releasing it into live production. As discussed in Suendermann *et al.* (2008), there are a number of techniques ("C[7]") aiming at ensuring quality of semantic annotation:

- **Completeness**: In industrial settings, the turn-around time for building semantic classifiers can be as short as a few days. Trusting in the well-known machine learning paradigm "there is no data like more data," workers could be tempted to focus on frequent and simple utterances first (also using the aforementioned acceleration features) and discard some of the rest, since, in terms of tokens, they account for the majority of data. Figure 10.7 shows an example of a recognition context where the easiest and most frequent utterances were annotated. If a classifier is trained using only this data, it is very likely that the distribution of semantic classes and underlying transcription tokens is skewed and does not well represent the situation in live production. This usually leads to production performance significantly worse than estimated in the development environment.
- **Correlation**: Since annotation requires knowledge about the domain and architecture of the application for which the semantic classifier is intended, it is a much more complex task than transcription. Even though the private cloud allows for workers to be thoroughly trained (as discussed in Section 10.1.3), there will always be situations in which workers will draw different conclusions about which semantic class fits an utterance best. By having

identical utterances processed by multiple workers, a correlation analysis can be carried out to identify workers with a high level of agreement and determine individuals who seem to require special training. A common way to do so is to use the kappa statistic (Rosenberg and Binkowski 2004).

- **Consistency**: Utterances that are identical or syntactically very similar are expected to convey the same meaning, that is, to belong to the same semantic class. Here, syntactic similarity means that if two utterances get transformed in some way, their transformation results are identical. A common transformation with little or no impact on semantics of utterances includes
  - removal of stop words;
  - word stemming (e.g., using the Porter algorithm, Porter 1980)
  - alphabetical sort of stems;
  - elimination of multiple occurrences of stems.

  Exceptions to the consistency check have to be manually marked.
- **Confusion**: This check investigates how confusable semantic classes are by running simulations on test data and analyzing the confusion matrix. Classes featuring high degrees of confusion should be screened more closely since they may not be semantically distinctive enough, or annotation guidelines are too fuzzy w.r.t. the concerned classes.
- **Congruence**: As discussed earlier in this section, originally, industrial spoken dialog systems primarily deployed rule-based grammars providing language models and semantic parsers. As described in Suendermann *et al.* (2009b), it is recommended to systematically replace such rule-based grammars by statistical language models and semantic classifiers in order to optimize performance. Despite this substitution, the original rule-based grammars should still return correct semantic parses for all the utterances they covered. These semantic parses need to be identical to the output of the semantic classifier unless changes to the annotation guidelines for the respective recognition context have taken effect. The congruence between output of rule-based grammars and semantic annotation can therefore be used for quality assurance.
- **Coverage**: One of the reasons why statistical language models and classifiers trained on data usually outperform their rule-based counterpart is the presence of a strong *no-match* rule. That is, all user utterances not matching any of the canonical classes expected in the call flow will explicitly trigger a no-match. A very simple example is the following context:

Do you want to pay by credit card or at a payment center?

Obviously, the two expected canonical classes are `creditCard` and `paymentCenter`. However, what if a user responds *I would like to pay by check*? In a rule-based grammar, this rule would simply be missing, possibly causing a false accept of one of the two canonical classes (see Suendermann *et al.* 2009a for a comprehensive overview on most common classification metrics). In case of a semantic classifier, if occurring in the training data, such an utterance would be annotated as `noMatch`, that is, a third semantic class. If a user responds with something matching a no-match rule, the call flow will usually respond with a retry such as

Sorry, I did not understand. Just say credit card *or* payment center.

Even though these types of correct rejects are much less disruptive than false accepts, it would be preferable if they could be avoided. In the above example, the user input should be mapped to an additional class, *check*. Generally, workers are permitted (and expected) to extend the semantic class set (see left side of Figure 10.6) by frequently observed classes reducing the overall fraction of no-matches. If this fraction exceeds a certain threshold (usually somewhere between 5% and 10%) annotations and annotation guidelines should be revisited.

- **Corpus size**: Before using semantic annotations for analysis or training purposes, a minimum number of utterances has to be accumulated. This is to make sure that analysis results are statistically significant and classifiers actually achieve the expected performance.

In order to facilitate all the aforementioned quality checks to be carried out systematically for thousands of distinct language models and classifiers, we designed an annotation portal where workers can monitor all the involved error metrics. Figure 10.8 shows a screenshot of the portal in which every table row represents a distinct family of language models and classifiers. Only if all involved quality criteria are fulfilled, will the status field of the respective row turn 'ready' for training, testing, and potential deployment in case of performance increase.

10.5   Subjective Evaluation of Spoken Dialog Systems

Transcription and semantic annotation are not only for training statistical language models and semantic parsers but, equally importantly, for monitoring performance of production systems, a process crucial for the success of an industrial speech application. As explained in more detail, for example, in Suendermann *et al.* (2010c), word error rate or multiple types of concept error rate can be used to evaluate speech recognition and understanding behavior. Even though these metrics provide a powerful insight into the interaction behavior of a speech application, there are certain effects they are unable to capture. Examples include the following:

- Flaws of the application (e.g., loops causing questions to be repeatedly asked).
- Redundancy or irrelevance (does the system ask questions whose answers are already known, can be inferred, or are irrelevant altogether for the task at hand?).
- How cooperative the user was.
- How well the system treated the user.

Most of the effects listed above can best be answered by listening to the entire conversation between user and system. In fact, a regular exercise of revisiting batches of full calls recorded in industrial production systems is important to ensure quality of the systems and catch flaws as early as possible. To revisit significant numbers of calls (100 or more per application, customer, and month is necessary to produce a relatively balanced picture) is a very expensive endeavor ("a 19 minute call takes 19 minutes to listen to" Suendermann *et al.* 2009b) recommending the use of crowdsourcing in this discipline as well.

Similar to the assignment of batches of utterances for transcription and annotation, also in the case of subjective evaluation of full calls, a subset of calls needs to be selected and distributed among participating workers. This can be done by flagging calls in the data warehouse layer

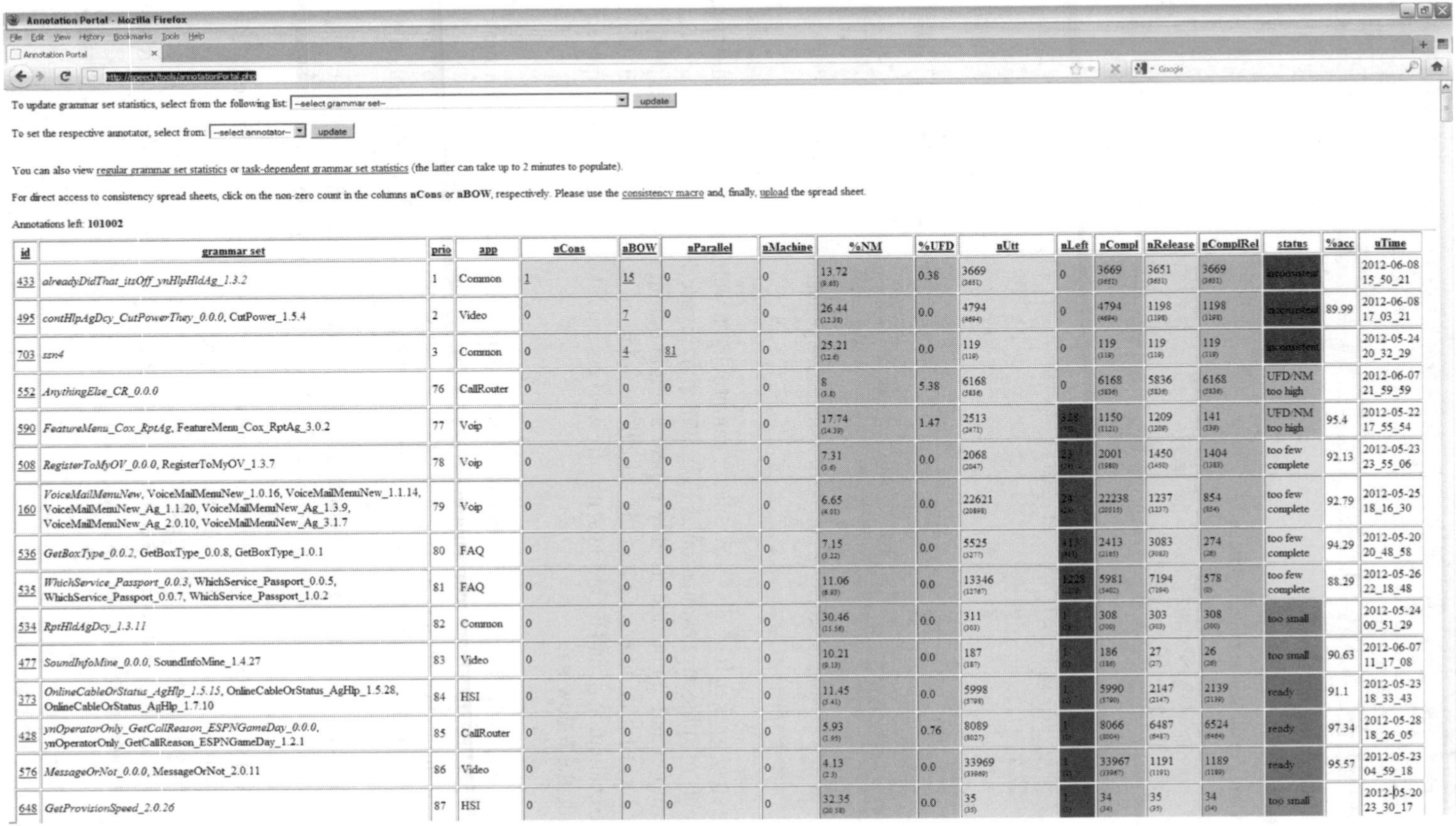

Figure 10.8 Example of an annotation portal.

Annotation Portal - Mozilla Firefox
File Edit View History Bookmarks Tools Help

To update grammar set statistics, select from the following list: --select grammar set--   update

To set the respective annotator, select from: --select annotator-- update

You can also view regular grammar set statistics or task-dependent grammar set statistics (the latter can take up to 2 minutes to populate).

For direct access to consistency spread sheets, click on the non-zero count in the columns nCons or nBOW, respectively. Please use the consistency macro and, finally, upload the spread sheet.

Annotations left: 101002

id	grammar set	prio	app	nCons	nBOW	nParallel	nMachine	%NM	%UFD	nUtt	nLeft	nCompl	nRelease	nComplRel	status	%acc	uTime
433	alreadyDidThat_inOff_yesHpHldAg_1.3.2	1	Common	1	15	0	0	13.72 (9.68)	0.38	3669 (3451)	0	3669 (3452)	3651 (3451)	3669 (3451)	uncooopsms		2012-06-08 15_50_21
495	contdHpAgDoy_CutPowerTuey_0.0.0_CutPower_1.5.4	2	Video	0	7	0	0	26.44 (12.28)	0.0	4794 (4494)	0	4794 (4494)	1198 (1198)	1198 (1198)	uncooopsms	89.99	2012-06-08 17_03_21
703	zzv4	3	Common	0	4	81	0	25.21 (12.4)	0.0	119 (119)	0	119 (119)	119 (119)	119 (119)	uncooopsms		2012-05-24 20_32_29
552	AnythingElse_CR_0.0.0	76	CallRouter	0	0	0	0	8 (0.8)	5.38	6168 (5836)	0	6168 (5836)	5836 (5836)	6168 (5836)			2012-06-07 21_59_59
590	FeatureMenu_Cox_RptAg_FeatureMenu_Cox_RptAg_3.0.2	77	Voip	0	0	0	0	17.74 (2439)	1.47	2513 (2471)	245 (10%)	1150 (1123)	1209 (1209)	141 (139)	UFD-NM too high	95.4	2012-05-22 17_55_54
508	RegisterToMyOV_0.0.0_RegisterToMyOV_1.3.7	78	Voip	0	0	0	0	7.31 (0.4)	0.0	2068 (2047)	37	2001 (2001)	1450 (1450)	1404 (1350)	too few complete	92.13	2012-05-23 23_55_06
160	VoiceMailMenuNew_1.0.16, VoiceMailMenuNew_1.1.14, VoiceMailMenuNew_Ag_1.1.20, VoiceMailMenuNew_Ag_1.3.9, VoiceMailMenuNew_Ag_2.0.10, VoiceMailMenuNew_Ag_3.1.7	79	Voip	0	0	0	0	6.65 (4.03)	0.0	22621 (20880)	24	22238 (20510)	1237 (1237)	854 (854)	too few complete	92.79	2012-05-25 18_16_30
536	GetBoxType_0.0.2, GetBoxType_0.0.8, GetBoxType_10.1	80	FAQ	0	0	0	0	7.15 (3.22)	0.0	5525 (5377)	412	2413 (2181)	3083 (3083)	274 (274)	too few complete	94.29	2012-05-20 20_48_58
535	WhichService_Passport_0.0.3, WhichService_Passport_0.0.5, WhichService_Passport_0.0.7, WhichService_Passport_1.0.2	81	FAQ	0	0	0	0	11.06 (8.03)	0.0	13346 (13787)	1223	5981 (5402)	7194 (7194)	578 (0)	too few complete	88.29	2012-05-26 22_18_48
534	RptHldAgDoy_1.3.11	82	Common	0	0	0	0	30.46 (13.96)	0.0	311 (303)	3	308 (308)	303 (303)	308 (308)	too small		2012-05-24 00_51_29
477	SoundInfoMine_0.0.0_SoundInfoMine_1.4.27	83	Video	0	0	0	0	10.21 (9.13)	0.0	187 (187)	1	186 (186)	27 (27)	26 (27)	too small	90.63	2012-06-07 11_17_08
373	OnlineCableOrStatus_AgHp_1.5.15, OnlineCableOrStatus_AgHp_1.5.28, OnlineCableOrStatus_AgHp_1.7.10	84	HSI	0	0	0	0	11.45 (8.41)	0.0	5998 (5748)	1	5990 (5740)	2147 (2147)	2139 (2139)	ready	91.1	2012-05-23 18_33_43
428	ynOperatorOnly_GetCallReason_ESPNGameDay_0.0.0, ynOperatorOnly_GetCallReason_ESPNGameDay_12.1	85	CallRouter	0	0	0	0	5.93 (3.69)	0.76	8089 (8027)	1	8066 (8004)	6487 (6487)	6524 (6485)	ready	97.34	2012-05-28 18_26_05
576	MessageOrNot_0.0.0_MessageOrNot_2.0.11	86	Video	0	0	0	0	4.13 (2.1)	0.0	33969 (33688)	1	33967 (33987)	1191 (1191)	1189 (1189)	ready	95.57	2012-05-23 04_59_18
648	GetProvisionSpeed_2.0.26	87	HSI	0	0	0	0	32.35 (25)	0.0	35 (35)	7	34 (34)	35 (35)	34 (34)	too small		2012-05-20 23_30_17

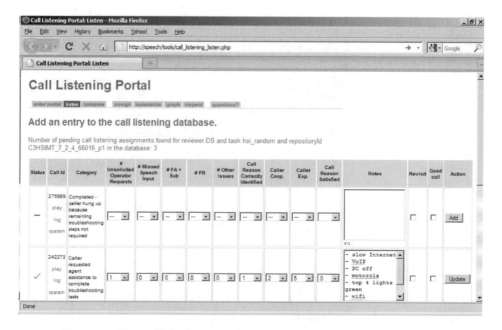

Figure 10.9   Example of a subjective evaluation portal.

with respective worker IDs. The crowd layer can provide a web portal tailored to the specific requirements of subjective evaluation as examplified in Figure 10.9.

Transcription and annotation usually expect a single output per analyzed unit, that is, the transcribed text for an utterance and the sematic annotation, respectively. When it comes to subjective evaluation of call recordings, usually workers are asked to evaluate multiple criteria that include the following:

- **Count of unsolicited operator requests**: This counter gets incremented when the caller requests speaking to a live agent without being offered to do so. In the opinion of the authors, it makes a difference on how to rate a user's cooperativeness when users accept an agent offer as compared to when they ask for one to end the conversation with the spoken dialog system.
- **Count of missed legitimate speech input**: This gets incremented when a speech input did not trigger a recognition event or was ignored due to system settings such as surpressed barge-in or command-word-dependent barge-in.
- **Count of other issues**: This gets incremented when an issue is observed in the call that is not accounted for by any of the other criteria. Examples include
  - strange timing or wording of system prompts;
  - backend integration (such as account lookup, and device reboot) failing or returning unexpected results;
  - latency;
  - lying to the user;
  - redundancy or irrelevance.

An additional notes field can be used to provide further detail on the encountered issues:

- **Correct identification of call reason**: The correct identification of a call reason is of importance for many task-oriented applications serving multiple call reasons including call routers (Williams and Witt 2004), troubleshooters (Acomb *et al.* 2007), FAQ systems (Dybkjær and Dybkjær 2004), and so on.
- **Call reason satisfied**: Did the application fulfill the caller's request?
- **Caller cooperation**: How did the caller treat the application—a mean opinion score (ITU 1996) describing the willingness of the caller to engage with the spoken dialog system. Cursing, unsolicited agent requests, side conversations, inattentiveness, and so on warrant lowering the score. Caller cooperation can depend on the dialog system's behavior (see *caller experience* below) but it is often outside of the system's control.
- **Caller experience**: How did the application treat the caller—a mean opinion score describing the appropriateness of system actions to satisfy the user's needs. Missed speech input, speech errors, and the aforementioned other issues warrant lowering this score.

In order to be able to draw correct conclusions on some of the above criteria, it is helpful for workers to have access not only to call recordings but also to the call logs. Thanks to the integration in the data warehouse layer, call logs are available and can be made accessible in the web interface (links "log" and "rparam" in Figure 10.9).

As concluded earlier for applying crowdsourcing to semantic annotation, quality assurance of the workers' output is equally important in the domain of subjective evaluation. While the mapping between transcription of an utterance and its respective semantic annotation can be defined more or less successfully by annotation guides and by providing numerous examples (practically the entire body of formerly annotated utterances), consistency of subjective evaluation of spoken dialog systems is much harder to ensure. This is mainly due to

- the sheer duration of the analyzed data (the average duration of one of the 25,000 calls evaluated in the portal described above is 3 minutes 36 seconds as opposed to the average length of an utterance being 4 seconds);
- multiple criteria to be tracked (not just one as in the case of semantic annotation);
- some of the criteria being subjective by definition (caller cooperation and experience among others).

Consequently, in order for the crowdsourcing endeavor to make sense and produce useful scores, the private crowd of workers has to be constantly monitored and trained. As proposed for both transcription and annotation, this can be done by assigning identical tasks to multiple workers. Figure 10.10 shows an example of an interface listing evaluation results per worker.

For all of the criteria assessed in the subjective evaluation of this specific call, the ratings and side nodes can be associated with a worker. In regular meetings, the involved workers of the subset of the private crowd whose discrepancies turn out to exceed a certain threshold (as measured by a kappa statistic or, as in the example, by the average MOS difference) join a short training session where they jointly listen to the affected calls and discuss the observed discrepancies. In the example of Figure 10.10, one can see that there is a clear disagreement about how many unsolicited operator requests occurred in the call. This in turn has a direct impact on the rating for caller cooperation.

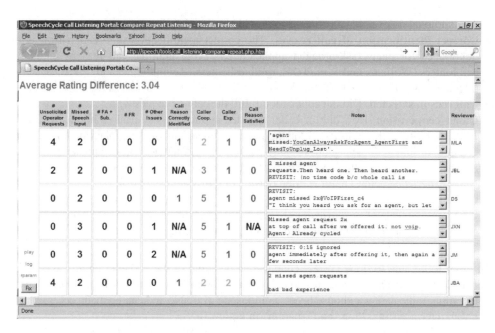

Average Rating Difference: 3.04

# Unsolicited Operator Requests	# Missed Speech Input	# FA + Sub.	# FR	# Other Issues	Call Reason Correctly Identified	Caller Coop.	Caller Exp.	Call Reason Satisfied	Notes	Reviewer
4	2	0	0	0	1	2	1	0	'agent missed:YouCanAlwaysAskForAgent_AgentFirst and NeedToUnplug_Lost'.	MLA
2	2	0	0	1	N/A	3	1	0	2 missed agent requests.Then heard one. Then heard another. REVISIT: (no time code b/c whole call is	JBL
0	2	0	0	0	1	5	1	0	REVISIT: agent missed 2x@VoIPFirst_c4 "I think you heard you ask for an agent, but let	DS
0	3	0	0	1	N/A	5	1	N/A	Missed agent request 2x at top of call after we offered it. not voip. Agent. Already cycled	JXN
0	3	0	0	2	N/A	5	1	0	REVISIT: 0:15 ignored agent immediately after offering it, then again a few seconds later	JM
4	2	0	0	0	1	2	2	0	2 missed agent requests bad bad experience	JBA

Figure 10.10   Discrepancy view of a subjective evaluation portal.

10.6   Conclusion

Speech processing in industrial settings often requires substantial manual data processing. The sheer amount of involved data (millions of calls) suggests the use of crowdsourcing which, however, has to respond to certain constraints dictated by the privacy policy of industrial companies. In particular, the crowd must not consist of anonymous workers leading to what the authors call *private crowd*, that is, a crowd of workers whose identity is known to the company. A private crowd comes along with a number of advantages compared to a public crowd including privacy of data, quality standards, and the ability to assign more complex tasks as well as to control workers' location, conditions, and throughput. To examplify crowdsourcing tasks for industrial speech applications, we discussed transcription of speech utterances, semantic annotation of utterances, and subjective evaluation of spoken dialog systems. We have seen how automation can significantly reduce the manual workload of transcription and annotation tasks and accordingly increase the workers' throughput. Independent of the task at hand, it is essential to implement quality assurance mechanisms that can identify workers or bodies of data not adhering to well-defined quality standards. These mechanisms can be used for workers to analyze and correct their own work (C^7) or, in the scope of crosstraining sessions, to achieve agreement within (parts of) the crowd.

References

Acomb K, Bloom J, Dayanidhi K, Hunter P, Krogh P, Levin E and Pieraccini R (2007) Technical Support Dialog Systems: Issues, Problems, and Solutions. *Proceedings of the HLT-NAACL*, Rochester, USA.

Basile V, Bos J, Evang K and Venhuizen N (2012) A Platform for Collaborative Semantic Annotation. *Proceedings of the EACL*, Avignon, France.

Burke D (2007) *Speech Processing for IP Networks: Media Resource Control Protocol (MRCP)*. John Wiley & Sons, Inc., New York.

Draxler C (1997) WWWTranscribe—A Modular Transcription System Based on the World Wide Web. *Proceedings of the Eurospeech*, Rhodes, Greece.

Dybkjær H and Dybkjær L (2004) Modeling complex spoken dialog. *IEEE Computer* **37**(8), 32–40.

Evanini K, Suendermann D and Pieraccini R (2007) Call classification for automated troubleshooting on large corpora. *Proceedings of the ASRU*, Kyoto, Japan.

Evermann G, Chan H, Gales M, Jia B, Mrva D, Woodland P and Yu K (2005) Training LVCSR systems on thousands of hours of data. *Proceedings of the ICASSP*, Philadelphia, PA.

Fernandez R, Bakis R, Eide E, Hamza W, Pitrelli J and Picheny M (2006) The 2006 TC-STAR evaluation of the IBM text-to-speech synthesis system. *Proceedings of the TC-Star Workshop*, Barcelona, Spain.

Hoffmann R, Jokisch O, Hirschfeld D, Strecha G, Kruschke H and Kordon U (2003) A multilingual TTS system with less than 1 megabyte footprint for embedded applications. *Proceedings of the ICASSP*, Hong Kong, China.

Hunt A and McGlashan S (2004) `http://www.w3.org/TR/2004/REC-speech-grammar-20040316` (accessed 17 October 2012).

ITU (1996) Methods for subjective determination of transmission quality. *Technical Report ITU-T Recommendation P.800*, International Telecommunication Union, Geneva, Switzerland.

Lee C and Pieraccini R (2002) Combining acoustic and language information for emotion recognition. *Proceedings of the ICSLP*, Denver, CO.

Marge M, Banerjee S and Rudnicky A (2010) Using the Amazon Mechanical Turk to transcribe and annotate meeting speech for extractive summarization. *Proceedings of the NAACL Workshop on Creating Speech and Language Data with Amazon's Mechanical Turk*, Los Angeles, CA.

McGlashan S, Burnett D, Carter J, Danielsen P, Ferrans J, Hunt A, Lucas B, Porter B, Rehor K and Tryphonas S (2004) VoiceXML 2.0. W3C Recommendation. `http://www.w3.org/TR/2004/REC-voicexml20-20040316` (accessed 18 October 2012).

Minker W and Bennacef S (2004) *Speech and Human-Machine Dialog*. Springer, New York.

Mohr D, Turner D, Pond G, Kamath J, de Vos C, and Carpenter P (2003) Speech recognition as a transcription aid: a randomized comparison with standard transcription. *Journal of the American Medical Informatics Association* **10**(1), 85–93.

Nuance (2009) Cost Savings with Computer Aided Medical Transcription: Three Case Studies. White paper.

Och F (2006) Challenges in Machine Translation. *Proceedings of the TC-Star Workshop*, Barcelona, Spain.

Oh A and Rudnicky A (2000) Stochastic language generation for spoken dialogue systems. *Proceedings of the ANLP/NAACL Workshop on Conversational Systems*, Seattle, WA.

Pieraccini R and Huerta J (2005) Where do we go from here? Research and commercial spoken dialog systems *Proceedings of the SIGdial Workshop on Discourse and Dialogue*, Lisbon, Portugal.

Pieraccini R, Caskey S, Dayanidhi K, Carpenter B and Phillips M (2001) ETUDE, a recursive dialog manager with embedded user interface patterns. *Proceedings of the ASRU*, Madonna di Campiglio, Italy.

Pieraccini R, Suendermann D, Dayanidhi K and Liscombe J (2009) Are we there yet? Research in commercial spoken dialog systems. *Proceedings of the TSD*, Pilsen, Czech Republic.

Porter M (1980) An algorithm for suffix stripping. *Program*.

Rochery M, Schapire R, Rahim M and Gupta N (2001) BoosTexter for text categorization in spoken language dialogue. *Proceedings of the ASRU*, Madonna di Campiglio, Italy.

Rose R (1991) Techniques for information retrieval from speech messages. *Lincoln Laboratory Journal* **4**(1), 45–60.

Rosenberg A and Binkowski E (2004) Augmenting the kappa statistic to determine interannotator reliability for multiply labeled data points. *Proceedings of the HLT/NAACL*, Boston, MA.

Schalkwyk J, Beeferman D, Beaufays F, Byrne B, Chelba C, Cohen M, Kamvar M and Strope B (2010) "Your Word Is My Command": Google Search by Voice: A Case Study, in *Advances in Speech Recognition: Mobile Environments, Call Centers and Clinics* (ed. A Neustein). Springer, New York.

SpeechCycle (2008a) SpeechCycle and Tellme Power Next Generation Enterprise Speech Self-Service. Press release.

SpeechCycle (2008b) SpeechCycle Rich Phone Applications Process 50 Million Automated Calls. Press release.

Suendermann D and Pieraccini R (2011) SLU in commercial and research spoken dialog systems, in *Spoken Language Understanding: Systems for Extracting Semantic Information from Speech* (ed. G. Tur and R de Mori). John Wiley & Sons, Inc., Hoboken, NJ.

Suendermann D, Liscombe J and Pieraccini R (2010a) How to drink from a fire hose: one person can annoscribe 693 thousand utterances in one month. *Proceedings of the SIGdial Workshop on Discourse and Dialogue*, Tokyo, Japan.

Suendermann D, Liscombe J and Pieraccini R (2010b) Optimize the obvious: automatic call flow generation. *Proceedings of the ICASSP*, Dallas, TX.

Suendermann D, Liscombe J, Dayanidhi K and Pieraccini R (2009a) A Handsome set of metrics to measure utterance classification performance in spoken dialog systems. *Proceedings of the SIGdial Workshop on Discourse and Dialogue*, London, UK.

Suendermann D, Liscombe J, Evanini K, Dayanidhi K and Pieraccini R (2008) C^5. *Proceedings of the SLT*, Goa, India.

Suendermann D, Liscombe J, Evanini K, Dayanidhi K and Pieraccini R (2009b) From rule-based to statistical grammars: continuous improvement of large-scale spoken dialog systems. *Proceedings of the ICASSP*, Taipei, Taiwan.

Suendermann D, Liscombe J, Pieraccini R and Evanini K (2010c) "How am I Doing?" A new framework to effectively measure the performance of automated customer care contact centers, in *Advances in Speech Recognition: Mobile Environments, Call Centers and Clinics* (ed. A Neustein). Springer, New York

Sun Microsystems (1998) Java Speech Grammar Format Specification Version 1.0. http://java.sun.com/products/ja va-media/speech/forDevelopers/JSGF/ (accessed 18 October 2012).

Tur G and de Mori R (2011) *Spoken Language Understanding: Systems for Extracting Semantic Information from Speech*. John Wiley & Sons, Inc., Hoboken, NJ.

Tur G, Schapire R and Hakkani-Tür D (2003) Active learning for spoken language understanding. *Proceedings of the ICASSP'03*, Hong Kong, China.

van Tichelen L and Burke D (2007) Semantic Interpretation for Speech Recognition Version 1.0. W3C Recommendation. http://www.w3.org/TR/semantic-interpretation/.

Williams J and Witt S (2004) A comparison of dialog strategies for call routing. *Speech Technology*.

Young S (2002) Talking to machines (statistically speaking). *Proceedings of the ICSLP*, Denver, CO.

Zissman M (1993) Automatic Language Identification Using Gaussian Mixture and Hidden Markov Models. *Proceedings of the ICASSP*, Minneapolis, MN.

11

Economic and Ethical Background of Crowdsourcing for Speech

Gilles Adda[1], Joseph J. Mariani[1,2], Laurent Besacier[3], and Hadrien Gelas[3,4]

[1]*LIMSI-CNRS, France*
[2]*IMMI-CNRS, France*
[3]*LIG-CNRS, France*
[4]*DDL-CNRS, France*

11.1 Introduction

With respect to spoken language resource production, crowdsourcing—the process of distributing tasks to an open, unspecified population via the Internet—offers a wide range of opportunities. Populations with specific skills are potentially instantaneously accessible somewhere on the globe for any spoken language. As is the case for most newly introduced high-tech services, crowdsourcing raises both hopes and doubts, and certainties and questions. This chapter focuses on ethical, legal, and economic issues of crowdsourcing in general (Zittrain 2008b) and of crowdsourcing services such as Amazon Mechanical Turk (Fort *et al.* 2011; Adda *et al.* 2011), a major platform for multilingual language resources (LR) production. These issues include labor versus leisure, spare time versus working time, labor organization and protection, payment and rewards, and so on. Given the multifaceted aspects of the subject, separating the wheat from the chaff might require an entire book, likely in a sociological or political science book series. This is clearly not the objective of the present contribution. However, given both the emerging role of crowdsourcing services as scientific tools and the ethical demands of science and research, a few issues of particular importance will be examined in order for researchers to sharpen their analysis and judgment. Some, such as the legal problems, are off-putting, and others are extraordinarily complex, as is the case of the economic models, but all are facets of crowdsourcing.

Crowdsourcing for Speech Processing: Applications to Data Collection, Transcription and Assessment, First Edition.
Edited by Maxine Eskénazi, Gina-Anne Levow, Helen Meng, Gabriel Parent and David Suendermann.
© 2013 John Wiley & Sons, Ltd. Published 2013 by John Wiley & Sons, Ltd.

Crowdsourcing is a neologism designed to summarize a complex process within a single word. To examine how ethics and economy are intertwined in crowdsourcing, the concept is dissected and a short review of the different crowdsourcing services (detailed in Chapter 9) is presented. We describe the major ethical and economic issues raised by representative crowdsourcing and microworking services, with a focus on MTurk, the main crowdsourcing, microworking service used nowadays by researchers in speech. In the context of this chapter, Microworking refers to the division of tasks into multiple parts and Crowdsourcing refers to the fact that the job is outsourced via the web and done by many people (paid or not). In particular, the issue of compensation (monetary or otherwise) for the completed tasks is addressed, as are the ethical and legal problems raised when considering this work as labor in the legal sense. This is particularly relevant when the tasks are in competition with activities performed by salaried employees. The proposed debate has to be considered in relation to both the economic models of the various crowdsourcing services and the task to be performed. The use of crowdsourcing for under-resourced languages is presented as a case study to exemplify the different issues exposed beforehand. Finally, this contribution aims to propose some specific solutions for researchers who wish to use crowdsourcing in an ethical way. Some general solutions to the problem of ethical crowdsourced linguistic resources are outlined.

11.2   The Crowdsourcing Fauna

11.2.1   The Crowdsourcing Services Landscape

The Crowdsourcing concept arose from the evidence that some tasks could be completed by Internet users, thus relying on the advantages of Internet, namely, instantaneous access to a huge number of people all over the world. Internet users may be compensated for their contribution, and this compensation can be monetary or not, depending on the crowdsourcing system and the tasks performed. A first phase of crowdsourcing development relied on the specific competences of some Internet users. WIKIPEDIA (http://www.wikipedia.org/) is one of the earliest and probably one of the most famous representatives of crowdsourcing systems relying on volunteer work. Besides WIKIPEDIA, there were many projects of collaborative science, such as THE GALAXY ZOO (http://www.galaxyzoo.org/). The first paid crowdsourcing systems involved users with special (professional) abilities such as programming, with TOPCODER (http://www.topcoder.com/) or designing, with 99DESIGNS (http://99designs.com/). In the speech domain, early attempts at collaborative annotation, such as Draxler (1997) should be highlighted.

More recently the concept of *Human computing*, in which the only required abilities are to be a human and to have some spare time, has appeared. This is a transposition of the Grid Computing concept to humans. The idea is to harness advantage of humans' "spare cycles" in order to develop a virtual computer of unlimited power, as the population potentially involved is no longer limited to a subset of Internet users with some special skills, but instead includes *any* Internet user. According to this concept, each user, like a processor in a grid, is assigned a basic subtask and only has access to the minimal information required to perform his or her subtask. If the task is straightforward and easy to explain (for instance, good-quality orthographic transcription of monologues as in Parent and Eskénazi 2010), then defining subtasks consists simply of splitting the data into small pieces. If the task is complex, it could be divided into successive, easier tasks, which in turn could be cut into small, simple elementary subtasks; for instance, Parent and Eskénazi (2010) have adopted a corrective strategy with successive teams of workers.

This type of crowdsourcing services, called *Microworking*, can be further classified depending on whether or not monetary compensation is provided. In this chapter, monetary rewards refer only to rewards in cash. Many sites of cloud labor provide monetary compensations, such as MTurk or CLICKWORKER (http://www.clickworker.com/), but GWAPS (*Games with a purpose*), which also make use of the concept of microworking, usually do not offer monetary compensations. The GWAP is another strategy of attracting large numbers of nonexperts through online games. It was initiated by the ESP online game (von Ahn 2006) for image tagging. Many projects of collaborative science are set up as a GWAP, for instance, FOLDIT (http://fold.it/) in the domain of protein folding. GWAPS provide entertaining or stimulating activities that are interesting enough to attract people willing to perform volunteer work, sometimes with nonmonetary rewards (e.g., SWAG BUCKS, http://www.swagbucks.com/). In the speech domain, collaborative experiments have been set up, such as (Gruenstein *et al.* 2009) about collecting and transcribing with an online educational game, or (Draxler and Steffen 2005) about the recording of 1000 adolescent speakers. Finally, some microworking systems are in an ambiguous situation. For instance, RECAPTCHA (http://www.google.com/recaptcha/learnmore) uses CAPTCHAs of words that optical character recognition software failed to read. RECAPTCHA aims to contribute to the digitization of difficult texts in the Google book project. RECAPTCHA does not offer compensation for the work done. However, this work cannot be considered as "voluntary work" as the user fills CAPTCHAs to get access to a service and not to help Google. This latest form of crowdsourcing, usually unbeknownst to the user, is described as "epiphenomenal" (Zittrain 2008b).

MTurk, introduced in 2005, is a precursor to and a leader of the myriad of paid microworking systems that exist today (see Chapters 1 and 9 for more details about the different alternatives).

The boundary between GWAPS and other microworking crowdsourcing systems is not precise. It is not the case that microworking systems propose tedious tasks, whereas other approaches are purely "for fun": entertaining tasks do exist on MTurk (see, for instance, The Sheep Market www.thesheepmarket.com/). GWAP and MTurk cannot be distinguished by the fact that MTurk provides remuneration, as some GWAPS do propose nonmonetary rewards (e.g., Amazon vouchers for PHRASEDETECTIVE, Chamberlain *et al.* 2008). Finally, collaborative and GWAP-based techniques are not the only "ethical alternatives," since ethical crowdsourcing platforms such as SAMASOURCE (http://samasource.org/) do exist.

To classify paid crowdsourcing services, Frei (2009) proposed four categories based on their complexity. At the lower end of the scale, there is the category of *micro tasks*, which includes MTurk. Here, the tasks are small and easy, require no skills, and offer very low rewards. Next comes the category of *macro tasks* with low pay and a substantial number of propositions, as in micro tasks, which, however, require some specific skills (e.g., writing a product reviews). Then come the *simple projects*, such as basic website design. Simple projects involve higher pay and fewer propositions while requiring more skills and time. At the highest level of the ranking scale are the *complex tasks*, those which require specialized skills and a significant time commitment (for instance, the tasks available in INNOCENTIVE, http://www.innocentive.com/). The two latter categories resemble tasks that can be encountered in the "real" world, unlike the first two categories. For instance, there is no direct communication between requesters and workers for the first two categories, while for the latter, communication is required. Other interesting taxonomies exist such as the one presented in Quinn and Bederson (2011) that uses six distinguishing factors to classify the human computation systems. Moreover, Frei's taxonomy of crowdsourcing services has been rendered oversimplified by the appearance of macro tasks or simple project services built upon micro tasks, such as CROWDFORGE (Kittur *et al.* 2011) (http://smus.com/crowdforge/) or TURKIT (Little *et al.* 2010) (http://groups.csail.mit.edu/uid/turkit/). Frei's taxonomy is, however, still useful for defining some targeted solutions, including fair compensation for the work done (see Section 11.5.1).

11.2.2   Who Are the Workers?

The backbone of the crowdsourcing system, workers constitute a population with rapidly evolving characteristics. This section gives some sociological details with recent facts and figures.

Country of Origin
The country of origin is not a selection criterion for most crowdsourcing services. For instance, ODESK's (https://www.odesk.com/) active workers are coming from (in decreasing order) the Philippines, India, the United States, Ukraine, Russia, Pakistan, Bangladesh, noting especially the Philippines' workers who seem to work 24/7 (Ipeirotis 2012c). Some services, such as MTurk, do impose restrictions: MTurk limits monetary remuneration (cash incentives) to workers with a valid US bank account (payment in dollars) or to workers from India (payment in rupees). Recently, requesters tend to a priori reject Indian workers, as they are more likely to be spammers or be less proficient in certain tasks involving language use; this change has led some Indian workers to lie about their location (Ipeirotis 2011a). MTurk also requires that requesters provide a US billing address and a credit card, debit card, Amazon Payments account, or US bank account in order to publish tasks. Some crowdsourcing services exclusively use underprivileged workers such as MOBILEWORKS (http://www.mobileworks.com/) or SAMASOURCE (http://samasource.org/). MOBILEWORKS has a team of workers from India and Pakistan ready to receive jobs via their mobile phone or computer and claims to be "socially responsible," suggesting that its workers are paid a fair wage to encourage higher quality work; SAMASOURCE is a nonprofit organization that establishes contracts with enterprise customers or other crowdsourcing services (such as CROWDFLOWER, http://crowdflower.com/) in order to provide crowdsourcing microworking services to people living in poverty around the world.

MTurk seems to be quite particular, given its bimodal distribution of workers: the ones from India and the ones from United States. It is difficult to obtain accurate figures concerning these workers, given their anonymity in MTurk. There is some evidence that the exact number of workers actually working in MTurk is much smaller than the official figure of 500,000 registered workers in 2011 (see Section 11.3.3).

Relying on surveys submitted within MTurk, studies in social sciences (Ross *et al.* 2010; Ipeirotis 2010b) may provide some insight into workers' socioeconomic profiles (country, age, . . .), the way they use MTurk (number of tasks per week, total income in MTurk, . . .), and how they qualify their activity. For instance, these studies enable one to estimate the number of Indian workers in MTurk: Indian workers represented 5% in 2008, 36% in

December 2009 (Ross *et al.* 2010), 50% in May 2010 (http://blog.crowdflower
.com/2010/05/amazon-mechanical-turk-survey/) and have generated more
than 60% of the activity in MTurk (Biewald 2010).

Sociological Facts
As for determining the number of workers, it is difficult to present an exact picture of who
the workers in crowdsourcing services are. The studies in social sciences mentioned above
revealed that, in MTurk, 91% of the workers expressed their desire to make money (Silberman
et al. 2010), even if the observed wage was very low: $1.25/h according to Ross *et al.* (2009)
and $1.38/h according to Chilton *et al.* (2010). If 60% of the workers think that MTurk is a
fairly profitable way of spending free time and earning cash, only 30% mentioned their interest
in the tasks, and 20% (only 5% of the Indian workers) said that they were using MTurk to kill
time. Finally, 20% (30% of the Indian workers) declared that they were using MTurk to make
basic ends meet, and about the same proportion stated that MTurk was their primary source
of income. Furthermore, the 20% of the most active workers who spend more than 15 hours
per week with MTurk (Adda and Mariani 2010) produce 80% of the overall activity.

The population and the motivations of the workers are heterogeneous. Nevertheless, those
20% of the workers for whom crowdsourcing is a primary income generate an activity that
should be considered as a labor, even if the actual labor laws (see Section 11.3.2) are unable to
clearly qualify this activity as such. Moreover, there is a huge difference between good workers
who have direct and regular connections with some requesters and know how to maneuver
between the tasks to avoid scams, and naive workers for whom the crowdsourcing platforms
blatantly lack a robust regulatory framework (see Section 11.3.3). There is also a difference
between workers who desperately need the money and those who do not: people who need
money will also undertake low-paying tasks, as there is an insufficient number of high-paying
tasks to fill a working day for those looking to daily earn a maximum of cash incentives (Adda
and Mariani 2010).

11.2.3   Ethics and Economics in Crowdsourcing: How to Proceed?

Economic and ethical problems in crowdsourcing are related (among others) to the type of
crowdsourcing services, the nature of the task and of the workers' activity, as well as to
the place where this activity is located. Given this complex situation, connecting all these
parameters poses a very difficult multivariate problem.

In Section 11.3, an overview of the economic model of crowdsourcing services is presented,
together with a summary of the situation regarding labor laws. Based on the insight gained from
the case study of transcribing speech of under-resourced languages, some possible solutions
may be envisioned for speech science, and more generally language sciences, in order to deal
with crowdsourcing services in a more ethical and efficient way.

11.3   Economic and Ethical Issues

Beyond the previously mentioned opportunities, the rapid growth of crowdsourcing services
introduces many problems, some of which are not only philosophical or ethical, as those
mentioned in Zittrain (2008b), but also legal (Felstiner 2011), while others are economic
(Ipeirotis 2010c). This section presents an overview of all these problems.

The main ethical and economic problems concern the worker, his or her relation with the task, the requester and the crowdsourcing service. Technically, workers are usually independent contractors. They are not subject to minimum wage or overtime protection. Ethical problems may arise in two situations: if the task is comparable to a human experiment, or if it corresponds to real labor. In both cases, researchers have specific ethical obligations. For instance, speech corpora transcription tasks were being performed for years by employees of agencies like LDC (Linguistic Data Consortium, `http://www.ldc.upenn.edu/`) or ELDA (Evaluations and Language resources Distribution Agency, `http://www.elda.org/`), while the collection of speech data was carried out on a more volunteer basis. For tasks which were performed by salaried employees, crowd labor could be viewed as offshoring on the web.

If the task corresponds to a human experiment, some experiments that would be legal for a private organization would not be approved by a university Institutional Review Board following the US National Research Act of 1974. For instance, to obtain the authorization of the Virginia Commonwealth University IRB (institutional review board; Virginia Commonwealth University 2009), there are the following guidelines about payment:

> Compensation for Research Participants: Payment for participation in research may not be offered to the subject as a means of coercive persuasion. Rather, it should be a form of recognition for the investment of the subject's time, loss of wages, or other inconvenience incurred.

Considering the statement in Section 11.2.2 that a significant fraction of the workers who are involved in microworking platforms seem to have no alternative way of earning, many experiments that are using these platforms are hardly compliant with most IRBs. Moreover, IRBs usually require that the study participants sign some charter or agreement to ensure their informed consent. Fortunately, the collection of anonymous speech or annotations is not considered to be a human experiment, and therefore does not really fall under the scope of IRB regulation. Nevertheless, it is always a good practice to explain the whys and wherefores of the study to the participants and, when possible, to obtain a signed agreement from them.

If the task corresponds to labor (as do for instance most of the tasks involving transcription or translation), the main question is the hourly wages paid in the crowdsourcing platforms, which are significantly lower than the minimum wage in many countries ($7.25/h in the United States, €9.22/h in France), and which may entail some ethical and economical problems. Defining a useful minimum hourly wage in crowdsourcing services is quite difficult (see Section 11.5.1). This minimum hourly wage should take into account the advantages of these tasks for the workers, such as self-assignment, the lack of time and money spent on commuting, and the fact that the crowd is not located in a single country, but defining a minimum hourly wage could help in addressing some of the problems uncovered in this chapter.

This question of labor within crowdsourcing services is situated in a more general framework. For instance in Albright (2009), it is noticed that the outsourcing of some jobs on crowdsourcing services is hardly avoidable: "Recognize that it is happening," advised Carl Esposti, founder of `crowdsourcing.org` (accessed 17 October 2012), which tracks the industry. "It's happening and an absolute inevitability that a new market for work is being created on both the supply and demand sides. An individual may not like this and may not want to participate, but they have no choice." Esposti, from `crowdsourcing.org` (accessed 17 October 2012), advises IT professionals who are currently employed to pay attention to this new business model and to the impact it could have on their careers. He suggests that individuals should try to determine whether crowdsourcing will be constructive or destructive for their organizations and how it might relate to their own particular jobs. He noted, for example, that companies do not crowdsource entire functions: rather they crowdsource work that can be broken up into manageable tasks. The more a person's job is activity based, therefore, the more his or her job could be at risk.

11.3.1 What Are the Problems for the Workers?

There are numbers of issues reported by the workers involved in crowdsourcing services. Some are general to all crowdsourcing platforms, while others are specific, but MTurk seems to manifest most of the problems and thus merits a closer examination.

Very low wages (below $2 an hour (Ross *et al.* 2009; Ipeirotis 2010b; Chilton *et al.* 2010) in MTurk) are a first point to be addressed. In Section 11.2.2, it is mentioned that a significant proportion of the workers use MTurk as their primary source of income, or to make basic ends meet (Ross *et al.* 2010; Ipeirotis 2010b). Below are some statements, from different sources, Ipeirotis (2008), Turkopticon (*supra*), Turker Nation (http://turkernation.com/), illustrating the economic situation of some MTurk workers in the United States:

> I realize I have a choice to work or not work on AMT, but that means I would also not need to make the choice to eat or not eat, pay bills or not pay bills, etc.
>
> How do you make ends meet on a dollar an hour? You don't. All you do is add to what you make with your regular job and hope it is enough to make a difference.
>
> I don't know about where you live, but around here even McDonald's and Walmart are NOT hiring. I have a degree in accounting and cannot find a real job, so to keep myself off of the street I work 60 hours or more a week here on mTurk just to make $150–$200. That is far below minimum wage, but it makes the difference between making my rent and living in a tent.
>
> I am currently unemployed and for some reason absolutely can not find a job. Every job I apply for either turns me down or I don't hear from them at all. I have been doing online surveys, freelance writing, and MTurk to try to make the most money I can. I don't make much but when you literally have no savings and no income you take what you can get.
>
> No available jobs in my area, have applied to over 40 jobs no calls so far been 3 months. Do it to pay my bills which includes rent and diapers for my kids until I find work again.

One may question whether or not the workers are free to choose or not this way of making money, especially considering the actual level of total or partial unemployment in United States and in Europe, and the living standards in Third World countries.

However, this conclusion must be tempered by other statements, from other workers, which illustrate the fact that the situation is not black or white:

> I have a high need for feedback and seeing my HITs get approved supplies me with that satisfaction.
>
> Mturk has given me a sense of conviction that 'I can'. I have started to believe in myself and the journey has been so enriching. I just love this place and when we get paid for something we love—nothing like it. Thanks to mturk.
>
> Mechanical Turk work is not only for money. This is an experience of the worldwide working methods.There are different kinds of hits. Every hit on this turk is challenge to our knowledge. So I like this job very much.
>
> I am a retired teacher who finds the more academic hits stimulating.

Another frequently mentioned problem (Silberman *et al.* 2010) is the fact that requesters pay late. In MTurk, there is an "autoapproval" delay in the permission of payment when requesters neglect to approve the task. It is very common for the requesters to choose the maximum delay, which is 30 days. This means that until the end of the delay, which is not visible to workers, the worker does not know if his work will be approved and paid or not.

To pay late is a problem, but to not pay at all could be a real issue for the workers. For example, many reported experiments dealing with speech crowdsourcing have implemented automatic filters in order to reject completed tasks that seem inadequate. To block a worker or to reject many tasks may result in the worker being banned from the crowdsourcing service, which could have real consequences, especially for workers for whom this money is essential. As misunderstandings of the guidelines or errors in automatic procedures are always possible, automatic procedures should be handled with great precaution.

A further point concerns the choice of anonymity by many crowdsourcing vendors in order to protect the workers and the requesters from e-mail spamming or from incorrect use of personal information. However, anonymity hides any explicit relationship between workers, and between workers and requesters. Even the basic workplace right of unionization is denied and workers have no recourse to any channels for redress against employers' wrongdoings, including the fact that they have no official guarantee of payment for properly performed work. They may complain to the site that the requester did not behave correctly, but without any guarantee.

Some regulation between requesters and workers exists through workers' Blogs or Forums, such as the Mechanical Turk Blog (`mechanicalturk.typepad.com`) or Mechanical Turk forum (`turkers.proboards.com`), or through the use of Turkopticon (`turkopticon.differenceengines.com`), a tool designed to help workers report bad requesters. However, all these solutions are unofficial, and nothing formally protects the workers, especially the new ones who are mostly unaware of these tools.

Given the anonymity, another concern is the nature of the task itself and the fact that the real task is cut into small pieces and presented to the workers in a way that may obscure the purpose of the work. An extreme case was provided by Zittrain (2008a) who mentioned the problem of matching photos of people, which could be used by an oppressive regime to identify demonstrators. However, more common cases include the task of solving a captcha that will give a spammer access to a protected site, and the task of "testing" if ads are working on a website, which will generate fake clicks but real money. Concerning speech science, it would be a good practice (see Section 11.3) to explain the purpose of the whole task to the workers in order to allow them the option of not participating in a study they do not agree with, such as an experiment in the domain of biometrics or a study funded by the army.

11.3.2   Crowdsourcing and Labor Laws

Given the ethical problems listed in previous sections, one may wonder why the law is not applied to the regulation of crowd labor. Some authors (Felstiner 2011; Wolfson and Lease 2011) looked at the possible extension of US labor laws to the crowdsourcing workplace. They found quite difficult to decide with precision how the laws could be applied to crowdsourcing. The first reason is the heterogeneity of the crowdsourcing. Section 11.2.1 already mentioned that crowdsourcing platforms could be very different, going from micro task platforms such as MTurk, to complex task platforms such as INNOCENTIVE (`http://www.innocentive.com/`) or ODESK (`https://www.odesk.com/`). The second and main reason is the inappropriateness of existing laws for dealing with the Internet, and especially with *work* on the Internet. In order to limit the complexity, Felstiner (2011) and Wolfson and Lease (2011) mainly looked at the application of US state and federal laws to MTurk, as it is one of the most used crowdsourcing platforms.

Worker Status in Participation Agreement

The main goal is to determine the exact status of workers: many crowdsourcing platforms (the vendors) include in their terms of use a statement that defines the workers as independent contractors. Workers are supposed to have accepted this term with the "clickwrap" participation agreement; for instance, MTurk Participation Agreement contains the statement: "As a Provider (worker), you are performing Services for a Requester in your personal capacity as an independent contractor and not as an employee of the Requester." The Amazon terms of use upon registration state that workers are not allowed "to use robots, scripts, or other automated methods to complete the Services," and that they should furnish the requester with "any information reasonably requested" and agree with not being entitled to any employee benefits or eligible for worker's compensation if injured. Amazon can cancel a worker's account at any time. When this happens, the worker loses all the earnings left in his Amazon account. As independent contractors, they have no protection of any sort and should arrange for their own insurance, pay self-employment taxes, and so on. By clickwraping participation agreements, requesters and workers of many crowdsourcing platforms are supposed to have accepted this contractual agreement. However, clickwrap participation agreements present two pitfalls:

1. Many requesters and workers do not really read the agreement and do not have a "clear" view of the contents of this contract.
2. The agreement has been drawn up by only one partner (the vendor), which calls into question the negotiated nature of the contract.

As stated by Felstiner (2011): "The vendors, in binding both workers and firms to their clickwrap, have, in essence, prospectively *filled in* the content of the worker-firm contract." What is very clear in the participation agreement is that its terms uniformly disclaim *any* vendor responsibility.

However, even though workers agreed with a click on the clickwrap participation agreement to classify themselves as "independent contractors," it is not a decisive determination of their status. First, this status is not always clear, even in participation agreements; for instance, in MTurk Participation Agreement: "Repeated and frequent performance of Services by the same Provider on your behalf could result in reclassification of independent contractor employment status." Second, the courts have already ruled that when the work is essentially done in the capacity of an employee, putting an "independent contractor" label on the worker does not exempt him or her from the protection of the act (Supreme Court in Tony and Susan Alamo Foundation v. Secretary of Labor, 471 US 290 (1985)).

Employee or Independent Contractor: Status of the Crowd Worker under FLSA

To decide if crowdsourcing workers are statutory employees or independent contractors, Felstiner (2011) and Wolfson and Lease (2011) use the Fair National Standard Acts (FLSA), which, given that the parties are "employers" and "employees," defines a federal minimum wage and overtime protection, and the National Labor Relation Act (NLRA). Courts have developed a series of tests to decide if someone is an employee under the FLSA or NLRA. In order to decide if, in the relations between the vendor, requesters and workers, some elements could be qualified as employer–employee relations, Felstiner (2011) and Wolfson and Lease (2011) look at the applicability of these tests on the crowdsourcing case. Under FLSA, courts use a multifactor "economic reality" test with seven factors. No single factor is determinative, thus all the factors are examined, with a different weight:

(i) **How integral the work is to the employer's business**: There is a large variety of requesters, some relying entirely upon crowd labor, others using crowdsourcing only sparsely. This factor is not decisive to determine a worker's status.

(ii) **The duration of relationship between worker and employer**: Some workers work repeatedly with the same requester, as in oDESK, but the relationship between requesters and workers could not be qualified as permanent. The sole long-term relationship is between these parties and the crowdsourcing service.

(iii) **If the worker had to invest in equipment or material himself to do the work**: This factor is often decisive, but the definition of "equipment" for the work on the Internet is vague. Is it the computer, which is basic equipment, or the specific web platform developed by the crowdsourcing service? This question could be debated, but courts have tended to be neutral on similar cases about tele-working.

(iv) **How much control the employer has over the worker**: The control of the requester over the worker is not direct, but (through the participation agreement) the requesters have a high level of control over how the work is done. In comparison with a contractual relation, the use of this control is not negotiated.

(v) **The worker's opportunity for profit and loss**: Crowdsourcing vendors did not structure their services such that workers may build and grow a business, and worker's opportunities for profit and loss are quite limited.

(vi) **How much skill and competition there is in the market for this type of work**: As in the preceding factor, crowdsourcing vendors leave very little room for initiative, judgment and foresight. For microworking services, it is clear that almost anyone of any skill level may perform the proposed tasks.

(vii) **If the worker is an independent business organization**: For microworking services, it would be surprising, especially given the very low observed compensation, for a worker to build an "independent business organization" devoted entirely to this activity. This could be different for complex tasks, such as the ones proposed in 99DESIGNS, a crowdsourced design contest marketplace (http://99designs.com/).

Concerning crowd labor, the last three factors weigh in favor of an employee status, while the first four do not decisively accord either employee or independent contractor status. Therefore, it is not clear if a crowd worker could be classified as employee under FLSA. However, there is uncertainty, which means that potential requesters must be aware of the possibility of regulation.

Many constraints listed in the license agreements of many crowdsourcing vendors are worded in order to *not* match the FLSA or NLRA tests' factors, and more generally to eliminate any explicit relationship of subordination. We may think that this is designed to limit the risk of a reclassification of the status of crowdsourcing workers as employees.

Crowdsourcing Vendor as Joint Employer

Workers in crowdsourcing services act as temporary employees, hired through a temporary staffing agency (here the vendor). In this case, the workers can be regarded as employees of the *vendor* instead of the requester. Usually, the vendor's participation agreement tries to reject this possibility, but if workers can show that the economic reality reflects an employee–employer relationship between the vendor and the pool of workers, courts could declare the vendor a *joint employer*. Felstiner (2011) argues that the vendors, and especially MTurk, could have difficulty escaping responsibility for the work rights of their workers as joint employers. Furthermore, any national or international regulation of the labor laws will be easier to apply to the vendor, who is relatively easy to identify and to locate. Regulation on the myriad of requesters, using only the informations provided by the requester to the vendor, may be difficult to apply efficiently.

Crowd Workers' Status in Other Countries

What about other countries? In France, the *Code du Travail* does not give an exact definition of who is a salaried employee. Instead, this status is accorded based on the jurisprudence, which lists three mandatory factors for deciding if a person is a salaried employee, linked to an employment contract:

1. A relationship of subordination with the employer
2. A monetary compensation
3. Completion of a task.

The relation between the point 4 of FLSA and the point 1 of the above list is clear: the subordination, which is conclusive in France for deciding if a worker is an employee (even if, in order to determine the status of a worker, all factors such as the points 1 and 3 of FLSA are taken into account in the French labor law jurisprudence). More generally, in many countries an individual will be considered an independent contractor if he or she independently carries out the job and if there is neither subordination nor exclusivity in the relationship between the parties. However, as in the US case, laws tend worldwide to elevate substance over form when examining the parties' actual relationship. Therefore, as some of the relations between vendors and workers or between requesters and workers could be defined as an employee–employer relationship, the outcome would be uncertain if any individual workers, or national or international labor agency, were to take legal action against either crowdsourcing platforms or requesters.

Crowdsourcing and Labor Laws: A Needed Regulation

Considering the existing labor laws, it is quite difficult to qualify the workers' status in crowdsourcing services as employees. At the moment, crowd labor is in a "gray area," because the current regulation is not adequate. It is likely that if the crowdsourcing market is still growing in terms of the number of workers and the size of the market, the national and international legislatures will take into account this innovation of the concept of work, and will amend the labor laws to regulate the market (Felstiner 2011; Wolfson and Lease 2011). Moreover, if crowd labor is growing at the expense of existing industries and jobs, instead of creating new activities with new types of workers, this will create a social pressure to address the deficiencies in the labor laws concerning crowd labor. The definition of a new type of temporary employment contract designed for crowdwork with monetary rewards could be beneficial to both the workers and the requesters. This contract, which could be drawn up between the vendor and the worker (see Section 11.3.2), should help to regulate the crowdwork and to assure a stability and a clear legal framework for the requesters.

11.3.3   Which Economic Model Is Sustainable for Crowdsourcing?

Pragmatically, making ethics a priority is feasible if the law enforces it and if the economic situation is compliant with it. Economics is indeed a major concern for obtaining the conditions for ethics in the real world. Highlighting some of the driving forces of its economic model will support our understanding of the crowdsourcing.

Low versus High Reward

The frequent assumption that the low rewards are a result of the classical law of supply-and-demand (large numbers of workers means more supply of labor and therefore lower acceptable salaries) is false. First of all, this assumption relies on the belief that the number of workers is huge, while Fort et al. (2011) observe that there are not too many active workers. Fort et al. (2011) looked at the number of tasks effectively completed by the 1000 MTurk workers (Ipeirotis 2010b) and compared it to the total number of tasks completed according to the Mechanical Turk Tracker (http://mturk-tracker.com). Taking into account the different factors, they found the number of effective workers to be below 50k, and the number of active workers below 10k. These figures are very far from the official figure of 500k registered workers. While only valid for MTurk, this ratio between the number of registered and active users/workers is compatible with other observations about the activity on the Internet, such as the "90–9–1" rule (Arthur 2006), and the number of active workers close to 2% observed in Ipeirotis (2012a). This calculation could explain the difficulty in finding workers with certain abilities, such as understanding a specific language (Novotney and Callison-Burch 2010), or speaking an under-resourced language: in Section 11.4, the expected number of Swahili speakers available on MTurk is lower than could be expected from the number of people speaking Swahili in the United States (3 instead of 32).

Many explanations could be provided to the fact that the rewards are so low. The first is that the low reward is a result of the requesters' view of the relation between quality and reward: many articles (see, for instance, Marge et al. 2010) observe that there is no correlation between reward and final quality. The reason is that increasing the price is believed to attract spammers (i.e., workers who cheat and not really perform the job, using robots or answering randomly instead). Spammers are numerous, for instance, in the MTurk system (Ipeirotis 2010c) due to a worker reputation system that makes it easy for a spammer to build a new account with 100% approval rate (Ipeirotis 2010a) (see also Chapter 9 for more details). This is a schema that is very close to what the 2001 economics Nobel prize winner George Akerlof calls "the market for lemons," where asymmetric information in a market results in "the bad driving out the good." He takes the market for used cars as an example (Akerlof 1970), where owners of good cars (here, good workers) will not place their cars on the used car market because of the existence of many cars in bad shape (here, the spammers), which encourage the buyer (here, the requester) to offer a low price (here, the reward) because he does not know the exact value of the car. After a period of time, the good workers leave the market because they are not able to earn enough money, given the work done (and sometimes they are not even paid), which in turn decreases the quality. At the moment, the crowdsourcing system is stable in terms of the number of workers, because workers leaving the system are replaced by new workers unaware of this situation (70% of the workers use MTurk for less than 6 months, Ross *et al.* 2009). A second explanation given in Bederson and Quinn (2011) uses the theory of moral hazard in economy (Holmstrom 1979), which explains that, given that the requesters do not incur the full cost of their actions because of the asymmetry of the relation, the anonymity, and the fact that they could reject the work done, . . . the cost for the other party (the workers) increases. In turn, workers tend to generate "just good enough" work or even cheat.

This lack of a well-designed reputation system is a stumbling block for many microworking services. For instance, Amazon's attitude toward reputational issues is passive. Amazon, as other crowdsourcing service vendors, maintains its position as a neutral clearinghouse for labor, in which all other responsibility falls of the two consenting parties (see Section 11.3.2).

As highlighted by Ipeirotis (2010c), without major developments, especially in financial rewards and reputation, flaws and a faulty economic model call into question medium term viability of microworking services such as MTurk.

Microworking crowdsourcing systems with low rewards should have difficulty keeping the "good" workers, because of the process described above. If work quality is an important aspect of the task (for transcription or annotation of speech for instance), this model is not adequate: relying on low-quality workers is not cost-effective, even if the standard redundancy method is used to improve the quality. For instance, Ipeirotis (2011b) shows that employing three high-quality master workers (the elite group of workers created in MTurk), and paying them 20% more than the usual price, results in the same quality as using 31 workers of 70% accuracy. It is far more beneficial to get around the system and retain good workers by paying them higher (but still modest) wages. In another example, Chen and Dolan (2011) note that given the incentive to maximize their rewards, workers often cheat; to solve this problem, Chen and Dolan (2011) used a two-tiered payment system to reward workers who submit good descriptions. This is a way to select the most qualified workers and to grant them with a bonus if they perform well. As the nature of the task becomes harder, Chen and Dolan (2011) report the benefits of long-standing relationships with the workers instead of the anonymous relation as proposed by MTurk: less quality control, ability to train the workers to improve their competence, and ability to correlate reward and quality (fair rewards result in worker loyalty).

It is indeed one of the reason many crowdsourcing services such as MTurk do not adhere to fair wages: a number of good workers operate within a separate framework built by some long-standing requesters who give higher financial rewards. These good workers are not available for other newly arrived requesters who offer the standard (very low) price.

Task versus Time Reward

As a requester, you can see the effective hourly rate along with the average completion time. However, MTurk workers do not have direct access to this information. In particular, they do not see the hourly rate, which is fundamental information for judging if the money received will be fair compensation for the work done and the time spent. A rational action of an experienced worker is to choose a large set of tasks, to use one unit task to test its real difficulty, and to determine his effective hourly rate. However, the workers are not all experienced or rational. This method of payment provokes behavior that is not always compatible with quality work, as the worker is not aware of the hourly rate before choosing the task. The same behavior may be observed in online games: a gamer is keeping track of his absolute score or level, and not of the time needed to obtain them. Similarly, the worker looks at the *absolute* level of funding rather than at hourly rate: "Today, I will work until I have made $10," which is certainly not the best way to optimize the overall reward. Moreover, Kochhar *et al.* (2010) reached the conclusion that an hourly payment was better (with some verification and time justification procedures), as task payment logically encourages one to place the number of performed tasks above the quality, regardless of payment. Our experiments described in Section 11.4 corroborate these observations.

Furthermore, piecework retribution is strictly regulated in developed countries in order to prevent a wage lower than the legal hourly minimum; for instance, piecework retribution, similar to other forms of variable remuneration, is possible in France only if it results in a wage above the legal minimum. However, determining an hourly rate is difficult in any remote workplace (see Section 11.5.1), as only all worked hours should be compensated; practical solutions should be learned from the telework/telecommuting case.

Which Is the Economic Model?

As pointed out in Ipeirotis (2011b), one may have the feeling that the visible flaws in some crowdsourcing services, such as MTurk concerning reputation system, anonymity, and very low rewards, are deliberate. This hypothesis has been put forward because these flaws induce an undue advantage for the first comers. They were able, using their own remuneration and reputation system, to catch the good workers and to subsequently keep them because the newcomers offering high rewards are overwhelmed by spammers and thus disappear or reduce their rewards. On the other hand, the low rewards proposed by the newcomers keep the remuneration at a sufficiently low level in order to present a very competitive cost. A visible effect of this is the growing number of specialized services which serve as interfaces between requesters and microworking platforms: CASTINGWORDS (http://castingwords.com/, speech transcription), SPEAKERTEXT (http://www.speakertext.com/, video transcription), SERV.IO TRANSLATE (http://www.serv.io/translation, translation)... The development of ethical crowdsourcing services enforces amendments and improvements to this model.

11.4 Under-Resourced Languages: A Case Study

It is difficult to discuss the ethical and economic aspects of crowdsourcing without experiencing the concept oneself. A case study is presented here in a domain where crowdsourcing seems to be a particularly hot topic: the processing of under-resourced languages. For these languages, data collection and annotation (for instance speech transcription) is a particularly difficult problem and crowdsourcing is a very attractive tool, especially for connecting speech technology developers and language experts. Moreover, since many under-resourced languages are spoken in developing countries, the potential workers (native speakers of the under-resourced language considered) tend to be the ones (mentioned in Section 11.2.2) who are more likely to rely on crowdsourcing for income, as do the Indian workers in the surveys from Ipeirotis (2010b) and Ross *et al.* (2010). In this section, the transcription of a speech corpora of two under-resourced languages from Africa using crowdsourcing is evaluated, and the main results of this experiment as well as the lessons learned are presented.

11.4.1   Under-Resourced Languages Definition and Issues

The term under-resourced languages introduced by Berment (2004) refers to a language characterized by some (if not all) of the following aspects: lack of a unique writing system or stable orthography; limited presence on the web; lack of linguistic expertise; lack of electronic resources for NLP (natural language processing) such as monolingual corpora, bilingual electronic dictionaries, and transcribed speech data. Developing an NLP system (e.g., a speech recognition system) for such a language requires techniques that go far beyond a basic retraining of the models. Indeed, processing a new language often leads to new challenges (special phonological systems, word segmentation problems, unwritten language, etc.). For its part, the lack of resources requires innovative data collection methodologies (crowdsourcing being one of them) or models in which information is shared between languages (e.g., multilingual acoustic models, Schultz and Kirchhoff (2006) and Le and Besacier (2009)). In addition, some social and cultural aspects related to the context of the targeted language bring additional problems: languages with many dialects in different regions; code-switching or code-mixing phenomena (switching from one language to another within the discourse); and massive presence of nonnative speakers (in vehicular languages such as Swahili).

11.4.2   Collecting Annotated Speech for African Languages Using Crowdsourcing

Recently MTurk has been studied as a means of reducing the cost of manual speech transcription. Most of the studies conducted on the use of MTurk for speech transcription have been done for the English language, which is one of the most well-resourced languages. The studies on English, including Snow *et al.* (2008) and McGraw *et al.* (2009), showed that MTurk can be used to cheaply create data for natural language processing applications. However, apart from a research conducted recently by Novotney and Callison-Burch (2010) on Korean, Hindi, and Tamil, MTurk has not yet been studied as a means to acquire useful data for under-resourced languages. As for as these languages are concerned, it is all the more important to collect data using highly ethical standards, as doing so usually involves people from developing countries who may suffer from extremely low standards of living.

The use of MTurk for speech transcription has been studied in the hopes of developing automatic speech recognition (ASR) for two under-resourced African languages without combining transcription outputs. The experimental setup, including the subject languages, is described in Section 11.4.3. Section 11.4.4 presents the result of the experiment, and a discussion is provided in Section 11.4.5.

11.4.3   Experiment Description

Languages

Amharic is a member of the Ethio-Semitic languages, which belong to the Semitic branch of the Afro-Asiatic superfamily. It is related to Hebrew, Arabic, and Syrian. According to the 1998 census, it is spoken by over 17 million people as a first language and by over 5 million as a second language throughout different regions of Ethiopia. The language is also spoken in other countries such as Egypt, Israel, and the United States. Amharic has its own writing system which is a syllabary. It is possible to transcribe Amharic speech using either isolated phoneme symbols or concatenated CV (Consonant Vowel) syllabary symbols.

Swahili is a Bantu language often used as a vehicular language in a wide area of East Africa. In addition to being the national language of Kenya and Tanzania, it is spoken in different parts of the Democratic Republic of Congo, Mozambique, Somalia, Uganda, Rwanda, and Burundi. Most estimations claim over 50 million speakers (with only less than 5 million native speakers). Structurally, Swahili is often considered to be an agglutinative language (Marten 2006). Despite being nontonal, it displays other typical Bantu features, such as noun class and agreement systems and complex verbal morphology. It was written with an Arabic-based orthography before it adopted the Roman script (standardized since 1930).

Corpora

Both Amharic and Swahili audio corpora were collected following the same protocol. Texts were first extracted from news websites and then segmented by sentence. Recordings were made by native speakers reading sentence-by-sentence with the possibility to rerecord mispronounced sentences. The whole Amharic speech corpus (Abate *et al.* 2005) contains 20 hours of training speech collected from 100 speakers who read a total of 10,850 sentences (28,666 tokens). The Swahili corpus used in this study corresponds to three and a half hours read by 5 speakers (3 male and 2 female). The sentences read by speakers serve as our gold standards and will be used to evaluate the transcriptions obtained by MTurk.

Transcription Task

For the transcription task, all 1183 of the audio files between 3 and 7 seconds (mean length 4.8 seconds and total one and a half hours) were selected from the Swahili corpus. The same number of files were selected from the Amharic corpus (mean length 5.9 seconds). These files were published (a task for a file) on MTurk. To avoid cheaters, task descriptions and instructions were given in the respective languages (Amharic and Swahili). The Swahili transcriptions did not require a special keyboard, but for the Amharic transcriptions, workers were given the address of an online unicode-based Amharic virtual keyboard (www.lexilogos.com/keyboard/amharic.htm), as the workers would not necessarily have access to Amharic keyboards.

11.4.4   Results

Analysis of the Workers' Contributions

After manual approval using the MTurk web interface, some experiments were conducted to evaluate a posteriori different automatic approval methods. Table 11.1 shows the proportion of approved and rejected tasks for both approval methods (manual and automatic). The higher rate of rejected tasks for Amharic can be explained by the much longer period of time during which the task was made available to workers. The tasks rejected with the manual process contained empty transcriptions, copies of instructions, nonsensical text, and tasks completed by workers without any knowledge of the language. This manual approval process is time consuming, thus an experiment involving automatic approval methods was also conducted.

Table 11.1 Submitted tasks approval.

	Number of workers		Number of tasks			
	AMH	SWH	AMH		SWH	
	(Man and Auto)		Man	Auto	Man	Auto
APP	12	3	589	584	1183	1185[a]
REJ	171	31	492	497	250	248
TOT	177[b]	34	1081		1433	

[a]Seven AMH transcriptions and four SWH transcriptions that were approved manually were rejected automatically, while two AMH and two SWH transcriptions that were rejected manually were approved automatically.

[b]It is the number of all the workers who submitted one or more Amharic tasks. It is not, therefore, the sum of the number of rejected and approved workers because there are workers who submitted some rejected tasks and some approved ones.

This was done a posteriori, and no worker was rejected using such an automatic procedure. As can be seen in Table 11.1, it is possible to obtain results equivalent to those of the manual approval with the following task filtering:

(i) Empty and short (shorter than four words) transcriptions
(ii) Transcriptions using non-Amharic writing system, including copy of URLs (for Amharic)
(iii) Transcriptions containing bigrams of instructions and descriptions from the tasks
(iv) Transcriptions that are outside the distribution space set by $Avg + 3 * Stdv(\log 2(ppl))$ (where *ppl* is the perplexity assigned by a language model developed with a different text).

The detailed completion rate per day was analyzed for both languages. Among the 1183 sentences requested, 54% of the Amharic tasks were approved in 73 days. On the other hand, Swahili was completed after 12 days, thus showing that there is a substantial variety in the rate of completion among different languages. This result is important since it shows that the Amharic transcription could not be achieved using MTurk with this setup.

One hypothesis for such a result could simply be the effective population having access to MTurk. A recent survey (Ipeirotis 2010b) shows that 47% of the workers were from the United States, 34% from India, and the last 19% were divided among 66 other nondetailed countries. However, US ENGLISH Foundation, Inc. (www.usefoundation.org/view/29) shows that Swahili speakers are less numerous than Amharic speakers in the United States (less than 40,000 Swahili speakers against more than 80,000 Amharic speakers).

Moreover, Table 11.1 shows that workers doing coherent work were more numerous for Amharic than for Swahili (12 and 3, respectively). A more probable explanation would thus be the input burden for Amharic language, considering the necessity to use an external virtual keyboard and to copy/paste from another web page. The difficulty to perform this task while managing and listening to the audio file may have complicated the task and therefore discouraged workers.

Nevertheless, the tasks' transcription productivity indicates similar mean worker productivities (15 and 17xRT for Amharic and Swahili, respectively). These numbers are close to the ones cited in (Novotney and Callison-Burch 2010) for transcriptions of English (estimated at 12xRT). However, these numbers must be considered with caution, since they do not include the time corresponding to the manual approval process or to the development of an *ad hoc* automatic approval procedure.

Evaluation of Workers' Transcriptions Quality

To evaluate workers' transcriptions (TRK) quality, the accuracy of the manually approved tasks was calculated based on our reference transcriptions (REF). As both Amharic and Swahili are morphologically rich languages, it was found relevant to calculate error rate at word level (WER), syllable level (SER) and character level (CER). Furthermore, real usefulness of such transcriptions must be evaluated in an ASR system. Some misspellings or differences of segmentation (which can be quite frequent in morphologically rich languages) will indeed not necessarily impact system performance but will still inflate WER (Novotney and Callison-Burch 2010). The CER is less affected and is, therefore, more reflective of the transcription quality than the WER. The reference transcriptions are the sentences read during corpora recordings, and reading errors may have occurred.

Table 11.2 presents error rates (ER) for each language depending on the computed level accuracy (five of the approved Amharic transcriptions and four of the Swahili ones were found unusable and were disregarded). As expected, WER is relatively high (16.0% for Amharic and 27.7% for Swahili), while CER is lower. It seems to approach disagreement among expert transcribers even if it was not possible to explicitly calculate such disagreement (because data was transcribed by only one worker without overlap). In the literature, it was found that the word-level disagreement for a nonagglutinative language with a well-normalized writing system ranges from 2% to 4% WER (www.itl.nist.gov/iad/mig/tests/rt). The gap between WER and SER may also be a good weight indicator of the different segmentation errors resulting from the rich morphology.

The low results for Swahili are clarified by providing per-worker ER. Among the three workers who completed approved tasks, two have similar disagreement with REF: 19.8% and 20.3% WER, and 3.8% and 4.6% CER. The last worker has a higher ER (28.5% WER and 6.3% CER) and was the most productive, performing 90.2% of the tasks. Looking more closely at error analysis, one could suggest that this worker is a second-language speaker with no difficulty listening and transcribing but with some variation in writing (see details below).

Table 11.2 Error rate (ER) of workers transcriptions.

Level	Amharic			Swahili		
	Number of Snt	Number of unit	ER (%)	Number of Snt	Number of unit	ER (%)
Wrd	584	4988	16.0	1,179	10,998	27.7
Syl	584	21,148	4.8	1,179	31,233	10.8
Chr	584	42,422	3.3	1,179	63,171	6.1

Table 11.3 Most frequent confusion pairs for Swahili.

Frq	REF	TRK	Frq	REF	TRK
15	serikali	serekali	6	nini	kwanini
13	kuwa	kwa	6	sababu	kwasababu
12	rais	raisi	6	suala	swala
11	hao	hawa	6	ufisadi	ofisadi
11	maiti	maiiti	5	dhidi	didi
9	ndio	ndiyo	5	fainali	finali
7	mkazi	mkasi	5	jaji	jadgi

Error analysis

Table 11.3 shows the most frequent confusion pairs for Swahili between REF transcriptions and TRK transcriptions. Most of the errors can be grouped into five categories that can also be found in Amharic:

- Incorrect morphological segmentation: see words *nini, sababu*, both preceded by *kwa* in REF.
- Common spelling variations of words such as *serikali* and *rais* (sometimes even found in newspapers article).
- Misspellings due to English influence in loanwords like *fainali* and *jaji* (meaning "final" and "judge").
- Misspellings based on pronunciation (see words *kuwa, ndio, suala*).
- Misspellings due to personal orthographic convention, which can be seen in words *maiti, mkazi, ufisadi, dhidi*.

Errors in the last two categories were all made by the same worker (the most productive one, having a high WER). Our assumption that this worker is a second-language speaker relies on the errors' frequency and regularity. One interesting conclusion of this analysis is that an accurate check of the origin of the workers (native / nonnative) is not easy to implement. For instance, in this latter case, qualification tasks would not have been efficient enough to detect this nonnative speaker.

In an experiment not reported here, an ASR system was trained using both REF and TRK transcriptions and nearly similar performances for both languages were observed. This suggests, therefore, that nonexpert transcriptions using crowdsourcing can be accurate enough for ASR. It also highlights the fact that even if most of the transcriptions are made by second-language speakers, it will not particularly affect ASR performances. This result is not particularly surprising: it demonstrates that ASR acoustic model training is rather robust to transcription errors. This result is in line with other works published on unsupervised and lightly supervised training where the machine (instead of the workers) transcribes speech data that will later be integrated into the training set (Wessel and Ney 2005).

11.4.5   Discussion and Lessons Learned

In this section, the use of Amazon's Mechanical Turk speech transcription for the development of acoustic models for two under-resourced African languages was investigated. The main results are the following:

- For a simple task (transcribing speech data), it is possible to collect usable data for ASR systems training; however, all languages are not equal in completion rate. The languages of this study clearly had a lower completion rate than English.
- Among the targeted languages, Amharic's task was incomplete after a period of 73 days; this may be due to a higher task difficulty (use of a virtual keyboard to handle Amharic scripts). This questions the use of Amazon's Mechanical Turk for less elementary tasks that require more of a worker's time or expertise.
- Analysis of the Swahili transcriptions shows that it is necessary to verify the workers' expertise (native / nonnative). However, designing a qualification test to filter out nonnative workers is not straightforward. Furthermore, the acoustic model training is rather robust to workers transcription errors. The use of MTurk in this context can be seen as another form of the lightly supervised scenario where machines are replaced by workers.

MTurk has proved to be a valuable tool for NLP domains, and some recommended practices were already proposed in (Callison-Burch and Dredze 2010), mainly concerning how to be productive with MTurk. However, one should be careful about the way in which the data are collected or the experiments are conducted in order to prevent any legal or ethical controversies. Due to the characteristics of MTurk discussed earlier in this chapter, it was decided, after that experiment, to work directly with a Kenyan institute (http://www.taji-institute.com/) to collaboratively transcribe 12 hours of our web broadcast news corpus. In order to reduce the repetitive and time-consuming transcription task, a collaborative transcription process was considered, based on the use of automatic preannotations (pretranscriptions) to increase productivity gains. Details on this procedure can be found in (Gelas *et al.* 2012). At $103 per transcribed hour, such collaboration is significantly more expensive than using MTurk ($37 per transcribed hour), but in this situation both employer and employee benefit from a more equitable relationship between the two. The price was set by the workers and corresponds to the task as well as to the reality of the local labor market (setting the fair price for work is important in such a context). This resulted in a positive and complete involvement of the workers; direct communication was a major benefit compared to MTurk. It allowed for both direct feedback on the experiment and a sufficient margin for adaptation. Such a direct collaboration is just one example of what can be done in order to carry out research along the highest ethical standards.

11.5   Toward Ethically Produced Language Resources

The preceding sections have illustrated the different economic, ethical, and legal problems of crowdsourcing. They are numerous and serious and, when paired with experiments such as the one described in Section 11.4, may lead to the adoption of a reserved stance on crowd labor use in speech science. However, given its huge potential, crowdsourcing will continue to develop even if some do not wish to participate. Solutions will be proposed for some, if not all, of the problems listed in this chapter in order to enable speech researchers or agencies to make use of crowdsourcing in an ethical way.

These solutions could be individual, namely guidelines for good practices. For instance, Wolfson and Lease (2011) provides some useful advice about the legal concerns that could be summarized in few points:

- **Be mindful of the law**: National and local legislatures and agencies may create new laws and administrative rules to protect the crowdsourcing workers and preserve the local labor. Anyone involved in crowdsourcing should consider all the potential legal ramifications, and weigh the costs and benefits.
- **Use contracts to clearly define your relationships**: The relationship between requester and worker, as defined by the clickwrap participation agreement provided by the crowdsourcing vendor, is not clear, and some crowdsourcing agreements may not stand up in court. Defining a clear contract between employer and worker could help resolve many problems in advance.
- **Be open and honest**: In order to prevent from possible problems, and especially to avoid legal problems, providers should be open and honest about their expectations so that workers can understand them and adjust their behavior.

Other solutions are general and involve the speech community as a whole in designing a more ethical framework.

After an outline of the various views concerning the difficult problem of the monetary compensation for the work done, a short overview of what could be the possible consequences of a "laissez-faire" attitude, especially for the development of LR, will be presented.

The different foreseeable individual and general solutions will be presented in the hopes of eliminating or at least reducing the ethical problems.

11.5.1 Defining a Fair Compensation for Work Done

There are evident solutions for the problem of payment to workers, including those pointed out by Sharon Chiarella, vice-president of MTurk (Chiarella 2011):

- **Pay well**: Do not be fooled into underpaying workers by comparing your HITs (tasks) to low-priced HITs that are not being completed.
- **Pay fairly**: Do not reject an assignment unless you are SURE it is the Worker who is wrong.
- **Pay quickly**: If you approve or reject Assignments once a week, Workers may do a few HITs and then wait to see if they are paid before doing more. This is especially true if you are a new requester and have not established your reputation yet.

These recommendations are a good starting point, but they do not address all the problems highlighted in this chapter.

Tasks could be subdivided (see Section 11.3) based on if they correspond to human experiments (speech acquisition for instance) or to a real labor (such as speech transcription). Furthermore, Section 11.2.1 lists some of the existing crowdsourcing services, the utility of which depends on the task to be accomplished. It should be said that it is easier to establish fair compensation for a given task if the chosen crowdsourcing service is adequately set up for doing so. Many crowdsourcing platforms look like a huge bazaar where tasks of different complexity, requiring workers with very different skills and therefore offering very different levels of compensation for the work done, coexist in an anarchic way. In the case of under-resourced languages transcription described in Section 11.4, a classical framework (for example using direct contact with a local university) has produced better results than the use of a microworking crowdsourcing platform.

In order to be able to establish fair compensation, it should be clear what the tasks are in the crowdsourcing platforms. However, even if the task is well defined, determining fair compensation is not only a question of ethics, but also a pragmatic question of efficiency, as it was already mentioned in Section 11.3. As it has been pointed out in (Ipeirotis 2011c), "Pay enough or not pay at all." If we want to set up an efficient sustainable framework, the only two stable solutions are:

1. Offer a fair reward, which could be modified in response to the quality delivered.
2. Do not to pay anything, as it is organized in most of the collaborative science or Game With A Purpose projects.

This quite radical assumption relies on the fact that the motivations underlying collaboration in a voluntary or a retributed work are drastically different for most people. One of the most counterintuitive result is that providing incentives for a task with no initial payment actually *reduces* performance (Gneezy and Rustichini 2000).

In addition to the problem of quality, completion time should also be considered. Frei (2009) shows a clear relation among the completion times of different tasks, depending on whether or not the task is interesting or involves payment. The conclusion is twofold: first of all, one should not ask that a tedious task be completed for free; and secondly, the level of incentives is clearly correlated to the obtention of a manageable completion time.

According to these different facts, a basic taxonomy could be defined. If the task could be performed through a traditional framework, or if some special ability is desired, a good strategy would be to attract and keep the "good" workers. In this context, use of a microworking platform such as MTurk is not useful. It is not a problem of ethics, but mainly a problem of quality and stability. A specialized service, such as CASTINGWORDS for transcription, or a crowdsourcing service that could provide a direct link with the workers (for instance oDESK) would be preferable. Designing a Game With a Purpose or building a collaborative science project is a good alternative. The task needs to be a large-scale one, as it requires significant development time, advertising, and so on, and interestingly enough; the incentive is not necessary in this context. The microworking services should only be used for tasks which do not call for high quality or special abilities but do require very rapid completion.

With project-based crowdsourcing (simple projects or complex tasks in the categories presented in Section 11.2.1), a requester usually hires a vendor that has access to a network of skilled professionals. The vendor is then responsible for recruiting people who can help with the work. The community can represent different categories of professionals, such as IT (information technology) experts, software developers, or CAD (computer-aided design) specialists, for example. Those selected to perform the work are compensated with cash prizes or other rewards or incentives. Here, a clear framework could be defined, one which resembles the classical relationship between employers and employees or client and individual contractors, as soon as the relations with the workers are clear and anonymity is discarded. In this framework, fair compensation is determined based on the classical balance between the difficulty of the task, the time spent, the amount of money available to perform it, and the negotiation with the workers. The fact that the task will be done through crowdsourcing will enable some cost reduction (streamlined recruitment and dismissal, no charge to equip the workers, etc.) but should not impose a lower wage on workers. The advantage for the workers, such as self-assignment and the lack of time or money spent on commuting, should compensate for the fact that they have to pay for their own insurance and equipment. In oDESK, for instance, requesters are connected with a team of skilled workers to complete a whole job. In this sense, oDESK is close to a traditional workplace: it allows requesters to distribute a task using an hourly wage; the communication between requesters and workers is direct; a requester could provide his team with relevant training and supervision; and the wages are substantially higher than the ones available on others platforms such as MTurk (between $10 to $25 per hour). However, paradoxically, this method which reduces many of the inherent crowdsourcing problems in regard to the workers raises the issue that workers in oDESK are much closer to being defined as employees under the FLSA (see Section 11.3.2) than MTurk workers. oDESK thus runs a higher risk of being taken to court by some of their "employees."

In contrast to project-based crowdsourcing, in most microworking services (micro and macro tasks in the categories presented in Section 11.2.1), and especially in MTurk, the situation is less clear, and establishing "fair" compensation is quite difficult. The payment should not be separated from the general economic model, and in a shaky economic model it is very difficult to determine a fair compensation. For instance, in the MTurk model it is not possible to establish a clear correlation between the reward and the final quality (see, for instance, Marge *et al.* 2010). This fact should also be seen in light of the article (Faridani *et al.* 2011), in which the authors show that increasing the reward decreases the demand for the task. The reason is that high rewards mean complex tasks, with higher risks for the worker.

Section 11.3.3 has pointed out that setting an hourly wage is better for quality and ethics. However, an hourly wage could be difficult to evaluate in a task-based environment, as there are individual variations among workers and the time spent working may decrease drastically as the workers learn how to perform the task efficiently, and so on. Nevertheless, an hourly based payment should be used (whenever possible) instead of a task-based one, as it is the common law for salaried employees in the majority of countries throughout the world. Moreover, hourly payment complies with the concept of *minimum* wage commonly found in developed countries which is not fulfilled in many crowdsourcing systems such as MTurk, which pays less than $2 per hour. Minimum wage has many ethical and practical advantages, but is quite difficult to settle in the unregulated crowdsourcing system. Minimum wage should be accompanied by regulation rules concerning both parties (requester and worker); the consequences of an minimum hourly wage without other regulations will be (among others):

- It is quite difficult to verify remotely how long a worker is actually working on a task. Online regulation tools should be designed to enable this verification while respecting privacy concerns.
- It will increase the "market for lemon' effect (see Section 11.3.3) by overpaying poor-quality workers and spammers. The relations between requesters and workers should be symmetrical, without anonymity and with an efficient reputation system.
- Minimum wages are country-specific, while crowd labor is spread across many countries in the crowdsourcing global marketplace. Based on classical laws of supply-and-demand, defining a minimum wage will effectively orient the work toward the places with the lowest minimum wage (see, for instance, the growth in the number of Indian workers in MTurk).

The minimum wage should be set in order to encourage the participation of workers from countries with higher minimum wages in order to preserve local work in these countries. One possible solution is to fix the reward according to the worker's country and to impose (by law, by social pressure, or through a quality label) on requesters a quota of unit tasks to be performed by workers from his own country; this quota could be adjusted based on the task to be performed (for instance, if the task could not be completed in his country, because of the language involved), or on the existing "ethical" foundation of a crowdsourcing platform (for instance, SAMASOURCE—http://samasource.org/).

11.5.2 Impact of Crowdsourcing on the Ecology of Linguistic Resources

What are the possible consequences of collecting, transcribing, or annotating speech with the help of the crowdsourcing services in their current state, given the ethical, economic, and legal problems related to these services?

For some of these crowdsourcing services, the future is insecure, given the flaws in their economic model (see Section 11.3.3). For most of them, national or international regulation of labor laws on the Internet may be foreseeable if the quantity of existing jobs outsourced on the Internet is sufficiently large enough to exert pressure on political decision makers (see Section 11.3.2). Until then, relying entirely on paid crowdsourcing services for the development of speech and Language Resources (LR) seems hazardous.

Beyond the present facts, some other problems could be considered foreseeable longer term consequences of the use of crowdsourcing for LR development. The main problem is derived from the fact that many researchers present the very low cost of crowdsourcing as its main advantage. If the language and resource community persists in claiming that with crowdsourcing it is now possible to produce any linguistic resource or perform any manual evaluation at a very low cost, funding agencies will come to expect just that. One can predict that in assessing projects involving language resource production or evaluation, funding agencies will prefer projects that propose to produce 10 or 100 times more data for the same amount of money. Costs such as the ones proposed in MTurk will then become the standard costs, and it will be very difficult to obtain funding for a project involving linguistic resource production at any level that would allow for more traditional, noncrowdsourced resource construction methodologies. The very low costs (available sometimes at the price of unreliable quality) would create a *de facto* standard for the development of LR detrimental to other development methods.

11.5.3 Defining an Ethical Framework: Some Solutions

The Situation

As is the case for many other implications of information and communication technologies (ICT) (Mariani *et al.* 2009), it is worth conducting a study on the ethical dimension of crowdsourcing, with ethics here meaning "the way to live well together," and following a precautionary principle: potentially harmful uses should be discouraged and beneficial uses should be encouraged (Rashid *et al.* 2009). And just as with many other consequences of ICT development, the population is faced with the problem once it has been largely deployed at the international level and has become a matter of concern even for the professionals in computer technology who created the problem (Albright 2009). Many researchers working in language science and technology still only see the positive aspects of crowdsourcing without apprehending the negative ones, and most papers on crowdsourcing simply ignore the ethical aspects. Large professional organizations such as the Institute for Electrical and Electronics Engineers (IEEE) (Rashid *et al.* 2009) and the Association for Computing Machinery (ACM) (Bederson and Quinn 2011) recently published papers warning the community about those ethical issues.

The scientific community working in the area of LR (Adda and Mariani 2010), as well as the one working on speech processing (Mariani 2011) or language processing (Fort *et al.* 2011), identified this problem and conducted discussions on the ethical dimension of crowdsourcing through conferences, journals, or forums. One researcher said she preferred using a machine to using a human crowd for evaluating a spoken dialog system, even if the human crowd may provide better results, because of the ethical problem attached to crowdsourcing (Scheffler *et al.* 2011). Another researcher remarked that crowdsourcing is in fact a way to identify specialists who were not known beforehand, and that this search for specialists came with its own costs (Karen Fort, personal communication, April 2012). The conclusion of a panel on crowdsourcing at the International World Wide Web WWW2011 conference revealed a similar orientation, stating that crowdsourcing is best for "parallel, scalable, automatic interviews" and for quickly finding good workers, as reported by Panos Ipeirotis (Ipeirotis 2011b). While domain independent crowdsourcing companies such as CROWDFLOWER or MTurk gather a taskforce of about 1 million workers, a more specialized company like TOPCODER also has a community of 400,000 specialized software engineers and computer scientists from more than 200 countries who develop software following a rigorous, standards-based methodology (Rashid *et al.* 2009).

Let us, therefore, consider the positive aspects of crowdsourcing, and explore how to encourage those positive aspects while avoiding the negative ones.

Toward Collaborative Solutions

Requesters should take into account principles of ethics when planning to use a crowdsourcing approach in the area of LR and language technologies (LT). Bederson and Quinn (2011) provide appropriate guidelines for requesters using a platform in the design of a crowdsourcing operation that can be summarized as follows:

Requester design guidelines

(i) **Hourly pay**: Price tasks based on time. The time to do tasks can be estimated in-house before posting tasks.

(ii) **Pay disclosure**: Disclose the expected hourly wage.

(iii) **Value worker's time**: Optimize tasks to use worker's time effectively.

(iv) **Objective quality metrics**: Decide to approve or reject work based on objective metrics that have been defined in advance and disclosed to workers.

(v) **Immediate Quality feedback**: Give immediate feedback to workers, showing whatever metrics are available.

(vi) **Longer term feedback**: Give warnings to problematic workers.

(vii) **Disclose payment terms**: Disclose in advance when payment will be made.

(viii) **Follow payment terms**: Pay as promptly as possible, and always within the disclosed time-frame.

(ix) **Provide task context**: Given the risk of doing objectionable work, tasks should be described in the context of why the work is being done.

System design guidelines

(i) **Limit anonymity**: Anonymity of requesters enables them to reject good work with near impunity. It also enables them to post unethical or illegal tasks with no public scrutiny. Anonymity for workers enables them to engage in large-scale cheating with nearly no risk since, as with requesters, if their reputation gets damaged, they can simply create a new account.

(ii) **Provide grievance process**: Provide a fair means for workers to request a review of work that was rejected.

It appears from the discussions within the scientific community that it is difficult for the researcher alone to determine the ethical way to use crowdsourcing. Scientific associations, specifically in the area of LR, such as the European Language Resource Association) (ELRA, http://www.elra.info), or in areas related to speech and language processing, such as the International Speech Communication Association (ISCA, http://www.isca-speech.org) or the Association for Computational Linguistics (ACL, http://www.aclweb.org), are expected to play a role in promoting and ensuring the ethical dimension of LR production and distribution in general, and of the use of crowdsourcing in particular.

Here is a list of tasks that those associations could take into consideration to that effect:

(i) **Promote an Open Data approach to LR in general and build up the LR ecosystem overall**:
- Promote a data sharing approach to the scientific community, encouraging all to share the resources they have developed for conducting research in order to allow others to verify the results, especially when the production of resources has been fully or partially supported by public funding.
- Attribute a persistent and unique identifier (LRID) to a language resource in order to facilitate its identification, use and tracking, in cooperation with all parties that are concerned worldwide.
- Compute a language resource impact factor (LRIF) in order to recognize the merits of LR producers.
- Attach a tag to a language resource that will accompany that resource for life by listing the contributors who participated in various aspects of its creation and improvement (design, specification, methodology, production, transcription, translation, annotation of various natures, validation, correction, addition, etc.).
- Assign a copyright status to a LR based on *creative commons* (CC) categories, or the like.

(ii) **Promote the Ethical dimension of crowdsourcing**:
- Make the community aware of the ethical dimension of crowdsourcing.

(iii) **Provide information in order to observe an Ethical approach for crowdsourcing**:
- Identify the requirements of the community, in terms of resources (data, tools, services) and of economic (price, payment), administrative (ordering, licensing), and ethical issues attached to resource production.

- Provide advice concerning the best approach to producing a specific resource (using a crowd, a set of specialists, automatic or semiautomatic systems, or a mixture of them).
- Identify the crowdsourcing platforms that exist and provide a description for each of them, including the pricing policy, the way it deals with the ethical aspects and the constraints of use.
- Provide information about the magnitude of the efforts (time spent, including time to learn the task, and so on) attached to various kinds of resource production and give an estimate of the corresponding salaries, with the aim of defining a fee schedule.

(iv) **Provide tools and services to facilitate and follow an ethical approach for crowdsourcing**:
- Provide a platform for producing the data:
  - Either its own platform, possibly involving a network of specialists and complying with ethics (fair trade principles, minimum guarantee of wages, autoapproval delay, etc.). However, this approach may not be convenient for communicating with a large community of speakers.
  - Or third-party generic crowdsourcing platforms addressing a large set of nonspecialists workers, either fully generic or tailored to a given purpose, after checking the ethical dimension of those platforms.
- Standardize simple tasks in order to facilitate reusability, trading commodities and true market pricing (Ipeirotis 2012b).
- Constitute and maintain a network of specialists for many different languages all over the world.
- Prequalify workers through ability tests, such as those concerning their proficiency in different languages:
  - Establish the means to conduct tests on worker ability through gold-standard data and ground truth.
  - Establish the means to provide public information about the reputation of workers and requesters.
- Help to define a network of local contacts for resource-needed languages; those local contacts might be nonprofit organizations that supervise the data annotation for one or several languages of a particular area and remunerate the workers in keeping with a minimum wage.

(v) **Provide recommendations and validation for Ethical approaches in crowdsourcing**:
- Write and distribute a charter for the ethical production of resources.
- Write and distribute best practices and guidelines for the ethical production of resources.
- Attribute an "ethically produced" label to LR that have been produced in an ethical way:
  - Such resources should be produced within the parameters of legal and ethical working conditions (fair trade principles, careful pricing of the tasks, minimum guarantee of wages, maximum number of working hours, compliance with the tax regulations, etc.) and should come with quality insurance (in terms of the technical quality of the resource, as well as of compliance with legal regulations (intellectual property rights, privacy, etc.)).
- Act so that third-party generic crowdsourcing platforms follow an ethical approach.
- Attribute an "ethically resource producer" label to such platforms.

11.6   Conclusion

As do many other topics, crowdsourcing can be considered from two angles.
  On the positive side, it

- allows to decrease the price of producing resources;
- may therefore increase the size of the data;
- allows one to address a large quantity and diversity of people;
- facilitates access to people who would be difficult to reach in other ways;
- establishes a direct link between the employers and the workers;
- offers a salary for those who have none;
- bypasses intermediaries.

It, therefore, seems to be an especially attractive option for **less-resourced languages** because it
is less costly, given that investments may be difficult to procure for economically uninteresting
languages for many reasons, including there may be fewer experts on those languages who
could intervene; access to native speakers of those languages who are abroad or who were
part of a diaspora may present different difficulties or have to be conducted via intermediaries
at a certain cost; the number of those native speakers may be low; and finding financial
support is complicated by difficult economic conditions. However, as it has been shown in
the related experience of using crowdsourcing for producing annotated corpus in the less-
resourced languages of Swahili and Amharic, the reality may be somewhat different and, in
some cases, may result in shifting back to a more traditional approach.
  On the negative side, it

- does not guarantee a proper salary for the workers;
- does not guarantee the quality of the result;
- may ultimately result in a more significant cost than traditional approaches;
- bypasses all legal aspects attached to social security, pensions, or union rights.

  The basic assumptions on quality, price, and motivation may, therefore, be discussed, as
well as the legal and scientific policy dimensions.

> **Quality**: The quality of the transcription of a speech corpus and/or of its trans-
> lation should be enough to train ASR or MT systems, for example, and teams
> participating in evaluation campaigns are highly sensitive to the quality of the
> training and testing material. However, some quality problems may appear, espe-
> cially if the task is complex. The task may then be subdivided into subtasks, but
> this increases the complexity of the organization, as it necessitates coordination
> and correlation. If people are primarily interested by the financial income, they
> may cheat in order to increase their productivity and thus their salary, and this
> also makes the quality checking mandatory. The initial detection of spammers
> is necessary and, in some cases, the task has to be duplicated or triplicated for
> crossvalidation. Final validation and postprocessing are also to be added.

Price: Salaries are usually rather low in crowdsourcing, and therefore, the production cost is supposed to be low. However, the development of interfaces for nonexperts, the detection of cheating, the spammer problem, and the quality issue necessitating the previously mentioned operations add extra costs. If the competence is hard to find, the salaries will also have to be higher.

Motivation: Some crowdsourcing actions, such as Wikipedia, are based on voluntary contributions, but most are conducted for money. Therefore, the workers may only be interested in the salary and not in the task. This may provoke those workers to consider efficiency first and to try to earn the maximum of money with the minimum of efforts.

Legal: The legal dimension must also be taken into account (Wolfson and Lease 2011). The employers may not pay taxes for the employees, while the employees may also not be taxed, as the action is conducted in an international framework which may escape national regulations. Wages are usually lower than the amount paid in the employer's country (while still sometimes being higher than the usual salary in the employee's country). There is no social or health security, no guarantee of payment and no support coming from unions (which may however be replaced on the Internet by blogs and forums such as Turker Nation (`http://www.turkernation.com/`) or Turkopticon (`http://turkopticon.differenceengines.com`)), and of course, there is no protection of IPR and copyright.

Science policy: Given that the use of crowdsourcing reduces production costs, funding agencies may reduce their support for resource production and therefore impose crowdsourcing as a *de facto* standard.

Disclaimer

The views and opinions expressed in this chapter are those of the authors and do not necessarily reflect the views and opinions of the authors of the other chapters, nor those of the editors of the book. None of these opinions were intended to be voluntarily defamatory or offensive with regard to any of the crowdsourcing sites cited here. Used in most of the studies mentioned in this chapter, Amazon Mechanical Turk (MTurk) serves in this context as a very famous example of a microtask site. We do not exercise any moral or political judgment on the general policy of this specific site.

References

Abate S, Menzel W and Tafila B (2005) An amharic speech corpus for large vocabulary continuous speech recognition. *Interspeech*, pp. 67–76, Lisbon.

Adda G and Mariani J (2010) Language resources and Amazon Mechanical Turk: legal, ethical and other issues. *LISLR2010, Legal Issues for Sharing Language Resources Workshop, LREC2010*, Malta.

Adda G, Sagot B, Fort K and Mariani JJ (2011) Crowdsourcing for language resource development: critical analysis of Amazon Mechanical Turk overpowering use. *Language & Technology Conference : Human Language*

Technologies as a Challenge for Computer Science and Linguistics (L&TC 2011), pp. 304–308, Poznan, Poland.

Akerlof GA (1970) The market for "lemons": quality uncertainty and the market mechanism. *Quarterly Journal of Economics* **84**(3), 488–500.

Albright P (2009) Is crowdsourcing an opportunity or threat? http://www.computer.org/portal/web/buildyourcareer/crowdsourcing (accessed 17 October 2012).

Arthur C (2006) What is the 1% rule? The Guardian.

Bederson BB and Quinn AJ (2011) Web workers unite! addressing challenges of online laborers. *Proceedings of the 2011 Annual Conference Extended Abstracts on Human Factors in Computing Systems (CHI EA '11)*, pp. 97–106. ACM.

Berment V (2004) *Méthodes pour informatiser les langues et les groupes de langues "peu dotées."* PhD thesis, Université Joseph Fourier, Grenoble.

Biewald L (2010) Better crowdsourcing through automated methods for quality control. *SIGIR 2010 Workshop on Crowdsourcing for Search Evaluation.*

Callison-Burch C and Dredze M (2010) Creating speech and language data with amazon's mechanical turk. *CSLDAMT '10: Proceedings of the NAACL HLT 2010 Workshop on Creating Speech and Language Data with Amazon's Mechanical Turk*, Los Angeles, CA.

Chamberlain J, Poesio M and Kruschwitz U (2008) Phrase detectives: a web-based collaborative annotation game. *Proceedings of the International Conference on Semantic Systems (I-Semantics'08)*, Graz.

Chen DL and Dolan WB (2011) Building a persistent workforce on mechanical turk for multilingual data collection. *Proceedings of The 3rd Human Computation Workshop (HCOMP 2011).*

Chiarella S (2011) Cooking with Sharon. http://mechanicalturk.typepad.com/blog/2011/07/cooking-with-sharon-tip-3-manage-your-reputation.html (accessed 17 October 20102).

Chilton LB, Horton JJ, Miller RC and Azenkot S (2010) Task search in a human computation market. *Proceedings of the ACM SIGKDD Workshop on Human Computation (HCOMP '10)*, pp. 1–9.

Draxler C (1997) WWWTranscribe—a modular transcription system based on the World Wide Web. *Proceedings of Eurospeech*, pp. 1691–1694, Rhodes.

Draxler C and Steffen A (2005) Ph@ttsessionz: recording 1000 adolescent speakers in schools in germany *Proceedings of Interspeech*, pp. 1597–1600, Lisbon.

Schultz T and Kirchhoff K (eds) (2006) *Multilingual Speech Processing*. Academic Press, Berlington, MA.

Faridani S, Hartmann B and Ipeirotis PG (2011) Whats the right price? pricing tasks for finishing on time. *Proceedings of the 3rd Human Computation Workshop (HCOMP 2011).*

Felstiner A (2011) Working the crowd: employment and labor law in the crowdsourcing industry. *Berkeley Journal of Employment and Labor Law* **32**(1), 143–204.

Fort K, Adda G and Cohen KB (2011) Amazon Mechanical Turk: gold mine or coal mine? *Computational Linguistics (editorial)* **37**(2), 413–420.

Frei B (2009) Paid crowdsourcing: current state & progress towards mainstream business use. Smartsheet white paper. http://www.smartsheet.com/files/haymaker/Paid CrowdsourcingSept2009-ReleaseVersion-Smartsheet.pdf (accessed 17 October 2012).

Gelas H, Besacier L and Pellegrino F (2012) Developments of swahili resources for an automatic speech recognition system *SLTU.*

Gneezy U and Rustichini A (2000) Pay enough or don't pay at all. *Quarterly Journal of Economics* **115**(3), 791–810.

Gruenstein E, Mcgraw I and Sutherl A (2009) A self-transcribing speech corpus: collecting continuous speech with an online educational game *the Speech and Language Technology in Education (SLaTE) Workshop*, Warwickshire, England.

Holmstrom B (1979) Moral hazard and observability. *Bell Journal of Economics* **10**(1), 74–91.

Ipeirotis P (2008) Why People Participate on Mechanical Turk, Now Tabulated. http://www.behind-the-enemy-lines.com/2008/09/why-people-participate-on-mechanical.html (accessed 17 October 2012).

Ipeirotis P (2010a) Be a Top Mechanical Turk Worker: You Need $5 and 5 Minutes. http://www.behind-the-enemy-lines.com/2010/10/be-top-mechanical-turk-worker-you-need.html (accessed 17 October 2012).

Ipeirotis P (2010b) Demographics of Mechanical Turk CeDER Working Papers http://hdl.handle.net/2451/29585. CeDER-10-01 (accessed 17 October 2012).

Ipeirotis P (2010c) A Plea to Amazon: Fix Mechanical Turk! `http://behind-the-enemy-lines.blogspot.com/2010/10/plea-to-amazon-fix-mechanical-turk.html` (accessed 17 October 2012).

Ipeirotis P (2011a) Do Mechanical Turk Workers Lie about Their Location? `http://www.behind-the-enemy-lines.com/2011/03/do-mechanical-turk-workers-lie-about.html` (accessed 17 October 2012).

Ipeirotis P (2011b) Does Lack of Reputation Help the Crowdsourcing Industry? `http://www.behind-the-enemy-lines.com/2011/11/does-lack-of-reputation-help.html` (accessed 17 October 2012).

Ipeirotis P (2011c) Pay Enough or Don't Pay at All. `http://www.behind-the-enemy-lines.com/2011/05/pay-enough-or-dont-pay-at-all.html` (accessed 17 October 2012).

Ipeirotis P (2012a) Mechanical Turk vs. Odesk: My Experiences. `http://www.behind-the-enemy-lines.com/2012/02/mturk-vs-odesk-my-experiences.html` (accessed 17 October 2012).

Ipeirotis P (2012b) The Need for Standardization in Crowdsourcing. `http://www.behind-the-enemy-lines.com/2012/02/need-for-standardization-in.html` (accessed 17 October 2012).

Ipeirotis P (2012c) Philippines: The Country That Never Sleeps. `http://www.behind-the-enemy-lines.com/2012/04/when-is-world-working-odesk-edition-or.html` (accessed 17 October 2012).

Kittur A, Smus B, Khamkar S and Kraut RE (2011) Crowdforge: crowdsourcing complex work. *Proceedings of the 24th Annual ACM Symposium on User Interface Software and Technology (UIST '11)*, pp. 43–52.

Kochhar S, Mazzocchi S and Paritosh P (2010) The anatomy of a large-scale human computation engine. *Proceedings of Human Computation Workshop at the 16th ACM SIKDD Conference on Knowledge Discovery and Data Mining, KDD 2010*, Washington, DC.

Le V and Besacier L (2009) Automatic speech recognition for under-resourced languages: application to vietnamese language. *Audio, Speech, and Language Processing, IEEE Transactions on* 17(8), 1471–1482.

Little G, Chilton LB, Goldman M and Miller RC (2010) Turkit: human computation algorithms on mechanical turk. *Proceedings of the 23nd Annual ACM Symposium on User Interface Software and Technology (UIST '10)*, pp. 57–66.

Marge M, Banerjee S and Rudnicky AI (2010) Using the Amazon Mechanical Turk for transcription of spoken language. *IEEE International Conference on Acoustics Speech and Signal Processing (ICASSP)*, pp. 5270–5273, Dallas, TX.

Mariani J (2011) Ethical dimension of crowdsourcing. *Special session on Crowdsourcing, Interspeech'2011*, Firenze.

Mariani J, Besnier JM, Bordé J, Cornu JM, Farge M, Ganascia JG, Haton JP and Serverin E (2009) Pour une éthique de la recherche en Sciences et Technologies de l'Information et de la Communication (STIC). *Technical report*, COMETS-CNRS. `http://www.cnrs.fr/fr/organisme/ethique/comets/avis.htm` (accessed 17 October 2012).

Marten L (2006) Swahili in *The Encyclopedia of Languages and Linguistics*, 2nd edn, vol. 12 (ed. K Brown). Elsevier, Oxford, pp. 304–308.

McGraw I, Gruenstein A and Sutherland A (2009) A self-labeling speech corpus: Collecting spoken words with an online educational game. *Proceedings of Interspeech*, pp. 3031–3034, Brighton, UK.

Novotney S and Callison-Burch C (2010) Cheap, fast and good enough: automatic speech recognition with non-expert transcription. *Human Language Technologies: The 2010 Annual Conference of the North American Chapter of the Association for Computational Linguistics (HLT '10)*, pp. 207–215, Los Angeles, CA.

Parent G and Eskenazi M (2010) Toward better crowdsourced transcription: transcription of a year of the let's go bus information system data. *Proceedings of IEEE Workshop on Spoken Language Technology*, pp. 312–317, Berkeley, CA.

Quinn AJ and Bederson BB (2011) Human computation: a survey and taxonomy of a growing field. *Proceedings of the 2011 Annual Conference on Human Factors in Computing Systems (CHI EA '11)*, pp. 1403–1412.

Rashid A, Weckers J and Lucas E (2009) Software engineering, ethics in a digital world. *IEEE Computing Now*. `http://www2.computer.org/portal/web/computingnow/0709/theme/col_softengethics` (accessed 17 October 2012).

Ross J, Irani L, Silberman MS, Zaldivar A and Tomlinson B (2010) Who are the crowdworkers?: shifting demographics in mechanical turk. *Proceedings of the 28th of the International Conference Extended Abstracts on Human Factors in Computing Systems (CHI EA '10)*. ACM, New York.

Ross J, Zaldivar A, Irani L and Tomlinson B (2009) Who Are the Turkers? Worker Demographics in Amazon Mechanical Turk Social Code Report 2009-01. http://www.ics.uci.edu/~jwross/pubs/SocialCode-2009-01.pdf (accessed 17 October 2012).

Scheffler T, Roller R and Reithinger N (2011) Speecheval: a domain-independent user simulation platform for spoken dialog system evaluation. *IWSDS 2011*, Granada.

Silberman MS, Ross J, Irani L and Tomlinson B (2010) Sellers' problems in human computation markets. *Proceedings of the ACM SIGKDD Workshop on Human Computation (HCOMP '10)*, pp. 18–21.

Snow R, O'Connor B, Jurafsky D and Ng AY (2008) Cheap and fast—but is it good? evaluating non-expert annotations for natural language tasks. *Proceedings of EMNLP 2008*.

Virginia Commonwealth University (2009) VCU Institutional Review Board Written Policies and Procedures. http://www.research.vcu.edu/irb/wpp/flash/XVII-2.htm (accessed 17 October 2012).

von Ahn L (2006) Games with a purpose. *IEEE Computer Magazine*, pp. 96–98.

Wessel F and Ney H (2005) Unsupervised training of acoustic models for large vocabulary continuous speech recognition. *IEEE Transactions on Speech and Audio Processing* **13**(1), 23–31.

Wolfson SM and Lease M (2011) Look before you leap: Legal pitfalls of crowdsourcing. *Proceedings of the American Society for Information Science and Technology* **48**(1), 1–10.

Zittrain J (2008a) *The Future of the Internet–And How to Stop It*. Yale University Press, New Haven, CT.

Zittrain J (2008b) Ubiquitous human computing. *Philosophical Transaction of the Royal Society A* **366**(1881), 3813–3821.

Index

Crowdsourcing for Speech Processing: Applications to Data Collection, Transcription and Assessment, First Edition.
Edited by Maxine Eskénazi, Gina-Anne Levow, Helen Meng, Gabriel Parent and David Suendermann.
© 2013 John Wiley & Sons, Ltd. Published 2013 by John Wiley & Sons, Ltd.